SP Pvt. Ltd.

Dedicated to My Son

SHREY SHEKHAR
(The Challenge of My Life)

Rocketry with Solid Propellants

Himanshu Shekhar
High Energy Materials Research Laboratory
(DRDO)
Pune – 411 021

2018

Studium Press (India) Pvt. Ltd.

Rocketry with Solid Propellants

ISBN: 978-93-85046-15-5

Published by:

Studium Press (India) Pvt. Ltd.
4735/22, 2nd Floor, Prakash Deep Building
(Near Delhi Medical Association),
Ansari Road, Darya Ganj, New Delhi-110 002
Tel.: + 91-11-43240200-15 (15 lines); Fax: 91-11-43240215
E-mail: studiumpress@gmail.com

Printed at India

About the Author

Dr. Himanshu Shekhar is a doctorate in Mechanical Engineering and is currently working as Scientist 'G', Associate Director in Defence Research and Development Organization (DRDO). He was instrumental in creation of facilities for processing propellants in large size case-bonded rocket motors in DRDO and also the development of production facility for case-bonded rocket motor. He has processed many tons of rocket propellants in rockets of different dimensions and has developed prediction and analysis tools for internal ballistics calculation of solid rocket propellants. He has also contributed significantly for performance analysis of gun propellants, high explosives and pyrotechnic composition, thus covering entire spectrum of High Energy Materials (HEMs). He is life member of High Energy Materials Society of India (HEMSI) and Aeronautical Society of India (AeSI). He has written 11 books and has one chapter in international multi volume set of nanotechnology. He has authored more than 120 research publications in international and national journals and seminars. Around 100 technical reports are published by him. He is on the editorial boards of international journals "Bioglobia" and "Central European Journal of Energetic Materials". He has delivered more than 80 invited talks and lectures in (inter-) national universities, academic institutions and research/production agencies of India.

Academics

- First class first with distinction during graduation in Mechanical Engineering
- GATE 91 score : 99.57
- M.Tech in Mechanical Engineering from IIT, Kanpur with CPI of 10.00 out of 10.00.
- Ph.D. in Mechanical Engineering from University of Pune, India in 2010.

Awards and Honours

- Agni Award for Excellence in Self-Reliance 2001 from DRDO.
- Mr. Engineer Award 2003 from Institution of Engineers (India), Pune Chapter.
- Young Scientist Award – 2004 from DRDO.
- Hindi Rajbhasha Pustak Puraskar – 2010 from DRDO.

Salient Books

- Solid Rocket Propellant Technology
- Mathematical Treatise on Interior Ballistics of Guns
- Explosive Science: A Mathematical Treatise
- Aircraft and Automobile Propulsion : A Textbook
- Hindi Books : Racket Rahasya, Bhartiya Vaigyanik, Adhunik Uchcha Visfotak

Preface

Design of Rocket for a particular mission is full of mathematical calculations and propellants being one of the major component share equal responsibility in performance, complexity, calculations and energy. Overall, rocket is a jet propulsion device and the working fluid for the generation of thrust is invariably hot exhaust product of combustion in gaseous forms. The formation of exhaust gases by the combustion of propellants, flow of these gases through the port of the propellant inside rocket motor, the burning characteristic of propellants, mass generation of combustion products, effect of environment are some of the effects, which are to be simulated for proper performance prediction of rockets in general and especially for solid rocket propellants. Although design of a rocket is a vast area, an attempt is made in this book to cover various design features of solid propellants for rockets. The book has total 15 chapter arranged evenly in three distinct sections –

- Theoretical Ballistics,
- Predictive Ballistics and
- Practical Ballistics.

The book covers theoretical aspects of internal ballistics, except thermo-chemical calculations in the first section. These theories are explained in a lucid way and numerical examples give an insight into derived equations of internal ballistics. Various parameters of internal ballistic calculations are also explained in this section of the book. The first chapter is introduction to rockets, their classification and their subsystems. Second chapter discusses introduction, classification, salient properties and manufacturing aspects of solid propellants for the rockets. Third chapter is devoted to terms used and derivations for internal ballistic calculations. Fourth chapter is about staging and features of multi-stage rockets are explained with numerical examples. Fifth chapter is summary of internal ballistic prediction for solid propellant rockets. Strategy for getting pressure time and thrust time profile from geometry of propellant, rocket nozzle and combustion gas properties are explained. Global analysis scheme, using conservation of mass flow rate at bulk-generation and discharge sections, is used to generate pressure from burning surface area. Appendix 1 is added to supplement theoretical calculations and numerical values of ballistic parameters for different operating conditions in the rocket are tabulated in this appendix.

The second section of the book is named predictive ballistics, because it deals about various geometrical configurations, where mathematical expressions for the evolution of burning area with web burnt is derived. This part gives solution to the most critical and major challenge faced by propellant designers. This section may also be treated as configuration ballistics or computational ballistics of solid rocket propellants. Mathematical formulations for axisymmetric cylindrical, tubular and multi-perforated configurations are derived under various chapters. Burning area is also evolved with web burnt for the internal star and fin configurations. Occasionally pressure time profile is also generated in this section, but major thrust of this section of the book is on generation of burning area *versus* web burnt profile for different geometries of solid propellants for rocket. This section of the book is supplemented with one Appendix. In Appendix 2, design of complete rocket from ab-initio design input is carried out in various configurations. It is clear that many configurations are possible for same input conditions and selection of a particular configuration is prerogative of designers.

In third section of the book, practical ballistics are discussed, where more attention is given to actual static firing of propellants. The data acquisition, data analysis and derivation of ballistic and performance parameters are main themes of this section. Calculation strategies from the static evaluation curves (pressure time profile or thrust time profile) are evolved. This type of analysis is rarely available in open literature. A short chapter on combustion instability is also incorporated in this section of the book. This section has one appendix. In appendix 3, neutral burning configurations are analyzed from practical considerations. Both end burning solid sustainer and end inhibited tubular propellant configurations are analyzed using practical values of ballistic parameters to arrive at the feasible or achievable values of thrust and burning time.

The title of the book is selected with lot of permutations to cover the entire content. The book concentrates on solid propellants, used in rockets and it discusses interior ballistic calculations predominantly. Compared to popular term "Solid Rocket Propellant", which is sometimes read and interpreted erroneously as "Propellants for Solid Rockets", the entire gamut of interior ballistics of rockets, illustrated in the book, is referred as rocketry in the title. The term "solid propellants", is clearly mentioned in the title of the book to broadly identify the chapter-contents. So the book is named "*Rocketry with Solid Propellants*".

The book is written in illustrative manner and each derived mathematical equation is followed by some numerical example for proper absorption, understanding and reuse by the readers. The book does not deviate from internal ballistics of rockets, but thermo-chemical calculations are excluded. Overall the book discusses various numerical and mathematical aspects of solid rocket propellants and it will be useful for students and amateurs for understanding basic principles of a solid rocket propellant design. The book will be equally useful for faculty, engineers, researchers and academic institutions for exploring the area of solid propellants as their field of research. The propellant production agencies, army and other users can peep inside the rocket through this book to understand the basic principles and philosophy of design of solid rocket propellants.

Although, journal articles are available for performance prediction with different solid propellant configurations, no book covers such vast majority of geometry at one place. The book is prepared on demand of young engineers, who face problem in understanding predictive and practical ballistics of rockets, on joining solid propellant development activities. The book can be treated as textbook for graduate/post-graduate students of engineering, rocket propellant designers, rocket propellant manufacturers (Industries, Factories), rocket users (Armed forces), rocket propellant developers, academicians (Faculties), freelance researchers (Central universities) and amateur readers. Feedback from readers may pave way for making this book more useful. Awaiting response.

Place: Pune **Dr. Himanshu Shekhar**
Date: 31 October 2016

Table of Contents

About the Author v
Preface vii
List of Tables xiii
List of Figures xiv

Section I:
Theoretical Rocket Propellant Ballistics

1. Rockets : Classification and Components 2
2. Solid Propellants : Processing and Properties 29
3. Internal Ballistics : Domain and Derivations 48
4. Staging of Rockets : Essence and Examples 80
5 Performance Prediction of Rockets 101

Section II:
Predictive Rocket Propellant Ballistics

6. Solid Cylindrical Configurations 120
7. Tubular and Multi-tubular Configurations 139
8. Multi-perforated Configurations 169
9. Internal Star and Fin Port Configurations 202
10. Simple Configurations and Numerical 226

Section III:
Practical Rocket Propellant Ballistics

11. Static Evaluation of Rockets 250
12. Ballistics from Pressure-time Curves 271
13. Ballistics from Thrust-time Curves 296
14. Advanced Evaluations of Firing Curves 308
15. Combustion Instability in Solid Rocket Propellants 324

Appendix 1: Tables of Internal Ballistic Parameters 340

Appendix 2 : Numerical Example for Design of a Rocket 348
Appendix 3: Feasibility of Neutral Configurations 358
Suggested Books for Further Reading 365
Publications by Author on Solid Rocket Propellants 365
Subject Index 373

List of Tables

1. Difference between Cartridge-loaded and Case-bonded Propellants for Rockets 19
2. Difference between Solid and Liquid Propellant Rockets 20
3. Difference between Head-end and Nozzle-end Ignition 24
4. Comparison of Types of Double-Base Propellants 34
5. Computation of Exit Plane Velocity 57
6. Maximum Vacuum Thrust Coefficient 66
7. Properties of Combustion Products 72
8. Final Combustion Product 74
9. Variation of Rocket velocity at ABP with different Mass ratio 83
10. Variation of Rocket Velocity at ABP due to staging 85
11. Parametric Study on End Inhibited Tubular Propellant Configuration 117
12. Salient Parameters for Dual Thrust Configuration 132
13. Web Transitions for Non-central Hole Multi-perforated Propellant Configuration 189
14. Calculated web Transitions for Concentric Multi-perforated Configurations 194
15. Calculation of Burning rate from Progressive Mode Firing 280
16. Generation of Burning Rate Data at Higher Pressures 281
17. Calculation of Burning Rate for Regressive Burning Tubular Propellant Configuration 286
18. Input Parameters for Pressure time Profile 288

List of Figures

1.	Thrust Generation in a Rocket	5
2.	Thermodynamic Cycle of Rocket Engine	10
3.	Pressure Variation in Rocket Engine	11
4.	Pressure Imbalance at Exit Plane of Rocket Nozzle	12
5.	Labeled Schematic of a Typical Rocket	16
6.	Different Types of Nozzles	26
7.	Complete Classification of Rockets	27
8.	Process Flow Chart of Manufacturing Double Base Propellant by Extrusion	40
9.	Flow Chart for Manufacture of Composite Propellant by Casting	44
10.	Variation of Mass Flow Rate with Pressure Difference	54
11.	Description of Nozzle Parameters	56
12.	Variation of Thermodynamic Parameter during Flow through Nozzle	58
13.	Thermodynamic Variations with Different g	59
14.	Variation of Area Ratio of Nozzle with Mach Number	60
15.	Variation of Area Ratio for Different γ	61
16.	Variation of Area ratio with Pressure ratio	62
17.	Variation of Vacuum Thrust Coefficient	65
18.	Comparison of Thrust Coefficients in Vacuum and under Optimum Expansion	66
19.	Variation of Thrust Coefficient under Optimum Expansion	67
20.	Variation of C* against Combustion Gas Parameters	69
21.	Variation of C* with g for Different Combustion Gas Parameters	70
22.	Variation of Specific Impulse of Rocket	76
23.	Operation of a Multi-Stage Rocket as Conceived by Tsiolkovsky	84
24.	Gain in Rocket Velocity at ABP with increasing number of Stages	86
25.	Percentage Gain in Rocket Velocity at ABP number of Stages	87
26.	Mass distribution in 2-stage rocket	88
27.	Variation of Different Rocket Parameters with Time	95

28.	Variation of Kinetic Parameters for Multi-Stage Rocket	98
29.	Variation of Altitude with Velocity	99
30.	Vieille's and Muraour's Law of Burning Rate	104
31.	Predicted Variation of Burning Area with Cigarette Mode Burning of Propellant Configuration	123
32.	Burning Area variation for Laterally Burning Solid Cylinder Configuration	124
33.	Pressure Time Curve for Laterally Burning Solid Cylinder Configuration	125
34.	Variation of Burning Area for Uninhibited Cylindrical Propellant Configuration	126
35.	Evolved Pressure-time profile for the Uninhibited Cylindrical Propellant Configuration	127
36.	Evolved Pressure-time profile for Alternate Uninhibited Cylindrical Propellant Configuration	128
37.	A Typical Dual-Thrust Configuration	129
38.	Salient Parameters for Burning Surface Evolution	130
39.	Surface of Revolution for Different Zones	131
40.	Variation of Volume Loading Fraction	133
41.	Effect of Varying Hole-Radius Fraction	133
42.	Variation in Performance Profile with Length Ratio	134
43.	Dimension of Dual Thrust Configuration	134
44.	Burning Area *Versus* Web Burn Profile	135
45.	Pressure-Time Profile for Dual Thrust Configuration	135
46.	Solid Sustainer with Blind Annular Hole	136
47.	Dual Thrust Solid Sustainer with Annular Hole	136
48.	Salient Features of Bare Tubular Propellant Grain	142
49.	Normalized variation of Burning Area with Consumed Web	142
50.	Parametric Variation for Burning to Port Area	144
51.	Parametric Variation for Burning to Port Area	145
52.	Effect of Inner Diameter of Propellant on Burning to Port Area Ratio	146
53.	Different Inhibited Tubular Configurations	146
54.	Variation of Non-dimensional Burning Area with non-dimensional Web	148
55.	Typical Burning Profiles for Externally Inhibited Tubular Propellants	149
56.	Burning Profiles for Tubular Configuration with One End and Outer Diameter Inhibited	151
57.	Burning Profiles with Variation of Inner Diameter	152
58.	Burning Profiles of Progressive Tubular Propellant Configuration	152

59. Propellant Configuration with Funnel Shaped Port 154
60. Burning area Variation for Taper Tubular Propellant Grain 155
61. Variation of 's' for Taper Tubular Propellant Grain 155
62. Variation of 'h' for Taper Tubular Propellant Grain 156
63. Moon Burner Configuration 157
64. Burning Area Variation for Offset Port Tubular Configuration 158
65. Burning Area Variation for Moon Burner Configuration 159
66. Cross-section of Multi-tubular Configurations 164
67. Comparison of Burning Surface Area Variation 165
68. Burning of Hepta-Perforated Uninhibited Propellant 172
69. Variation of Mass of Propellant with Web Consumed 175
70. Variation of Burning Surface Area of Propellant with Web Consumed 175
71. Two Hole Inhibited Propellant Configuration 177
72. Burning area *versus* web consumed profile for 2-hole Configuration 178
73. Parametric Study with 2-Hole Configuration 179
74. Burning Profile for Sun-Planet Type 2-Hole Configuration 182
75. Burning Profile for 2-Hole Configuration in Three Webs 183
76. Burning Profile for 2-Hole Configuration in Four Webs 185
77. Burning Profile for Four Web Configuration with Reduced Tail-off 186
78. Burning Profile for Generic 2-Hole Configuration 187
79. Burning Profile for Non-concentric Multi-hole Configuration 190
80. Burning Profile-2 for Non-concentric Multi-hole Configuration 191
81. Burning Profile for 4 hole Concentric Multi-perforated Configuration 194
82. Burning Profile for Concentric Multi-perforated Configuration 195
83. 10-Hole Multi-Perforated Configuration 197
84. Predicted Burning Area Variation for 10-hole Configuration 198
85. Predicted Pressure-time Profile for 10-Hole Multi-perforated Configuration 199
86. Convex-Concave Configuration 203
87. Burning Area Profile for Convex-Concave Lobed Star 205
88. Parametric Variation of Burning Area for Convex-Concave Lobed Star 206
89. Concave Lobed Star Configuration 207
90. Burning Area Variation for Concave Lobed Star 209
91. Comparison of Lobed Star Configurations 209

92. Fin Shaped Concave Lobed Star Configuration 210
93. Burning Area Variation for Fin-Shaped Star 211
94. Convex-Concave Fin Shaped Propellant Configuration 212
95. Variation of Burning Surface Area for Fin-2 Configuration 214
96. Variation of Burning Circumferences of Fin-Configuration 215
97. Effect of Number of Stars on Profile 216
98. Top-Land Star Port Configuration 217
99. Top Land Star Configuration 220
100. Burning Surface Area Variation for Top Land Star Configuration 220
101. Parametric Study for Change of Angular Fraction 222
102. Parametric Study for Change of Half Angle for Star Opening 223
103. Neutral Burning Star-Port Propellant 223
104. Propellant Configuration with Longitudinal Variations 227
105. Behavior of Ramped Tubular Inhibited Propellant Configuration 229
106. Burning Profile for End Uninhibited Ramped Tubular Configuration 229
107. Dendrite Configuration 231
108. Burning Surface Area for Dendrite Configuration 233
109. Special Case of Dendrite Configuration 234
110. Cross-Shaped Propellant Configuration 235
111. Length of Periphery in Half Limb of the Leg 237
112. Burning Surface Area of Cross-Shaped Grain 238
113. Burning Surface Area for Longitudinally Inhibited Cross-Shaped Propellant 239
114. Burning Surface Area with Straight Sections 241
115. Variation in Longitudinal Geometry 242
116. Curved Projections and Cavities in Configurations 243
117. A Typical Set-up for Horizontal Static Firing of Rockets 252
118. Vertical Rocket Firing Set-up 253
119. A General Instrumentation Schematic 261
120. Curve from Static Evaluation of Rockets 265
121. Area under Rocket Firing Curves 267
122. Analysis of a Pressure Time Profile 272
123. Flow Chart for Calculation of Burning Rate Law from Pressure Time Profile 277
124. Pressure Time Profile in Progressive Mode firing 279
125. Superimposed Ideal and Practical Firing Curves 280
126. Superimposition of Prediction with Two Burning Rate Laws 281
127. Pressure Time Profile from Static Firing of Regressive Tubular Propellant 284

128. Superimposed Pressure Time Profile for Regressive Propellant Configuration 286
129. Pressure Time Profile from Static Firing of Rocket 288
130. Variation of Burning Area with Time 288
131. Variation for Burning Area with Web Consumed for the Static Firing 289
132. Predicted Pressure Time Profiles for Same Configuration with Different Compositions 289
133. Superimposed Pressure Time Profile with New Composition 291
134. Prediction for Equal Total Burning time 293
135. Superimposed Pressure Time Profiles with Varied Burning rate 294
136. Typical Rocket Evaluation Curves 298
137. A Typical Pressure Time Profile 304
138. Generated Thrust Time Profile 305
139. Variation of Thrust Coefficient with Time 305
140. Superimposed Predicted and Actual Thrust Time Profiles 306
141. Prediction of Thrust Time Profile with Nozzle Efficiency (88%) 306
142. Pressure-Time Profile for Multi-Perforfated Configuration 310
143. Burning Rate Calculation for Multi-Perforated Configuration 311
144. Superimposed Pressure Time Profiles for Multi-perforated Configuration 311
145. Nozzleless Propellant Configuration 313
146. Recorded Pressure and Thrust against Time for Nozzleless Propellant Configuration 314
147. Calculation of Parameters at an Elapsed Time 315
148. Variation Throat Diameter in Nozzleless Propellant Configuration 316
149. Dual Thrust Propellant Configurations with Two Propellants 319
150. Firing Curves for Dual Thrust Rocket 322
151. Variation of Thrust Coefficient in Booster and Sustainer Phases 323
152. Nature of Combustion Instability 327
153. Longitudinal Mode Combustion Instability 331
154. Transverse Mode of Combustion Instability 332
155. Combustion Behaviour in different Operating Pressures and L* 333

Section I:

Theoretical Rocket Propellant Ballistics

This Section of the book discusses salient philosophies, principles, hypothesis and derivations in rocket design. It defines rocket in a number of ways and energetic are converted to ballistics through smooth mapping of physical, thermodynamic and chemical concepts. The staging of rockets is introduced as an exploration area and performance prediction of rocket by global analysis is presented after introduction of various critical ballistic parameters. This section has total five chapters, moving around themes like introduction to rocket, solid rocket propellants, rocket ballistics, staging and performance prediction. The rocket ballistic and performance prediction of rockets is supplemented with Appendix 1 for numerical values of ballistic parameters. This section also paves way for the second section on performance prediction for propellants with different geometrical shapes. This section contains five chapters.

- Rockets: Classification and Components
- Solid Propellants: Processing and Properties
- Ballistics: Domain and Derivations
- Staging of Rockets: Essence and Examples
- Performance Prediction of Rockets

1

Rockets: Classification and Components

CHAPTER SUMMARY

This chapter introduces rocket as a reaction device, as a propulsive device, as a thermal engine, as an energy conversion device, and as the fastest means of transportation. Various concepts are introduced, explained and illustrated through diagrams and examples. The components of rockets are introduced and based on each of the components, classification of rockets is elucidated. The basis of classification, types of rockets and their comparison are also introduced in this chapter. Overall, complete idea about the rocket is introduced in this chapter. The chapter introduces the following:

- Definition of rocket
- Purpose of rockets
- Propulsion of rockets
- Operation of rockets
- Thermodynamics behind rockets
- Components of rockets
- Classification of rockets

INTRODUCTION

Rocket is the fastest means to transport a payload from launch position to target. It is a *reaction device*, where kinetic energy of outgoing exhaust gases imparts propulsive force. Thermodynamically, basic function of a rocket is *energy transfer* where chemical energy of combustible fuel is converted into kinetic energy of exhaust combustion gases. So any rocket has *fuel*, which is power-house for chemical energy, *a combustion chamber*, which confines combustion gases to generate high pressure (with simultaneous temperature) and *a nozzle*, which converts thermal and pressure energy into kinetic energy of exhaust gases.

Type of payload affects nomenclature of rockets. If payload is a satellite, orbiter or other similar objects meant for projection into space, the rockets are called *launch-vehicles*. If payload is a warhead to destroy various targets, the rocket becomes part of a missile. If rocket uses liquid fuel, it is popularly known as *Engine*, while rocket is a generic term used to denote solid fuel based systems. The word engine is probably taken from energy conversion devices, where fuel storage and combustion chambers are different units. Sometimes, *Motor* is appended to rocket for solid fuel based systems. Although rocket based on solid propellants rarely contain any moving parts, but nomenclature as motor looks to be a misnomer.

For every rocket, two phases of motion exist; one is powered phase, during which rocket gains energy by combustion of fuel and second is non-powered phase, where it loses already gained velocity (in the first phase) due to air drag and other resistances. First phase is called *Boost* phase, while second phase is called *Coast* phase. Interface of boost and coast phase is called all burnt point (ABP) because fuel is completely consumed at this moment. Sometimes rockets are designed on the philosophy of boost-sustain where after booster phase low thrust rocket called sustainer, is operational to maintain velocity attained at ABP.

Propulsion Device

Propulsion is derived from Latin words: *pro* meaning *before or forward* and *pellere* meaning *to drive*. So, *propulsion means to push forward or to drive an object forward*. The forward movement is generally executed by generation of thrust. Propulsion includes all sorts of forward moving devices. By muscle energy bicycle, wheel chair, cart, etc are propelled. Similarly, in broad sense car, truck, bus, rail-bogeys, boats, ships, submarines, aircrafts, helicopters, spacecrafts etc are propelled by various energy sources. Even movement of food inside body is a kind of propulsion, generated by regular periodic smooth contraction of muscles called *peristalsis*.

Rocket is a propulsion device and its features in this perspective are important. *Propulsion is an act of changing motion of an object.* Propulsion in general is synonym of force, which imparts or tends to impart certain acceleration to a body. The acceleration may be observed as enhancement in velocity, reduction in velocity, stopping a moving object or moving a stationary object. Overall, propulsion has many facets, but rocket propulsion is a reaction device, which operates on the Jet Propulsion principles. *Jet propulsion is identified by fast ejection of stored mass from the object to generate thrust.* Jet is high velocity discharge of materials. In motor boats, water enters from one side and is ejected out in rearward direction at high velocity. This gives propulsive force for the forward movement of the motor boat. On similar lines, if jet of material is ejected, reaction force is automatically generated. *In jet propulsion, ejected mass is stored inside body*. In fact, in gravity-free drag-free vacuum environment (in space), throwing an object in one direction results in motion in opposite direction. Rocket operation is one way or the other, similar to this reaction generation principle.

Rocket is propulsion device, where thrust is generated by internal reaction of expanding gases.

Many a times, confusion is created by the word *reaction*. Our feet push ground backward and movement in forward direction is generated by reaction force. If ground fails to provide reaction (slippery or sandy), rather than moving forward, we tend to fall or get stuck. Similarly, a sailor pushes water in backward direction to move the boat in opposite direction. When we jump to shore from a boat, boat is pushed backward by reaction forces. When a gun is fired, bullet moves forward and backward pushes is experienced by gun due to reaction. *As an extrapolation to these, rocket is erroneously thought to be getting thrust because outgoing gases from nozzle, in rearward direction, push atmosphere and as a reaction to this outside atmosphere, rocket is pushed in forward direction*. There is no such reaction force in the rocket. It is *internal reaction*, which is responsible for propulsion of rocket and it is *independent of outside reaction*. The outside atmosphere does not contribute to generation of thrust by rocket and action of rockets is independent of external atmosphere. Once exhaust gases are out of rocket nozzle, they cannot contribute towards thrust generation. So, whether outside atmosphere exist or not is immaterial for the operation of a rocket.

In fact, when Robert Goddard, the first person to practically demonstrate lifting of payload by rocket, proposed operation of rocket in space (vacuum), The New York Times dismissed the notion in an editorial-page feature, on Jan 13, 1920 with the following comment.

"That Professor Goddard, with his 'Chair' in Clark College and the countenancing of the Smithsonian Institution, does not know the relation of action to reaction, and of the need to have something better than a vacuum against which to react – to say that would be absurd. Of course he only seems to lack the knowledge ladled out daily in high schools."

However, 49 years later (24 years after the death of Robert Goddard and during flight of the first moon mission with Apollo 11) on 17 July 1969, The New York Times corrected their mistake with the following statement.

"Further investigation and experimentation have confirmed the findings of Isaac Newton in the 17th century and it is now definitely established that a rocket can function in a vacuum as in an atmosphere. The Times regrets the error."

In fact in the definition of rocket, above, the word '*internal*' is very important and for generation of reaction no outside support is needed by rockets. *It must be noted that outside atmosphere does not pushes rocket in forward direction.*

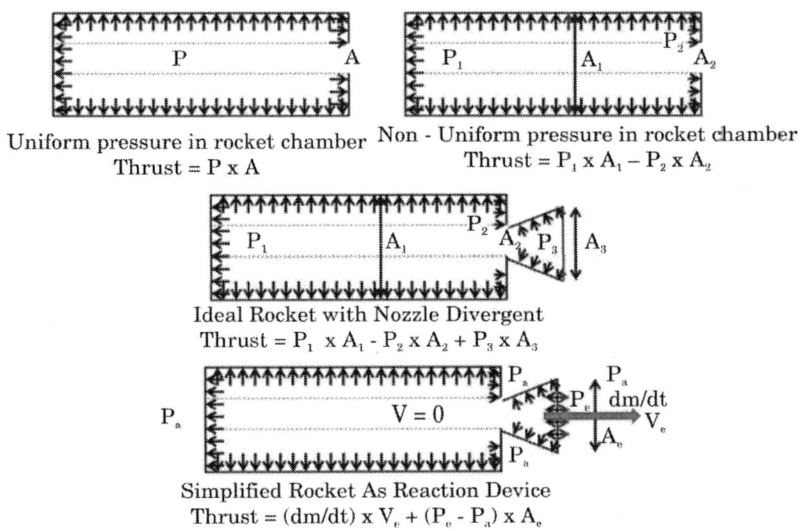

Fig. 1: Thrust generation in a rocket

Basically, exhaust gases are pushed out of rocket in rearward direction, as a reaction to which rocket moves in forward direction. Rather than concentrating on reaction and searching action of the reaction, the operation of rocket can be simply considered as a result of internal pressure imbalance. In case of rocket, which is a solid of revolution with one end closed and other end containing a small vent, pressure generated by combustion of gases is equally distributed. There is no pressure imbalance on the longitudinal surfaces, so lateral forces are not available in rockets. The pressure acting at the side lying directly opposite to the vent remains unbalanced. This unbalanced pressure multiplied by vent area gives (pressure) thrust, which pushes rocket in forward direction.

Additionally, near vent, the pressure levels are lower than homogeneous pressure inside rocket chamber because of conversion of some part of the pressure energy into kinetic energy there. This increases pressure difference and extra (momentum) thrust generated by the rocket. If a divergent section is added after the vent, there is further enhancement in pressure difference and thrust generated by rocket becomes higher further (Fig. 1). The augmentation of thrust over products of pressure difference and vent area is called thrust coefficient (to be discussed in Chapter 3), which is one of the important ballistic parameters in rocket operation.

Rocket is a propulsion device, where thrust is generated by pressure imbalance inside rocket.

Obviously, if pressure imbalance is so important, then treating rocket as a device working on reaction principles needs proper justification. During operation of rocket, propellant or fuel is consumed by combustion inside rocket motor chamber and hot combustion gases (low density) are generated. Because of confinement, high pressure is realized inside rocket motor chamber. However, this generated pressure is converted into kinetic energy of gas in the nozzle due to thermodynamic expansion. The gain in momentum (velocity in the nozzle) by the combustion gases lead to generation of thrust (force). This thrust is said to be occurring due to reaction. As per second law of motion by Newton, *product of mass flow rate and velocity gain gives thrust called momentum thrust.* So, it is desirable in rocket operation to increase mass flow rate as well as outflow velocity of combustion gases from the nozzle, so that higher thrust is obtained. In fact, *thrust generated per unit mass flow rate is treated as one of the important rocket performance parameters, called effective exhaust velocity and is also related to specific impulse* (Chapter 3). In case of rocketry, *momentum thrust surpasses pressure thrust in numerical value*.

The calculation for a rocket performance is derived on the basis of action by a rocket. Thrust produced by rocket is basically sum of actions shown by the rocket. A rocket (nozzle) changes momentum (velocity) of combustion gases (generated in combustion chamber) from zero in the combustion chamber to a value numerically equal to product of rate of mass flow and exhaust velocity. Since direction of both mass flow and flow velocity is along axis of rocket in rearward direction, an equivalent reaction force is developed. Similarly, at the exit plane of the nozzle of the rocket, there is pressure imbalance. Rocket exhaust pushes atmosphere with certain pressure, resulting in some thrust. The product of excess of rocket exhaust pressure over atmospheric pressure and exit plane diameter of the nozzle is a force in outward direction longitudinally, a reaction of which generates pressure thrust. Since reaction corresponding to actions of the rocket generates thrust, it is apt to call it a reaction device.

So rocket is generally explained to be a reaction device working on jet propulsion principle.

The main purpose of the rocket is the generation of thrust to transport payload from launch sites to target. It has capacity to lift objects, overcome drag (in fluid)/friction (in contact with solids) and fill chamber with hot combustion gases. The rockets can be utilized for a variety of purposes. Lifting of objects is utilized during warhead and satellite launches. During launch, it provides high acceleration to the payload. During flight, the drag and frictional forces are countered and rocket continues to accelerate during the powered phase. One of the major applications of rockets is as *thruster*, where it gives small pulses (time/ duration) of thrust. A large thrust is delivered for small duration. This type of devices is used for attitude correction of satellites. In a closed chamber, it can be used as *gas generator*, which has a variety of applications like displacement of themolytic liquid for battery operation, initiation of cartridge actuated devices, and ejection of large missiles from canisters/silos etc. These closed chamber applications in true sense is not a rocket operation but combustion of any high gas producing composition can demonstrate these actions.

> *Rocket is a propulsive device, meant to generate thrust by ejection (at high velocity to get momentum) of stored mass (Jet Propulsion principle) called propellant (fuel), for the quickest transportation of payload (warheads or satellites) from launch point to target.*

Thermal Engine

Operation of rocket is based on combustion of gases and the energy in the form of heat is generated, which is converted into useful work. The useful work is kinetic energy. In the operation of a rocket, it is attempted to get maximum velocity of outgoing gases. Since combustion gases are accelerated at the expense of pressure and temperature, reduction of both these parameters to raise kinetic energy is obvious. However, the pressure should not be reduced to a value lower than outside atmospheric pressure, else it reduces thrust produced. This is contrary to what is explained in previous section that rocket operation is free from external atmosphere. *In fact, if external atmosphere is absent (vacuum), rocket performance improves.* So, rocket propulsion may be treated to generate thrust by undergoing a thermodynamic cycle. A look at the thermodynamic principle of various cycles is necessary to understand rocket propulsion.

Thermodynamic cycle is a series of thermodynamic processes, transferring heat and work, while varying pressure, temperature and other state variables, and returning to initial state. State properties depend on thermodynamic state and cumulative variation of such properties add up to zero. However work and heat are path functions and they depend on the processes adopted. Their cumulative sum gives a non-zero number. However, as per first law of thermodynamics, if system and surrounding returns to initial state at the end of a thermodynamic cycle, heat and work transfer are equivalent. The operational systems based on thermodynamic cycle are used in propulsion of various transport vehicles by combustion of fuel.

Thermodynamic cycle for thrust production (in automobile) has six processes – intake, compression, combustion, expansion, blow down and exhaust. The first process in the cycle is *intake*, where fuel is sucked inside system. Sometimes fuel air mixture is also sucked, while at other occasions only air is sucked inside. Whatever may be the design requirement intake charges the system and makes it ready for doing certain work, later. Since for combustion of fuel, air is also needed, interaction of both must occur at some point of time in the cycle. If only air is sucked in intake stage, fuel will be added at the beginning of combustion. If air-fuel mixture is sucked in, then it is followed by homogenous mixing. If only fuel is sucked, then it is injected in the system waiting for availability of adequate air and temperature for combustion. The second process is *compression*, which is meant to raise temperature and density of the system. Only air is compressed to raise its density, so that proper air-fuel ratio can be maintained. Air fuel mixture is compressed to raise temperature to auto-ignition temperature to bring mixture to flammable limits. During compression system has do some work and work output is negative for this step. Once suitable combustion conditions are established in the system, combustion takes place. *Combustion* may be spontaneous or it may be actuated by some device called spark plug. Combustion is a heat addition step. Combustion is followed by *expansion* process, where work is extracted from the combustion gases. This results in cooling and pressure reduction of working fluid (combustion gases containing both fuel and air). This process is reverse of compression step and is work producing step. The work produced during this step is utilized for powering compression step. After expansion *blow down* occurs, where further extraction of work may take place. This is followed by *exhaust* step, where working fluid is discharge and system is returned to original state to start intake step afresh.

The entire process of thermodynamic cycle is represented by four processes on a thermodynamic plane. The intake and exhaust processes are not relevant and only energy interaction steps are included. The processes include compression, heat addition, expansion and heat rejection. For a simple diesel cycle (proposed by *Rudolf Christian Karl Diesel, France in 1892*), the thermodynamic cycle is shown as Fig. 2. This is valid for all compression ignition cycles, where air is compressed to auto-ignition temperature of fuel and fuel injection result in their immediate combustion. The first process is isentropic compression where air, alone is compressed (process 1-2) to attain high pressure, density and temperature. This is physically realized in a piston cylinder arrangement by inward movement of piston. At the end of compression, fuel is injected in air. This results in combustion giving a constant pressure heat addition process (2-3). This raises temperature of working fluid in the system to higher values at constant pressure, as shown on temperature-entropy plot of the cycle. This is followed by isentropic expansion. The expansion of gases presses piston and does mechanical work by outward movement of piston. Some part of mechanical work is stored for powering next compression stroke also. During this step both pressure and temperature of working fluid reduces. Work is extracted at the cost of these reductions. System is restored to initial state by the process called constant volume heat rejection, which brings both temperature and pressure to normal levels in the cycle. This cycle is repeated at a very fast speed in any automobile to power it.

As far as, rocket engine is concerned, fuel, in the form of solid propellant, is already stored in the system. So, intake stroke is dispensed with. However for liquid propellant based rockets, the intake stroke is executed in same way as done for the specified cycles. But the liquid rocket does not behave as reciprocating engines. They resembles in action with the aircraft propulsion. However, if solid propellant based rocket engine is considered, the process does not have any intake step. When propellant is ignited, heat addition starts and equilibrium pressure is attained. The heat addition process may be assumed to be occurring at constant pressure. In many cases, the heat addition process may be at varying pressures. This combustion process, increases pressure in the beginning and this transient phase may be treated as compression stroke. The process is not an isentropic one and cannot be represented on thermodynamic plane correctly. However, pressure and temperature rise in rocket chamber occurs in milliseconds. Simultaneously, produced combustion gases are expelled out giving a combination of both expansion and heat rejection step. The expansion in nozzle generates momentum thrust. So, rocket engine has only two valid processes, on the thermodynamic plane.

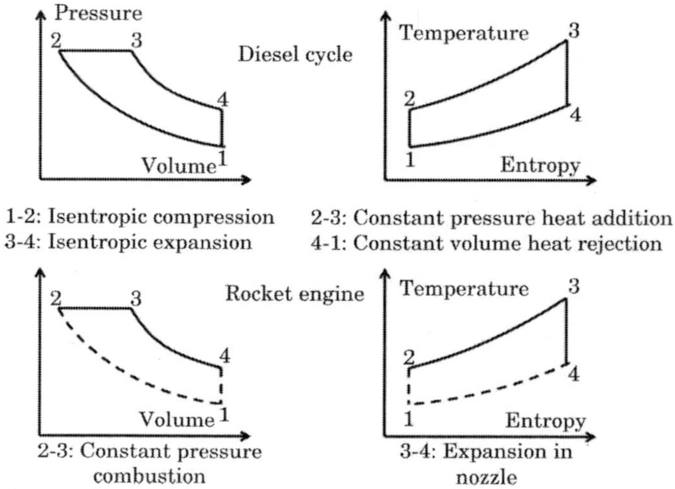

Fig. 2: Thermodynamic cycle of rocket engine

Another significant variation is that once solid propellant is initiated, there is no cycle. After complete consumption of propellant only the system is restored to original state. *This indicates that rocket engine operates as a single stroke thermodynamic cycle*. There is no periodicity of operation. There is continuous generation of work till complete propellant (fuel) is consumed.

Rocket is a single stroke two process (heat addition and expansion) heat engine.

For stable operation of a rocket engine, combustion chamber cum fuel storage space of rocket is pressurized by hot combustion gases. Ideally, the entire chamber till nozzle entrance is assumed to be experiencing same constant pressure (Fig. 3). In the nozzle, expansion reduces the pressure to exit plane pressure. The reduction of pressure in the nozzle imparts kinetic energy to the gases and maintains constant pressure inside rocket motor chamber. The nozzle is responsible for control of pressure inside rocket chamber and higher chamber pressure results in higher mass flow rate, resulting in reduction of chamber pressure. So, combustion chamber is gas producer to increase the pressure where as nozzle reduces pressure, generates thrust by utilizing the energy associated with pressure reduction. Since combustion gases are assumed to be stagnant (zero velocity) inside rocket chamber, the chamber pressure may be taken as stagnation pressure. Practically conversion of pressure to velocity does not start abruptly the moment nozzle inlet plane is identified. But velocity gain by the combustion gases continues in the combustion chamber also and pressure in the combustion chamber at the nozzle end side drops. The rate of drop of pressure in combustion chamber is small, which gets accelerated in the nozzle. The velocity variation is also following similar trend. It increases slowly at the tail end of combustion chamber and increases rapidly in the nozzle of the rocket. The pressure variation in the rocket engine is shown as Fig. 3.

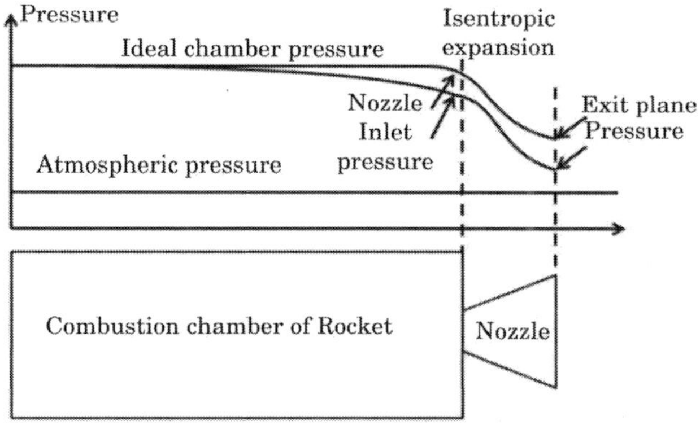

Fig. 3: Pressure variation in rocket engine

It is also worth noting that exit plane pressure from a rocket is more than atmospheric pressure in most of the rockets for smooth operation. Suppose exit plane pressure drops below atmospheric pressure, atmosphere will press inside the nozzle and some part of the exhaust plume will get compressed and rocket exhaust may not follow contours of the rocket, resulting in flow separation at the exit plane. This reduces thrust produced. Most of the rocket operation occurs with exit plane pressure of nozzle higher than atmospheric pressure. This condition is called under-expanded nozzle operation. The reverse situation is called over-expanded nozzle. Both the cases are depicted in Fig. 4. Till exit plane, nozzle guides the flowing stream of combustion gases. Once it comes out of exit plane, the nature changes. If exit plane pressure is equal to atmospheric pressure, the flow continues in streamline direction parallel to original flow velocity. When exit plane pressure is higher, outgoing gases pushes atmosphere and at the exit plane the gases suddenly expands. The outgoing angle with respect to longitudinal axis increases for the combustion gases. Reverse situation occurs for over-expanded condition.

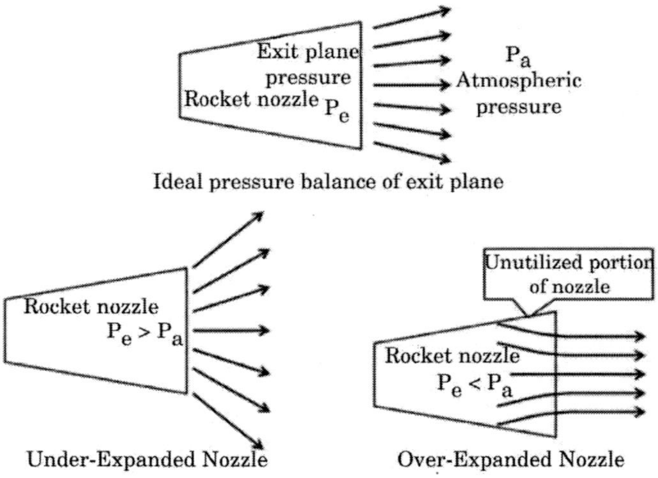

Fig. 4: Pressure imbalance at exit plane of rocket nozzle

Energy Conversion Device

Rocket is basically a thermodynamic system where conversion of energy from one form to the other takes place. The potential energy source in case of chemical propulsion is chemical bond energy of molecules and method to extract this energy for the generation of thrust is the main mechanism of rocket propulsion. The moment energy conversion is stated, comparison with other contemporary devices in terms of efficiency is genuine. It is better to compare rocket with various transport vehicles, where efficiency matters. The comparison may be in terms of power, maximum speed and fuel efficiency.

As far as power is concerned, it is expressed for transport vehicle in terms of horse power (hp). A normal scooter is around 7.5 hp, while a normal motorbike has 20 hp powers. A car exhibits 70 hp, a truck around 200 hp, a Volvo bus around 450 hp. In the heavy power category, railway engine demonstrates a massive 1500 hp to 4500 hp. This is again very meager as compared to aircraft power as 60000 hp. The power from small rocket containing around 100 kg propellant is 20000 (20 thousand) hp. The largest rocket is space shuttle booster, which delivers power of 44000000 (4.4 crore) hp. Clearly in terms of power, rocket engines are power packed systems. They are compact (small size), but very high power producing systems.

Another concern is average speed, which is relevant for any propulsion device. The winners of 100 meters race exhibit average speed of little over 10 m/s. A normal municipal transport vehicle moves at 7-

11 m/s. Speed bike can reach 25 m/s easily, while train cruises at above 50 m/s. Airplane travels at a speed of around 280 m/s and any large rocket travels at an enormous speed of 4500 m/s. Even smaller rocket travels at a speed of around 1000 m/s. This number itself is not sufficient, as escape velocity from the earth is 11400 m/s and rockets have achieved this speed, to transport satellites outside the earth's reach. Definitely maximum speed attained by a rocket engine is very high.

But obviously, the parameter, where rockets fail miserably, is the fuel economy. This feature is so poor that rocket engine as alternative for normal transport cannot be thought off. Truck, bus, car, railway engine etc give a mileage of around 10-15 kilometers per liter. Some scooter claims little above 45 kilometers per liter while some 4-stroke bikes claim 80-100 kilometers per liter. Compared to these systems an average rocket gives a fuel consumption rate of 2-5 liters per kilometers (0.25 - 0.50 kilometers per liter). Nobody will like to invest in rockets for high fuel consumption rate despite its top speed and high power.

In nutshell, speed is high, power is high for rockets but fuel consumption is also very high. For comparison, speed bike having mileage of 20 km/lit moving at a speed of 25 m/sec consume 1.25×10^{-3} lit /sec fuel. Train at 13 km/lit moving at a speed of 50 m/sec consumes 3.84×10^{-3} lit /sec fuel. Rockets at 0.25 km/lit moving at a speed of 4500 m/sec consume 18 lit/sec fuels. It is almost 10000 times higher than normal fuel consumption rates.

Rocket is the fastest and the most powerful means to transport payload from launch point to target at the highest fuel consumption rate.

The rocket operates as an inefficient energy conversion device. If efficiency of combustion and its conversion into useful work is numerically calculated, actual assessment of rocket performance can be obtained. For a typical example, a rocket has 100 kg propellant in 300 kg total system weight and it attains a peak speed of 1000 m/s. 100 kg propellant with calorific value of 1300 cal/g can propel a 200 kg payload to a top speed of 1000 m/s. On energy producing side, the energy imparted to the system = $100 \times 1000 \times 1300 \times 4.187$ J = 5.4431×10^8 J. The output is appearing as kinetic energy of the system. The kinetic energy of the system = $\frac{1}{2} \times 200 \times 1000^2$ J = 1.000×10^8 J. The efficiency of the system is 18.37%. This number is low because normal efficiency of mechanical systems is generally of the order of 30%. The rocket systems, which attains higher speeds of 4500 m/s, consumes so much propellant that efficiency of conversion drops to abysmally lower values. However, the actual end-output is transportation of warhead only and not the total empty hardware. If warhead is weighing 100 kg, then overall efficiency is half the calculated value *i.e.*, around 9%, only.

The power from a rocket is very high. The system described above for the calculation of efficiency may be consuming the propellant in 5-10 seconds. If this much energy is generated for 5 seconds, the power of the rocket can be stated as 20 Megawatt. This is comparable to capacities of some of the power generation plant of the world.

As chemical propulsion is energy limited device and maximum energy available is limited by quantity of fuel carried by the system, alternate means of propulsion is attempted. The non-chemical means include electrical propulsion, magnetic propulsion, solar propulsion, nuclear propulsion, electromagnetic propulsion, vacuum propulsion, tether propulsion, etc. However, for application in defence, rocket propulsion still relies mostly on chemical propulsion. The research on various new molecules is directed by the requirement of higher and higher energy from the breaking of chemical bond. At the same time, stability, safety, hazard and other compatibility and processing issues are also to be addressed. In light of this, the workhorse propellant is composite propellant for chemical rockets, which delivers around 250 kg thrust by mass flow rate of 1 kg/s at chamber to exit pressure ratio of 70. Overall, rocket is a poor energy conversion device. Energy balance in a rocket engine is explained in various ways.

The only source of energy in the rockets is chemical energy, released during combustion. Actually, each chemical has certain heat of formation and combination of molecules result in certain heat of formation of the propellant compositions. This is total energy available, which can be obtained, when all the constituting elements are released in its ground state. However, this is not possible and the energy from combustion of propellant is roughly depicted by *calorific value*. Since propellant carries its own oxygen, the heat released during combustion in absence of any oxygen (air) supply depicts correct energy release. The energy released by combustion of propellant has certain loss due to incomplete combustion of ingredients improper mixing of combustion gases etc. This is denoted by combustion efficiency and is of the order of around 98%. Heat loss to the rocket motor casing, thermal insulation layer and nozzle also consumes certain part of the released energy and a loss of around 5-8% is envisaged there. If proper thermal insulation is applied to restrict heat flow to walls of the rocket, increase in overall output from a rocket is envisaged. Out of the remaining energy, the exhaust gases carry around 40-60% of energy in the form of kinetic energy and heat energy.

This may result in only 30-50% of energy available for propulsion purpose. So, as energy conversion device the rockets may not be very impressive.

The main output of rocket is kinetic energy, which is half the product of mass of outgoing gases and square of velocity of outgoing gases. If velocity of outgoing gases is doubled, kinetic energy becomes four-fold higher. If velocity is tripled, kinetic energy becomes nine times of original. The velocity of outgoing gases at the exit plane of the nozzle is generally supersonic in nature. For supersonic flow through nozzle, velocity is proportional to cross-sectional area of throat. This indirectly indicates dependence of kinetic energy on nozzle throat area. If throat diameter is doubled, outgoing flow velocity of gases becomes 4 times and kinetic energy becomes 16 times. Similarly, if cross-sectional diameter of throat is enlarged to three times, kinetic energy increases by 81 times. The enlargement of throat diameter result in rise of mass exiting the nozzle also and rise in kinetic energy is further enhanced. However, nozzle losses restrict such exorbitant rise in kinetic energy.

Components of Rockets

Rocket is an assembly of components, with end purpose of converting chemical energy of propellants into kinetic energy. The major components of a rocket may be classified under energetic and non-energetic heads. Propellants and igniters form the energetic components of any rocket, while rocket casing, thermal insulation and nozzle form the non-energetic components of a rocket. The energetic components are consumed in combustion and are dispensed with during rocket propulsion. They generate high temperature product of combustion in gaseous form. Igniter gases initiate propellants and help in establishing stable self-sustaining flame front at the free surface of the propellant, while propellant gases generate pressure inside rocket motor chamber due to confinement.

The non-energetic components are used to handle the combustion gases generated by the energetic components. The rocket motor casing gives a closed chamber for the combustion gases. It collects them and restricts their expansion to generate high pressure. The thermal insulation layer protects rocket casing from hot combustion gases and restricts rise of their temperature beyond certain level. Sometimes inhibition layers are placed over propellant surfaces to selectively restrict the propellant surface from combustion and control the rate of combustion gas generation. The nozzle gives a passage for flow of combustion gases and limits abnormal rise in rocket chamber pressure due to confinement of rocket casing. Nozzle governs pressure generated during rocket operation and is also responsible for the accelerated expulsion of combustion gases in rearward direction from the rocket. A labeled sketch of simple rocket is shown in Fig. 5.

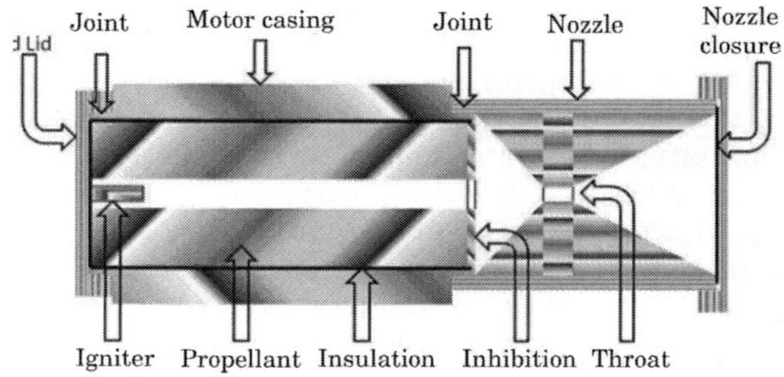

Fig. 5: Labeled schematic of a typical rocket

It is clear from the figure that for the most simple rocket configuration the number of components can be much more than normally assigned.

Classification of Rockets

Rockets are chemical propulsion devices with multiple applications. Hence, they are classified as per different norms. One of the obvious ways to classify them is purpose for which they are utilized. The rocket can be *main motor* used for lifting, transportation and delivery of payloads. Another is *booster* motor, which imparts high acceleration to the system by generating more thrust. The booster rocket motor may be available in the form of additional rocket, which may be functioning along with main motor for limited duration for providing initial high thrust. Such boosters are present in space shuttle in wrap around configuration along with the main liquid propellant based engine. The space shuttle has two wrap-round solid propellant rocket boosters (SRB) to work as booster for providing initial thrust. The boosters may be arranged in tandem with main rocket. Such boosters will be initiated before main rocket to generate high initial thrust, gets exhausted before main motor picks up. Sometimes boosters are provided in internal fashion also by suitable design of rocket propellants and/or nozzles. Solid propellant configurations may be tailored to give dual thrust configurations, where initial high thrust phase constitutes booster phases. Subsequent, low thrust phase forms sustainer. Tailoring of thrust through solid propellant configuration is possible with single propellant or two propellant configurations. Indian missile NAG uses multiple grain dual thrust configuration, while Konkurs and Milan uses single grain dual thrust profiles. Enlargement of throat diameter in-flight during propulsion phase reduces working pressure and thrust. Such arrangements may be utilized for generating booster and sustainer phases. The Indian missile AKASH has such arrangement of booster.

Yet another purpose of rocket is of *sustainer*, where thrust sufficient to counter drag and other retarding forces are only generated by the rocket. No extra thrust is generated to accelerate or change velocity of the system. One more purpose of rocket is auxiliary propulsion. Under this, rockets act as *attitude correction motor, retro-motor, ullage motor, gas generator motors* etc. The *attitude correction motor* may be working simultaneously to correct yaw, pitch or roll. Yaw is nose-left or nose-right movement, pitch is nose-up or nose-down movement and roll is left-up or left-down movement. Small rockets are used in pairs to correct path of the big missiles. Similarly *roll stabilization motors* are also provided in bigger missiles for auxiliary purpose. For separation, rockets are used and after stage separation, the burnt out stage is provided rearward thrust by firing small *retro motors*. Similarly small rockets are used for nose-cone separation of missile front (protective dome). Despite this general classification on the bases of utility of rockets, there are several basis for classification of rockets, which are given in detail in subsequent paragraphs.

First and foremost method of classification of rockets is on the basis of source of energy. Generally, three forms of rockets are conceived from source of energy. The first is *chemical rocket*, which gets energy from disintegration and reassembly of chemical bonds during self-sustained exothermic chemical reaction (oxidation). For such rockets, heat of formation of chemicals used becomes one of the important governing parameter and they are energy limited systems. The usual principal of chemistry including Gibbs free energy minimization, heat balance, heat of reaction, Hess law etc are applicable for thermo-chemical understanding of the heat release process. The heat released by chemical reaction is converted into kinetic energy for subsequent use in propulsion. The second type of rocket is *nuclear rocket*, which gets energy from manipulation of subatomic particles by fission or fusion reaction. Law of mass action, intra-particle interactions and formation of new molecules in the process are responsible for release of tremendous amount of heat in accompanying nuclear processes. Such rockets have very high energy release and the amount of applied thermal insulation restricts the process of tapping and harnessing such tremendous release of energy. Another concern in nuclear rockets is protection from radioactive radiations, which are inadvertently present. Yet another form of rocket on the bases of source of energy may be operating on *solar energy*, where heat energy is directly utilized for the process. In such rockets heat energy is directly available and is stored in fuel cells and other energy storing devices and is released on demand. This device operates on efficient storage of solar energy and their subsequent release on demand. Apart from this, *enthalpy rockets, vacuum rockets, hall thrusters, electro-magnetic propulsion, tether propulsion, electrical propulsion* etc are also being explored for use in rocketry. However, by far chemical propulsion is the most versatile form of energy release process used in rocket propulsion.

The main components (discussed in next section) of any standard chemical rocket are propellant, igniter, casing, nozzle and thermal insulation. Based on each one, rockets are classified. Each component has several possible materials of construction. On the basis of propellants, rocket may be classified in several manners. First is physical state of the propellant. The propellant used in rockets may be *solid, liquid, hybrid or gelled propellants*. Solid propellants may be *heterogeneous* (chemically cross-linked) or *homogenous* (mixed and gelling). *Homogenous solid propellants* has ingredients containing both fuel and oxidizer moieties in single molecule. They are gelled and solidified under elevated temperature. Double base propellants (DBP) based on Nitrocellulose (NC) and nitroglycerin (NG) constitutes homogenous propellants. They can be manufactured by casting (cast double base: CDB) or extrusion (Extruded double base: EDB). *Heterogeneous propellants* need a chemical cross-linking reaction, for solidification. Composite propellants are based on Ammonium perchlorate (Oxidizer), aluminium powder (metal fuel) and binder (Hydroxy terminated polybutadiene: HTPB). They are manufactured by casting route, generally. Combination of double base propellants and composite propellants, *composite modified double base propellants* (CMDB) are conceived. CMDB propellants have ammonium perchlorate (Oxidizer) particles loaded in the matrix of double base propellants (NC and NG). Sometimes nitramines (RDX, HMX) are loaded in double base matrix to give *nitramine propellants*. *Fuel rich propellants (FRP)* are another variety of propellants, used in rockets. They contain more metal powder and fuel components. They are used in air-breathing mode, where burning takes place in presence of sucked air from outside atmosphere. They operate in ramjet mode and deliver high energy. With search for new ingredients, advanced propellants based on new binders, oxidizers, plasticizers and other additive are also manufactured and used. They all can be put under advanced propellants.

The loading of solid propellants in rocket motor chamber may be basis for classification of rockets. Generally rockets are classified as *cartridge-*

loaded or case-bonded. In case of *cartridge loaded propellants*, solid propellants are manufactured separately and qualified. They are assembled in rocket motor casing with the help of grids and are not an integral parts of the rocket motor casing. Like cartridge (bullets) of a gun, they are filled and fired in rocket motor casing. Such propellants have different production line for propellants and rocket motor casing. Contrary to this, *case-bonded solid propellants* are processed inside rocket motor casing and they become inseparable from the rocket casing, later on. Such propellants have integrated production line and full rocket comes out as final product. This sometimes causes problem with defects and entire rocket motor has to be discarded for such processing. Since casing and propellant has different coefficient of thermal expansion, propellant is likely to break incase of temperature change and interface of propellant and casing has to be designed with high accuracy level and tailored mechanical properties. The main difference between cartridge loaded and case-bonded propellants are given in Table 1.

Table 1: Difference between cartridge-loaded and case-bonded propellants for rockets

Basis from differentiation	***Cartridge-loaded propellants***	***Case bonded propellants***
Manufacture of propellant	Separate in molds	Along with casing
Manufacturing process	Casting, pressing or extrusion	Casting alone
Radiography of propellant	Low kV source, as only defect in propellant is to be investigated.	High kV source as penetration of metallic casing needed
Rejection	Only propellant	Manufactured motor
Reuse of casing	Possible with limited rework	Possible with major rework
Casing thickness	High for prolonged exposure of hot gases	Low as exposure to hot gases is less
Utilization of case-volume	Spare space. Grids to hold propellant.	Optimum volume utilization
Mechanical properties of propellant	Low percentage elongation, high tensile strength and modulus.	High percentage elongation, moderate tensile strength and modulus.
Structural integrity	Not critical as propellant is free to expand	Very critical due to restriction on propellants
Thermal insulation	For protection from hot gases alone	For structural integrity also
Grain configuration	Any configuration with suitable inhibition	Radial outward burning or cigarette mode

Another type of propellants used in rockets is in liquid phase. Such rockets are called *liquid propellant rockets*. They operate in same manner as any automobile operates. This contains separate propellant storage tanks and combustion chamber. In the flow passage from storage tank to combustion chamber, the rocket contains flow regulating valves to control mass flow rates. The system becomes complicated, but this results in higher specific impulse. The main difference between, solid and liquid propellant rockets are enumerated in Table 2.

Table 2: Difference between solid and liquid propellant rockets

Basis for differentiation	*Solid propellant rockets*	*Liquid propellant rockets*
Density of propellant	High	Low
Specific impulse	Low	High
Number of parts	Less	More
Construction	Simple	Complicated
Ready to use system	Yes	No
Combustion chamber and tank	Same	Separate
Thrust values	Predetermined	Tailorable
Thrust variation	Limited	Range 10^8
Thrust termination and restart	Difficult	Possible
Effect of storage temperature	Severe	Negligible
Structural integrity	Needed	Not critical
Chance of sloshing, spillage	No	Yes
Chances of accident	More	Less
Condensate deposition on inner surface	Yes	No
Exhaust products	Toxic	Clear
Suitability	Small size	Large size
Utilization of complete propellant	Possible	Not possible
Shelf life of propellants	20 years	Limited

The liquid propellant based systems were used in sounding rockets for meteorological investigations during 1925 to 1975. There were used in *Jet Assisted Take-Off* (JATO) for making heavier aircrafts, airborne at marginal airports. First time JATO was used by Germany in 1935 and the use continued till 1970. Similarly liquid propellants were used for rocket propelled aircrafts, aircraft assisted rocket engines, guided tactical missiles, space launch vehicles, reaction controls and minor airborne maneuvers. The selection of liquid propellants for rockets takes into account economic, performance, physical hazards, corrosion and compatibility factors. The desirable properties of liquid propellants are enlisted below.

- Low freezing point
- High boiling point

- High decomposition temperature
- Lower vapour pressure
- High specific gravity
- Low viscosity
- High specific heat
- High thermal conductivity
- Stability in operating temperature ranges
- Less variation in properties with temperature

The liquid propellants for rockets can be classified in various ways. First is *monopropellant or bipropellant* on availability of oxidizer and fuel in the same molecule. In bipropellants, oxidizer and fuel are stored separately and are not mixed outside combustion chamber. These bipropellants may get ignited by coming in contact with their counterparts in combustion chamber. Such propellants lead to hypergolic ignition. If hypergolic ignition is not possible, then addition source for ignition of propellant is needed. Contrary to this, monopropellant may be a mixture or homogeneous materials, stable at ordinary temperatures. Liquid propellant in oxidizer category includes oxygen, fluorine, nitrogen tetra-oxide, hydrogen peroxide, red fuming nitric acid (RFNA) etc. The fuel components include hydrazine, hydrogen, RP-1, Unsymmetrical DiMethyl Hydrazine (UDMH), MonoMethyl Hydrazine (MMH) etc. Another way of classification is on the basis of boiling point of liquid propellant and classification includes *earth-storable liquid propellants and cryogenic propellants*. RFNA and gasoline are earth storable liquid propellants. They can be stored in sealed tanks for prolonged duration. Cryogenic propellants are liquid gases at very low temperatures. Liquid oxygen (LOX: bp: -183°C) and liquid hydrogen (LOH: bp: -253°C) are examples of cryogen propellants. These classifications include almost all forms of liquid propellants.

To reduce risk of propellant spillage and sloshing, *gelled propellants* are conceived. In this case, liquid propellants are made highly viscous by suitable gelling agents. Selection of gelling agent as well as liquid propellants is critical for such systems. The liquid propellants in this case, are generally thixotropic in nature. Such gelling agents reduce energy of liquid propellants, but they give almost solid like storage ease with liquid propellants. The chances of sloshing and spillage are completely eliminated. On application of pressure, such gelled liquid propellants flow easily through the valves to combustion chamber.

Hybrid propellant rockets are conceived to take advantage of solid and liquid propellant based systems. In this case physical form of propellant is divided into solid and liquid forms. Either liquid oxidizer with solid fuel or solid oxidizer with liquid fuel is used in hybrid propellant based rockets. Most of the systems tried, include solid fuel (rubber, plastic, metals) with liquid oxidizers (Oxygen, fluorine, hydrogen peroxide, nitric acid, nitrogen oxides). The system is slightly less complicated than solid propellants rockets, but has higher specific impulse than solid propellant rockets. The complexity of valve and pumps needed for pumping of liquid ingredients is reduced, as only one ingredient is in liquid form. Since solid part of propellant contains only fuel or oxidizer, the product is insensitive and manufacturing is safe. The system has high reliability and performance is less affected by temperature of the propellant. The cost of propellant is also less than solid propellants. Compared to solid propellants, where propellant combustion takes place in the direction perpendicular to the initial burning surface, in case of hybrid propellants, burning takes place in the direction perpendicular to oxidizer flow direction. The burning process is through macroscopic turbulent diffusion flame, where oxidizer to fuel ratio changes down-stream due to consumption of oxidizer. Presence of any crack in propellant is less severe in case of hybrid propellants than they are, in solid propellants. The propellant has less burning rate and low bulk density. The ignition transience is also sluggish. The combustion efficiency of the system is lower and chances of left-out residue after combustion is possible. In 1970, a target missile based on nitrogen oxide (N_2O_4/NO: liquid) and hydrocarbon (PMMA or plexiglass + Magnesium: Solid) was tested to give 2200 N thrust at throttling range of 8:1. Several tests followed at isolated work-centers all over the globe. However, operational systems with hybrid propellants are not made fully operational yet. Inline with hybrid propellants, Tribid systems for rocket propulsion is also conceived. In one such system, the liquid part may contain a mixture of both fuel and oxidizer (O/H_2 system) impinging on solid fuel (Beryllium loaded Polybutadiene) block.

Ignition means for a chemical rocket is also a basis for classification. There are specific high heats releasing gas/solid producing pyrotechnic compositions, which are used for ignition of rocket propellant. Although energy for initiation of igniter compositions is electrical energy, but the pyrotechnic compositions, whose combustion products impinge on the solid propellant surface for initiation, comes under purview of igniters. The igniter compositions are characterized by their calorific value and gas content. They are solid mixture of metal with an oxidizer manufactured by pressing in tablet shapes. Although manufacturing in the form of dry powder is also preferred for quicker ignition. Gun powder (potassium nitrate + sulphur + charcoal) is probably the oldest pyrotechnic igniter composition. It has calorific value of 730 cal/g and has only 40% gaseous (rest 60% solid) products of combustion. Metal (like Magnesium or Boron) along with potassium nitrate also forms an igniter composition with very high gaseous content (~ 97%) in combustion product. Magnesium based composition has calorific value of 2100 cal/g, and Boron based composition has a value around 1700 cal/g. Another popular combination is Magnesium Teflon and Viton mixture (MTV), which has calorific value of 2200 cal/g and gaseous content of 93% in combustion products. In fact, different pyrotechnic compositions are development as igniter for different propellants.

Rockets can be classified on the basis of igniters as *pyrotechnic ignition or pyrogen ignition*. For smaller rockets, combustion products of igniter composition is sufficient to ignite entire exposed surface of the propellant and establishment of stable self-sustained combustion wave front without any peak pressure. The quantity of igniter is fixed using several variables like igniter must develop a pressure in the free volume, so that low pressure combustion limit (LPCL) of propellant is realized. This is needed for stability of subsequent combustion of propellant. For achievement of initial port pressurization, generally a nozzle closure is designed, which blocks initial flow of gases. They are designed to break at certain pressure, after which smooth one-direction flow of combustion gases takes place through nozzle of the rocket. Since thrust from a rocket is delivered only when flow velocity of combustion gases is established, in practical situations, chamber pressure rises before any appreciable thrust is generated by the rocket. Other parameters are temperature of propellant, gas velocity, propellant type and surrounding gas filled in the propellant port. One of the most important requirements is bringing certain fraction of exposed propellant surface to auto-ignition temperature. This requires sufficient heat content (calorific value) in the igniter composition. Pyrotechnic ignition has very small ignition transience (duration) and it burns very fast, develops pressure and brings propellant surface to auto-ignition temperature. However, when such conditions are applied to larger rockets (large surface area, large empty space), high pressure is developed near the igniter surface leading to excessive burning of propellant in that area. Pressure is not uniform throughout the available free volume in the rocket motor. This may sometimes lead to failure of rockets. Pyrogen igniters are conceived to overcome this problem. Pyrogen igniters are actually another small rocket motors with high burning rate (high mass flux) propellants, placed inside rocket to initiate main propellant piece. A pyrotechnic composition initiates pyrogen rocket, whose combustion products initiate main rocket propellant.

In the rocket motor, position of igniter is also a basis for classification of rockets. Conventionally igniters are generally placed in the head end side (closed end) opposite to the nozzle. On supply of electrical pulse, ignition gases travel the entire surface of propellant, heating the propellant surface and pressurizing the available empty free volume. This is called head end ignition. Another variant may be *nozzle end ignition*, where igniter is placed in the nozzle itself. On initiation, combustion gases execute all the actions enumerated above for nozzle end ignition. Additionally, it has to execute ejection of igniter cleanly before stable flame front is established at the propellant surface and flow of gases across rocket motor is established. The difference between head end ignition and nozzle end ignition is listed in Table 3.

Table 3: Difference between head-end and nozzle-end ignition

Basis of differentiation	*Nozzle end ignition*	*Head end ignition*
Location of igniter	Nozzle end	Head end
Igniter canister during propellant combustion	Detached from rocket	Inside rocket
Quantity of igniter for same rocket	More	Less
Loss of igniter gases	More	Less
Damage of propellant and nozzle by Igniter canister pieces	Not possible	Possible
Chances of combustion instability due to igniter sub-components	Less	More
Attachment of electric supply leads	Easy	Pre-attached
Criticality in design	Pressurization and discharge of igniter sub-components	Pressure kink and igniter pieces hitting propellant surfaces
Gas flow versus flame spread	Opposite	Same direction

Nozzle in rocket engines converts thermal energy into kinetic energy. Based on their construction, shape and orientation, rockets are classified. Nozzles are inert non-energetic component and minimum length or weight is preferred for this component. They are manufactured with three sub-components mainly. From rocket end side, first sub-component is called *convergent section*, which collects the gases generated inside rocket casing by combustion of propellants. This section should be long enough to give a streamline flow to the combustion gases. Mostly it is contoured in ellipsoidal shape. The second sub-component houses the smallest opening for flow of gases. This is called *throat* and is virtually a straight section. Again, length has to be restricted but very small length leads to throat erosion due to flow of gases. The opening diameter of throat section is responsible for pressure inside the combustion chamber. The third section is called *divergent section*, which discharges gases into outside atmosphere. The half-angle of divergent is generally kept around 15^0 to avoid flow separation. Since this section is attached to enhance thrust from rocket, higher than mentioned half-angle leads to flow separation and gain in thrust is not achieved. Such nozzles with three distinct sections is called *De-Laval nozzle*.

The rocket motors are classified as *single nozzle or multi-nozzle motor*. As the name suggests, this method of classification is based on number of nozzles. If for a given condition, single nozzle is replaced with multiple nozzles, the throat diameter for multiple nozzles to maintain same pressure inside rocket chamber is lower. Suppose 7 MPa pressure is maintained in a rocket motor with a single nozzle of 20 mm throat diameter. If same pressure is to be maintained with two nozzles, the throat diameter will be ($20/\sqrt{2}$ mm =) 14.14 mm. As a thumb rule, if single nozzle is replaced with n-nozzles, the throat diameter will be one by n time throat of the single nozzle to maintain same chamber pressure and same flow velocity. This type of adjustments is made to meet certain project constraints. Although pressure inside rocket chamber is equalized, but reduction in throat diameter, gives higher nozzle throat erosion, practically. However, 2 or 4 nozzle rocket motors are prevalent. If more than one nozzle is used, the nozzles may not discharge on rocket axis but at certain offset from the rocket axis to accommodate the nozzles. The additional issue of thrust misalignment has to be addressed in such rocket motors.

It is not number of nozzles alone, but on the bases of orientation of nozzle, with respect to axis of the main rocket motor, rocket can be classified as *straight nozzle and canted nozzle*. It is obvious that multiple nozzles can discharge at certain angle from the rocket axis, so as to accommodate all the nozzles. However, symmetry has to be maintained. Canting of nozzle leads to thrust loss. For auxiliary rockets like roll correction or roll stabilization motors, cant angle of 90^0 is also practically employed. Canting is alone not sufficient for typical cases and to avoid protrusion of divergent section of such canted nozzle out of the rocket surface, the divergent section is truncated to give *scarfed nozzle*. The scarfed nozzle gives lower thrust than other shapes of the nozzles. Rockets with different types of nozzles are given in Fig. 6.

Rocket propulsion can be classified on various basis and the summary of classification is given as Fig. 7. The entire classification procedure is given in text form with description earlier. The major four criterion may be identified for classification. As per source of energy, chemical form of rocket propulsion is more prevalent in the world due to less complexity, less hazard and quick operations, but in advanced systems,

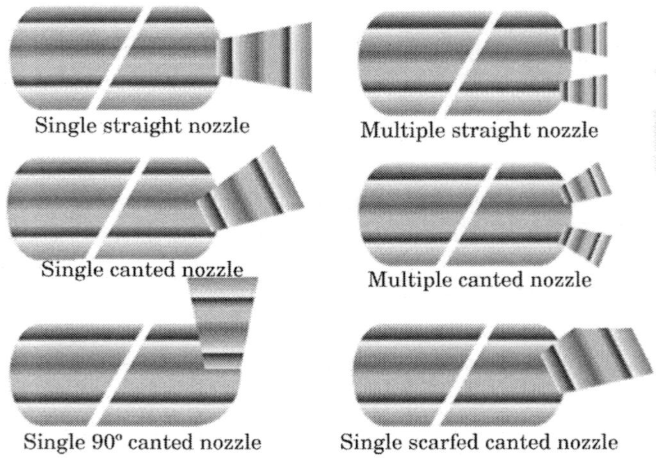

Fig. 6: Different types of nozzles

other forms of rocket propulsion is resorted to. The components of rockets are basis for classification and based on propellants, igniters, nozzles and hardware, rockets can be classified. The stage of rocket is also one basis for classification, where rockets are classified as per their altitude of operation. The design criteria, nozzle area ratio, exit plane pressures are different for different stages. Rockets are employed for many activities other than transportation of payload from launch point to target. Such different actions may be one basis for classification of rockets. Fig. 7 covers all such aspects of classification of rockets in general, with special emphasis on solid propellant based systems.

REVIEW QUESTIONS

1. How is rocket a reaction device?
2. Explain the term 'Propulsion'.
3. What is Jet Propulsion? Why motor-car is not a jet propelled device?
4. Rocket does not push atmosphere to move forward. Support or contradict this claim.
5. Explain pressure imbalance in the rocket for the generation of thrust.
6. What are similarities between a rocket and a heat engine?
7. How is power in a rocket calculated?
8. What is the role of kinetic energy in rocket operation?
9. What are main components of a rocket?
10. What is difference between booster and sustainer rockets?

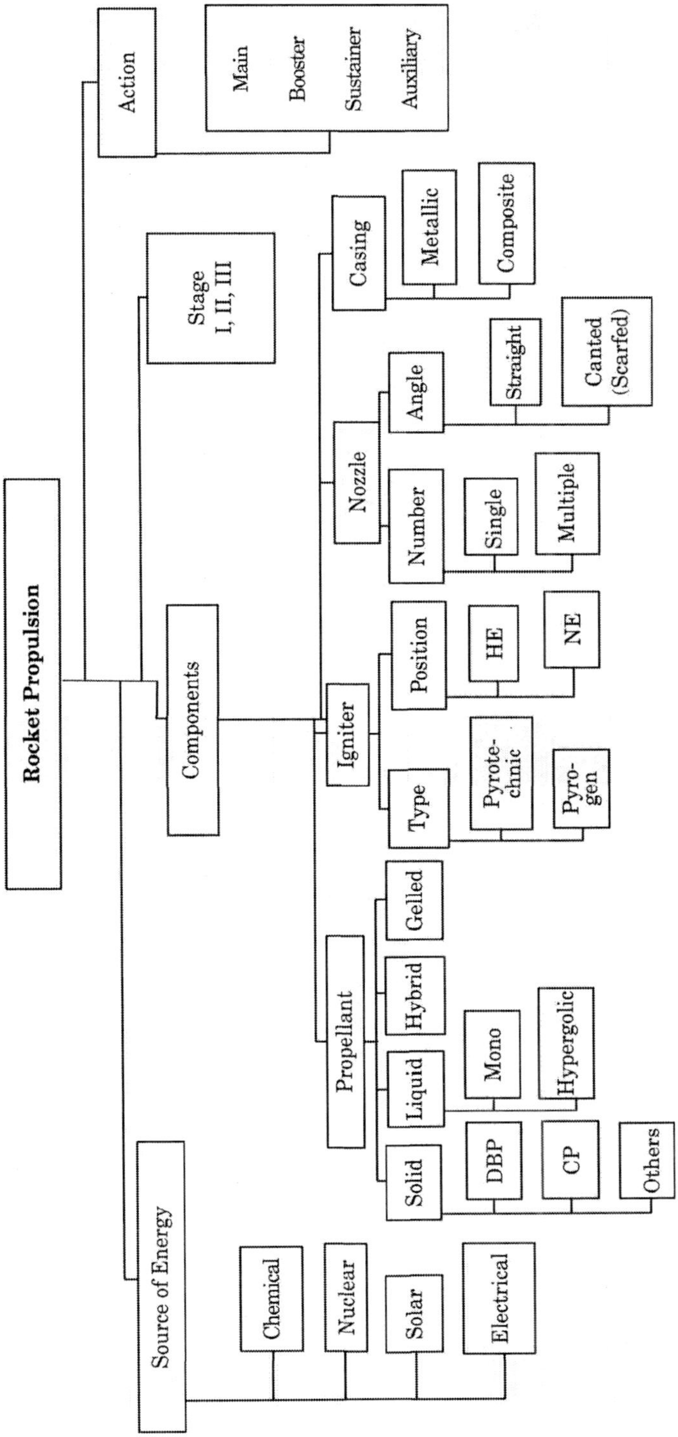

Fig. 7: Complete classification of rockets

11. What are usages of rocket as an auxiliary propulsion device?
12. Compare chemical and nuclear rockets.
13. What are various types of solid propellants for rocket?
14. Compare cartridge-loaded and case-bonded rocket configurations.
15. What are advantages of solid propellant rockets over liquid propellant rockets?
16. What are favorable conditions for use of liquid propellants in a rocket?
17. What are main properties needed for liquid propellants?
18. Why are gelled and hybrid propellant based rockets evolved?
19. What are various pyrotechnic compositions used for the ignition of solid rocket propellants?
20. How is pyrogen igniter different from pyrotechnic igniter?
21. What are advantages of nozzle end ignition over head end ignition in rockets?
22. How nozzle converts thermal energy into kinetic energy?
23. What are various types of rockets of the bases of nozzle configuration?
24. Write short notes on (i) Igniter (ii) Nozzle (iii) gelled Propellants (iv) Hybrid propellants (v) Solar propulsion (vi) Thrust in rocket.
25. Differentiate between the following
    a. Solid Propellant Rockets and Liquid Propellant Rockets
    b. Cartridge-loaded propellant and Case-bonded propellant
    c. Head-end ignition and nozzle-end ignition
    d. Straight nozzle and canted nozzle
    e. Pyrogen igniter and pyrotechnic igniters

2

Solid Propellants: Processing and Properties

CHAPTER SUMMARY

For the performance prediction and ballistic analysis, properties of Solid rocket propellants are needed. As properties are a function of raw materials and processing techniques, the popular varieties of solid rocket propellants, namely Extruded Double Base (EDB) propellants, Cast Double Base (CDB) propellants and Composite Propellants (CP) are introduced in this chapter in more detail. Although other varieties of solid rocket propellants are also in use, but these three varieties supersedes all other varieties, in terms of versatility, bulk production capacity, operational systems and consistent ballistic properties. Following points are illustrated in this chapter:

- Characteristics of Homogeneous propellants
- Characteristics of Heterogeneous propellants
- Raw materials for Homogeneous propellants
- Raw materials for Heterogeneous propellants
- Manufacture of Extruded Double Base (EDB) propellants
- Manufacture of Cast Double Base (CDB) propellants
- Manufacture of Composite propellants (CP)

INTRODUCTION

Solid propellants are used as fuels for chemical rockets for various applications like missiles, launch vehicles, cartridge actuated devices, gas generators, control motors, nose-cap motors, etc. Liquid propellants lead history of rocket propulsion, but ease of operation, simple design, readiness of system and higher density makes solid propellants the most popular choice for rocket propulsion. In this chapter, solid rocket propellants are discussed for their general properties and processing.

Those properties, which affect their performance, are deliberated in more detail. Propellants undergo exothermic combustion reactions to generate high pressure exhaust gases. This requires two main types of ingredients – fuel and oxidizer. On elemental level, fuel (hydrogen, lithium, beryllium, boron, carbon, sodium, magnesium, aluminium, silicon etc) fall in first to fourth columns of the periodic tables, while oxidizers (oxygen, fluorine, chlorine etc) fall in group six to seven of the periodic-table. The fuel and oxidizers are defined for various compounds also. *A compound, which oxidizes or accepts electrons are called oxidizers*. Fuels generally donate electrons. Based on nature of fuel and oxidizer, this chapter is divided into four sections, each discussing one type of propellant. First section discusses homogenous propellants and second section discusses heterogeneous propellants. A final section is added, where discussion about advanced propellant ingredients, their processing and properties are discussed. Propellant manufacturing methodologies and salient properties are elaborated in the chapter.

HOMOGENOUS PROPELLANTS

Solid Propellants for rocket applications invariably contain polymeric macro-molecules. These macromolecules act as binders. *When size of non-homogeneities in the propellant does exceed that of the largest macromolecules, the resulting propellant is called homogeneous propellant*. Pure substances, in which fuel and oxidizer are present in the same molecule, are called BASE for propellants. This has nothing to do with the alkaline nature of the substance. For example Nitro-Glycerin ($C_3H_5N_3O_9$) has both fuel (C, H) and oxidizer (O) moieties in its single molecules. This is one of the main components (BASE) of homogenous propellant. Another main ingredient (BASE) of homogenous propellant is nitro-cellulose (NC), which is obtained by esterification or O-nitration of cellulose [$C_6H_7O_2(OH)_3$]. There is possibility of replacement of any one or more of the three hydroxyl groups (- OH) by nitro (- ONO_2) groups. Most of the homogenous propellants of homogeneous type contain NC and NG as main ingredients. Hence they are called Double (two) Base Propellants (DBP). Although both the bases of DBP (NC and NG) are explosive in nature with velocity of detonation of the order of 6-7 km/s, but in combination and with proper compounding, they undergo slow deflagration and give a burning rate of the order of few mm/s at standard atmospheric conditions and in rocket operating conditions. *These propellants are thermoplastic in nature* and they can be softened on application of high temperature. *NC forms colloidal solution of very high viscosity with NG and hence it is named colloidal propellants*. Since, there are no metallic constituents in any of the main ingredients, homogenous propellants do not produce any visible signature on combustion. Absence of ions, condensed materials (solid or liquid), soot and other opaque agents in combustion products makes flame of these propellants almost transparent and hence *they are called smokeless propellants*. They are employed in tactical missiles/rockets, where launch sites could not be traced from the visible exhaust gas plumes by enemy.

Homogenous propellants have many other synonyms, which are enlisted below. These alternate nomenclatures are given to indicate one of the significant properties of homogenous propellants.

- No non-homogeneity bigger than the sizes of macromolecules gives *homogeneous propellants*.
- *Double base propellants*, containing NC and NG (two main ingredients or bases) are conventional homogeneous propellants.
- Double base propellants are also called *colloidal propellants* as they form colloidal mixture of components.
- Homogenous propellants are also called *smokeless propellants* because their combustion products contain water vapour, oxides of nitrogen and carbon and are *free from* solids, soot, coloured gas, liquids or condensable gases.

Nitroglycerin (NG) is a liquid ingredient of Homogenous propellants. It is a plasticizer and is obtained by nitration of glycerin. Its molecular weight is 227 ($C_3H_5N_3O_9$) and has equivalence ratio (defined as ratio of oxidizing valences to fuel valences) of [valences of (O/C & H) = 9 x 2/ (3x4 + 5x1) =] 18/17. So, this is oxygen rich. Sometimes this is expressed as ratio of deficiency in available oxygen to amount of oxygen needed for complete oxidation of fuel moieties. Three carbon-atoms need 6 oxygen atoms for oxidation to CO_2 and 5 hydrogen-atoms require 2.5 oxygen-atoms for complete oxidation to H_2O. So, net requirement of oxygen for complete oxidation of fuel moieties is 8.5. Available oxygen (9) is 0.5 more than requirement for complete oxidation. So oxygen balance (defined in percentage as molecular weight of extra oxygen to molecular weight of the molecule) of NG is (0.5x16x100/227 =) +3.524%. NG is a colourless transparent oily liquid. It is a very sensitive explosive and needs proper handling system. NG starts boiling at 180°C and decomposes violently at 215-218°C. It is a good solvent and has density of 1600 kg/m^3 at 15°C. For NG, explosion temperature is 3185°C, volume of gas evolved is 715 liter/kg and heat of formation is -85 kcal/mol (or - 85x1000/227 kcal/kg = -374.4 kcal/kg). It was prepared for the first time by Ascanio Sobrero in 1847.

Nitro-cellulose [mono-nitrate, $C_6H_7O_2(OH)_2(ONO_2)_1$; di-nitrate, $C_6H_7O_2(OH)_1(ONO_2)_2$; tri-nitrate, $C_6H_7O_2(ONO_2)_3$] is a solid nitrate ester, where degree of nitration of cellulose determines heat of formation, density, nitrogen content, reactivity and percentage nitrogen in the NC molecules. In fact, NC is not a pure macro-molecule, but it contains many macro-molecules with different degree of nitration (denoted by x in this section). NC is assumed to have 4 molecules and Molecular formula of NC is generally given by $[C_6H_7O_2(OH)_{3-x}(ONO_2)_x=]\ C_6H_{10-x}N_xO_{5+2x}$. Molecular weight of NC is given by 162 + 45x. Oxygen balance of NC is given by 1600x (2.5x - 12)/(162 + 45x). For mononitrate, nitrogen content is (1400/207%=) 6.76%. For di and tri-nitrates, nitrogen contents are (2 x 14 x 100/252% =) 11.11% and (3 x 14 x 100/297% =) 14.14%, respectively. Fully nitrated NC is unstable and it undergoes slow decomposition with reduction in nitrogen percentage (and degree of nitration). In propellants, percentage nitrogen lies between 12.4% to 13.2%. NC is a white, amorphous, fluffy powder for any degree of nitration and has density of around 1660 kg/m^3. If nitrogen percentage is given by P, then formula for calculating x is given by 162P/(1400-45P). If percentage nitrogen in NC used in a propellant is 13.1% then, P = 13.1, then x = 2.618. For this type of NC, molecular weight is 279.8, oxygen balance is (-5.454x1600/279.8% =) -31.18%, explosion temperature is 3100°C, volume of gas evolved is 765 liter/kg and heat of formation is -156.5 kcal/mol (or -156.5 x 1000/279.8 kcal/kg = - 559.3 kcal/kg). NC dissolves in polymeric solvents and acts as binder with non-volatile solvents.

If oxygen balance of homogenous propellant is considered, ratio of NG to NC for balanced composition is (31.18/3.524 =) 8.85:1 (or 89.85% NG). However, propellant with NG mass fraction more than 47% is difficult to process and hence *DBP is invariably fuel rich* in nature. So, NG (oxygen rich part) is less than stoichiometric balanced composition. Any standard DBP contain around 50% NC and 40-45% NG. Generally NG/NC ratio is from 0.7 to 1.1, in propellant formulations.

Other ingredients are also part of DBP composition. They are added into composition to impart certain specific properties like sensitivity, processing ease, stability, opacity, flash reduction, ballistic modification, combustion stability suppression, etc. one of the main prerequisites for such addition is compatibility of added ingredients to the existing ingredients. Although NG is main plasticizer in DBP, sometime inert plasticizer like camphor, centralite, dinitro toluene, triacetin, diethyl phthalate, dibutyl phthalate, tricrysyl phosphate, methyl methacrylate, etc are added as inert plasticizer. The main purposes of these modifications to DBP compositions are reduction in sensitivity of NG, improvement in mechanical properties, burning rate adjustments, processing ease etc. Stabilizers are added to DBP compositions to capture oxides of nitrogen generated by slow reaction (decomposition) of NC and NG. These oxides of nitrogen has tendency to catalyze the decomposition and reduce performance and life of DBP composition. The presence of moisture may enhance the decomposition also. The common stabilizers in DBP are diphenyl amine, orthonitro diphenyl amine, symmetric dimethyl diphenyl urea (centralite), carbazole, diphenyl benzamide, nitronaphthaline, aniline, etc. DBP is translucent and falling radiation (light) can lead to sub-surface heating and decomposition. Darkening agent like carbon black (0.1-0.5%) is added in propellant formulation as darkening agent to make the propellants opaque.

Burning rate modification of DBP is another area of concern, which includes several ingredients in the propellant formulations. Although burning rate of DBP are not affected by ingredients and mostly processing parameters are tailored to give desired burning rates, some transition metal salts are found to change burning rates of DBP. Burning rate of DBP has high dependence on pressure as depicted by high value of burning rate index. Some copper and lead salts are capable of imparting plateau burning (burning rate independent of pressure) and mesa (burning rate reducing with enhancement in temperature) effects. These effects are observed in certain given pressure ranges, only.

DBP has density of around 1600 kg/m^3 and has heat of combustion of 900-1200 cal/g. The combustion temperature of DBP is about 2600 K to 2800 K. The combustion gases of DBP contain (by volume) CO (10-13%), CO_2 (13-25%), H_2O (23-25%), H_2 (5-8%) and N_2 (12-15%). The temperature sensitivity coefficient of burning rate for the double base propellant is between 0-0.3%/°C. DBP are found to be insensitive to spark upto 5 Joules. The friction and impact sensitivity of double base propellants is around 20-28 kg and 30-60 cm. The temperature of spontaneous ignition of propellant on progressive heating is 170-180°C.

Manufacturing of DBP is carried out by two methods (i) Extrusion (ii) Casting. The detail of manufacturing is covered later in this chapter. Extrusion gives homogeneous composition and certain properties are improved through this method of manufacturing. The properties and effect of both the methods on properties of propellant formulation is depicted in Table 4.

Table 4: Comparison of types of double-base propellants

Properties	***Extruded double-base (EDB) propellant***	***Cast double base (CDB) propellant***
Homogeneity	Better	Moderate
Burning rate	High	Low
Specific Impulse (s)	High	Low
Calorimetric value (cal/g)	High (more NG Absorption)	Low
Ultimate Tensile Strength (MPa)	10-14	5-8
Percentage Elongation (%)	30-50	10-20
Manufacturing hazard	Adiabatic compression of entrapped air	Relatively safe
Manufacturing of large web grain	Difficult	No such limitation
Manufacturing non-uniform cross-section	Difficult	No such limitation
Processing time	Less	More
Ease of bulk production	Exists	Relatively less

DBP is translucent in appearance and has smooth surface finish with excellent mechanical properties (high tensile strength, high compressive strength and high modulus). It can be configured in cartridge-loaded configuration and the propellant grains has sufficient strength to display very little sagging. The propellant is non-hygroscopic and machining with water as coolant is possible. Storage in humid atmosphere does not degrade performance of DBP. The thermoplastic nature of polymer gives flexibility to reuse the polymer mix, leading to less wastage.

DBP has lower energy in terms of specific impulse (thrust produced per unit mass flow rate), which is maximum around 200 seconds. The higher strength and lower elongation makes DBP unsuitable for case-bonding. So, maximum utilization of chamber volume is not possible with DBP and additional payloads like grids are also encountered. Double base propellants have higher value of low pressure combustion limit (LPCL). This is minimum pressure at which propellant burning is smooth. LPCL for DBP is around 25-30 bar. Attempts are made to reduce LPCL, so that stable combustion of propellant and predicted performance of rocket with DBP can be practically realized. Because of thermoplastic nature of propellant, it softens at high temperature and looses strength. Contrary to this at low temperatures, it becomes brittle and crack development is possible. As far as, self-life is concerned, DBP, on prolonged storage looses NG by diffusion and by slow chemical reaction with NC. This restricts life of DBP.

HETEROGENEOUS PROPELLANTS

Heterogeneous propellants have heterogeneity of size of ingredients, types of ingredients and also in flame structures. *They are also called composite propellants to signify role of ingredients in the final properties of propellants.* The ingredients of composite propellants are having particle sizes bigger than largest macro-molecules. Each ingredient may be classified as either fuel or oxidizer and presence of both the moieties in single molecule/ingredient of CP is rarely found. In composite materials, solid oxidizer and metallic fuel are dotted in the liquid binder matrix, which is chemically cured by polymerization and cross-linking to give solidified binder. The particles of solid ingredients are present as it is, in the final propellant, also. Ammonium perchlorate (AP) is universally used oxidizer and HTPB (Hydroxy Terminated Poly Butadiene) is workhorse binder for CP. Ammonium nitrate, potassium perchlorate, lithium perchlorate etc are also mentioned as possible oxidizers for CP. Glycidyl Azide Polymer (GAP) is an alternative binder for CP. Metallic fuel are used in composite propellants for enhancement in density, suppression of combustion instability, raising temperature of flame, enhancement of energy and improvement in performance. Aluminium is generally used as metallic fuel in composite propellants, due to matching density, high energy of combustion, non-toxic nature, ease of manufacture and good combustion characteristics.

Chlorate, perchlorate, nitrate, peroxides are commonly used oxidizers and composite propellants also uses one such oxidizer. Compared to metallic oxidizers, ammonium salts are preferred oxidizers to give more and more gaseous product of combustion. Although potassium nitrate (AN) is advocated as eco-friendly oxidizer, with adequate energy content, but it undergoes multiple crystal phase changes in the operational ranges. Specially, at 32°C, it undergoes a phase change with significant change in the crystal density. Such changes may result in generation of stresses and sometimes cracks, in the propellant formulations. Propellant with ammonium nitrate has low burning rate and low specific impulse also and can be at most used for gas generating applications. Another potential oxidizer potassium perchlorate (KP) has metallic component and propellant made out of this ingredient gives very high burning rate exponent, leading to high variations in pressures in actual rocket motor operations. Ammonium perchlorate is used in composite propellants as oxidizer due to following main reasons.

- It is white crystalline powdery free-flowing solid, which can be loaded on binder matrix adequately.

- It undergoes crystal phase transition from orthorhombic to cubic at 240°C, which is beyond normal storage range of propellants on the Earth. So, it does not show any thermally unstable behavior in the operating temperature range.
- It can be commercially produced in plants, without any significant health or environmental hazard.
- It is safe to handle and even grinding of particles to get finer sizes are easily executed.
- It affects burning rate through particle size and burning rate of composite propellants can be tailored by control of particle size, with finer sizes giving higher burning rates.
- It has high oxygen-content (~32%).

However, there are certain limitations of using AP. First and foremost is its hygroscopic nature, which is forwarded to manufactured propellants also. The absorption of moisture leads to formation of lumps, which cannot be dispersed uniformly into binder during propellant manufacturing. AP before use is always dried and stored in oven at elevated temperatures and anti-caking agents (tri-calcium phosphate, preosil, furtflow) are dispersed in AP-powder. The purity of AP is another concern and more that 89-99% purity is difficult to attain. However, more than purity, it is type of impurity, which plays a vital role in performance of composite propellants. With AP, generally potassium and sodium are present as impurities.

Although metals are classified as fuel in the ingredients of composite propellants, but they do not burn like ammonium perchlorate (oxidizer). They get ignited at the surface of the propellant and burn in propellant cavity, thus increasing temperature of the combustion gases. Oxides and fluorides of metals have very high negative heat of formation and metals react with oxygen and fluorine in exothermic manner. From energy point of view, metals suitable for use in composite propellants are aluminium, magnesium, boron, beryllium etc. Magnesium and boron are useful in fuel rich propellants, but in composite propellants, their use is restricted. Boron has problem with ignition and magnesium is too fast to ignite. Energy content with magnesium is lower than that with aluminium. Beryllium gives toxic fumes. Overall, aluminium remains only suitable choice as metal powder ingredient in composite propellants due to following reasons.

- It gives higher energy per unit weight of the metal.
- It is available in abundance at cheaper rates.
- It burns relatively ease with respect to other metals.

- It does not emit any toxic or hazardous fumes on combustion.
- It suppresses combustion instability by damping.

However, use of aluminium is also restricted by various limitations. Aluminium can be loaded effectively up to a level of maximum 30%, only. Higher percentage of aluminium makes propellant slurry very viscous and processing of slurry becomes difficult. Higher percentage of aluminium leads to deposition on nozzle throat and convergent. Higher percentage also leads to reduced performance in terms of specific impulse. More aluminium creates more alumina on combustion and molecular weight of combustion product increases. More aluminium also increases two-phase flow in the rocket chamber combustion gases and performance is further degraded. So, overall, percentage of aluminium in composite propellant is restricted to less than 20%. It emits hot luminous exhaust plume, which can be detected easily and launch point of rockets/missiles could be traced.

Although binders or pre-polymers are discussed in detail in many available literature, but HTPB (Hydroxyl terminated polybutadiene) has established itself as the best binder for composite propellants due to following reasons.

- It is liquid with workable viscosity in operating temperature ranges.
- The viscosity changes with temperature and propellant slurry can be processed easily.
- It can take adequate solid loading (as high as 88%).
- It has functionality of around 2 (number of functional groups present per molecule sufficient to bond with curing agent).
- It has adequate density.
- It imparts good structural integrity and mechanical properties to composite propellants.
- It can be easily cross-linked with different isocyanate curing agents without any release of gases or other by-products.
- It gives high flame temperature and specific impulse composite propellants.
- It is compatible with other ingredients of propellants, like plasticizer, metallic fuel, oxidizer, curing agents etc.
- It has low glass transition temperature (-80°C).

Although binder takes solid loading, but for reduction of viscosity during processing, improvement in elongation of propellant and enhanced pot-life of premixed propellant, plasticizers are added.

Plasticizers are low-viscosity organic liquid added to propellant at about 3-4% level for ease of processing, but at the cost of performance and shelf-life of propellant. Di-octyl adepate (DOA), Di-octyl Phthalate (DOP) etc are some of the popular plasticizers. Plasticizers are uncross-linked liquids and are responsible for migration, evaporation and crack formation in composite propellants.

In composite propellant formulations, several means for better cross-linking of NCO and OH groups are adopted. Pyrogallol, a tri-functional chemical is added for create 3-D network of cross-linking. Isocyanate is added as curing agent. Curing catalysts like Ferric Acetyl Acetonate (FeAA), is also added for mentioned purpose. N-Butane Diol (nBD) and tri-methylol propane (TMP) is also added for better cross-linking in the form of adduct.

Burning rate of propellant is a function of particle size of oxidizers, but there are special ingredients added to composite propellant formulations to change the burning rate. These are called burning rate modifiers. They affect burning rate in two ways (i) by changing burning rate in a given pressure range and (ii) by changing burning rate exponent to alter dependence of burning rate on pressure. These burning rate modifiers are added in very small proportion (2-3%). Those burning rate modifiers, which increases burning rate are called burning rate catalysts. Copper chromite, iron oxide, butacene, ferrocene, catocene etc are popular burning rate catalysts (enhancers). Just opposite to this, if addition of an ingredient reduces burning rate of composite propellants, it is called burning rate suppressant or burning rate poison. Typical burning rate poisons are oxamide, lithium fluoride, nitroguanodine, ammonium nitrate, calcium carbonate etc.

The oxidizer to fuel (binder and metal) ratio for stoichiometrically balanced composition is 9.2:1, but such propellants with so low binder percentage cannot be processed, easily. As more binders are added to CP, the propellant formulations become fuel-rich in nature. In fact, this can be expressed as extent of solid loading, permitted for getting more or less homogeneous mixture of propellant ingredients in the form of liquid slurry of adequate viscosity. Higher quantity of binder lowers solid loading. Composite propellants with solid loading as high as 92% (by weight) are reported, but binder matrix and even other ingredients are changed. Generally composite propellants with solid loading of 85-86% are processed.

The composite propellants have higher energy and practical specific impulse of 240 s is generally reported with 86% solid loading. However, with change of binder and oxidizer, specific impulse, as high as 260s, is also practically demonstrated. Composite propellants exhibit higher density (1760 kg/m^3) than DBP. It gives a flame temperature of 3400 K and the combustion products contain (by volume) H_2O (~35%), HCl (18%), N_2 (10%), CO_2 (6%), CO (6%), Al_2O_3 (5%), H_2 (4%) and Cl (3%). Another advantage of composite propellant is lower value of Low Pressure Combustion Limit (LPCL) than DBP. The LPCL for composite propellant is 12-15 bars and this makes composite propellants a better choice than DBP in low pressure operating consitions of rockets.

Since oxidizer of CP is hygroscopic in nature, the properties are reflected in the properties of propellants, also. Composite propellants are hygroscopic in nature and their performance degrades in humid atmosphere. Processing also gets affected in humid environment and creation of blow-holes is expected in presence of moisture. The combustion gases produce higher temperature and it invariably needs thermal insulation at the inner surface of rocket motor even for small burning duration. The combustion gases are corrosive in nature, they produce smoke in moist environment and they contain greenhouse gases. The high temperature and corrosive nature of combustion gases are responsible for the nozzle erosion. The throat diameter changes during rocket operation and for around 10 seconds of firing, nozzle throat enlargement by around 0.5 mm is practically observed. Throat enlargement leads to reduced pressure and characteristic velocity. The alumina particles of the combustion gases, is liquid due to high temperature but as it passes through nozzle, there are chances of its deposition on the nozzle throat. The alumina also results in two-phase flow thus reducing momentum and heat exchange between gases. This leads to a loss of 2-5% in performance.

Propellant Manufacturing

Homogenous and heterogeneous solid rocket propellants are manufactured through casting and extrusion routes, but there are other classes of propellants, which are excluded deliberately from the text. Fuel rich propellants are one variety of propellant where high metallic loading require processing through pressing route, only. The cross-linked double base propellants, composite modified double base propellants, nitramine propellants, extruded composite propellants are some of the varieties, which have specific use, but they are not very popular for universal applications. So, in manufacturing section, only double base propellants and composite propellants are covered in details.

Manufacture of Extruded Double Base (EDB) Propellants

Double base propellants are manufactured by extrusion and casting routes. The main process step in the manufacturing is gelatinization of NC by NG. Before combination of both the main ingredients, NC is processed separately. The processing of NC may include drying of NC or preparation of NC-NG paste or manufacturing of spheroidal/globular/dense NC. Extrusion can be carried out by solvent or solventless route. Solvent method of manufacturing is preferred for smaller web size propellant. NG makes propellant surface moisture impermeable. This is suitable for composition containing moisture sensitive ingredients. In this method water-drying can be dispensed with.

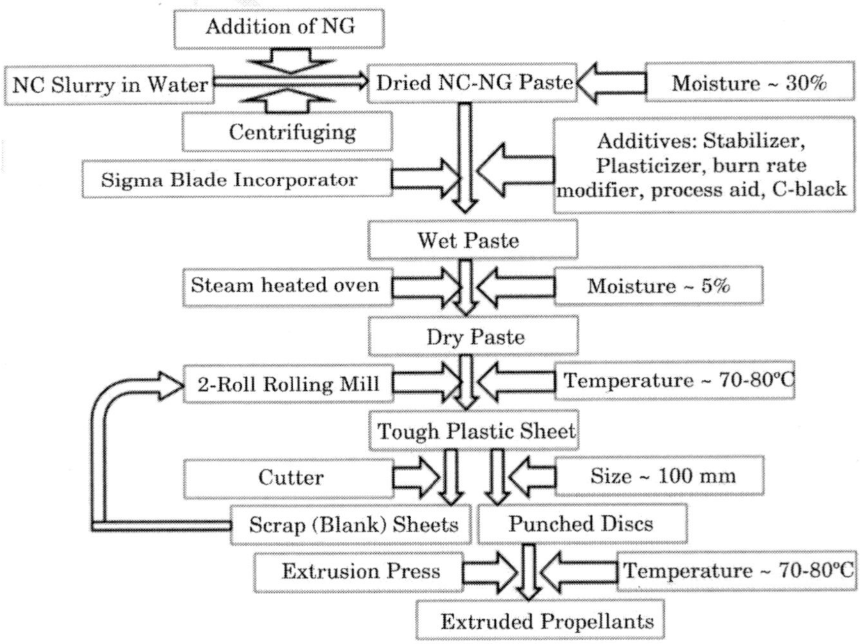

Fig. 8: Process flow chart of manufacturing double base propellant by extrusion

DBP are mostly processed by solventless extrusion process (Fig. 8). Rocket propellants are generally manufactured by this method. The process begins with addition of NG in NC (required nitrogen content: 12.0-12.3%), soaked with water. The slurry thus formed is agitated so that NG is absorbed by fibrous NC. Water percentage is restricted to 30% in the slurry by centrifuging. This is formation of NC-NG paste, which is kept for 3-4 weeks for maturation (absorption of NG in NC). Such NC-NG paste of varying NG content and nitrogen content of NC is dewatered at the time of taking a batch. The mix is taken for kneading in a sigma blade incorporator. During needing stabilizer, plasticizer, burn rate modifiers, carbon-black, process aids etc are incorporated in the paste and mixed for their homogeneous mixing. The paste thus formed is dried to reduce moisture level to 5%. The processed paste is rolled in two-roll rolling mill for gelatinization. The rolling mill has horizontal even speed heated (70-80°C) rollers and paste is given certain minimum number of passes (40-50) for good mixing. The process of rolling has inherent fire hazard and water quenching system is mandatory for safety of such process. When sheet is ready, small discs are punched out of the sheets.

Since paste of DBP can be recycled, the scrap blanked pieces generated after punching the discs are fed again in the roller for next round of rolling. The punched discs are fed to the basket (cylinder) of extrusion press. The stacked discs in basket are pressed through die of predefined size in hot condition to get DBP propellant grain. The manufactured DBP is cooled through defined cycle for proper annealing and reduced stress concentration. The extruded DBP called extruded double base (EDB) propellant is taken for characterization. The characterization may include evaluation of density, burning rate, calorific value, chemical stability, sensitivity, mechanical properties, radiography, ballistic evaluation etc for lot clearance.

Although extrusion process of manufacturing DBP is suitable for bulk production and has homogenous propellant properties, it has following limitations in performance, manufacturing and properties:

- EDB cannot be manufactured in large sizes due to improper annealing, large extrusion press requirements and adiabatic heating during pressing.
- EDB manufacturing process has roll-fire hazard.
- EDB propellants with RDX, AP and other sensitive ingredients have manufacturing issues.
- EDB propellant cannot be manufactured in all shapes and sizes.
- EDB propellants cannot be prepared in bi-composition formulation.
- EDB propellants cannot be made case-bonded except by separate bonding to casing through resin impregnated cloth or other additional aids.

Manufacture of Cast Double Base (CDB) Propellants

In view of above mentioned limitations, another route of manufacturing DBP is preferred, which is casting route. This process is more complex than EDB manufacturing routes, but it takes care of certain limitations of EDB propellants. DBP made by casting route are called cast double base (CDB) propellant. Manufacturing of CDB has two steps – (i) Preparation of Casting Powder and Casting Liquid (ii) Casting and Curing.

Preparation of casting powder starts with the preparation of *casting powder*, in which solid ingredients along with plasticizers and stabilizers are added in an incorporator. Sequence of addition of ingredients, temperature during incorporation and duration of agitation after addition control the final properties of casting powder. This incorporation is carried out in presence of volatile solvents like mixtures of ether and alcohol or acetone and alcohol. Type of solvent, their internal mix-ratio, and quantity for given quantity of NC-NG paste, govern the processing and end properties. Solvent, plasticizer and mechanical action in the incorporator help in gelatinization of NC. After incorporation, thick dough is obtained. This is extruded in two steps – first is pre-pressing for removal of entrapped gases and second is extrusion through multi-perforated die in the form of strands. The strands are dried and cut in 1 mm length. Cut strand is called casting powder. It is sieved and dried (volatile matter < 2%). The casting powder is coated with 0.05% graphite for getting free-flowing conducting casting powder.

Since casting powder is basic raw material for processing of CDB propellant and propellant properties are dictated by this intermediate, characterization of casting powder is very important. The characterization of casting powder includes assessment of density and comparison with theoretical density. Density is direct indication of presence or absence of voids and volatile matters. CDB propellants are manufactured by filling casting powder in a mould and filling it with casting liquid. So, it is not density alone, but screen loading density (SLD), which is packing efficiency of casting powder in a given volume, is also assessed. This is indirect indication of uniformity of grain sizes and extent of graphiting.

Parallel to manufacturing of casting powder, *casting liquid* manufacturing is also undertaken. Casting liquid is a mixture of NG, inert plasticizer (triacetin, diethyl phthalate etc) and stabilizer (carbamite, 2-nitro diphenyl amine etc). Inert plasticizer desensitizes NG and helps in safe processing and handling with NG. Inert plasticizer is maintained in the casting liquid to a level of 18-25% level. The casting liquid is evacuated before use, which help in drying also.

For the preparation of CDB propellant, dried casting powder is filled in moulds containing base plate, central core, top plate etc. Attempt to achieve maximum screen loading density (SLD) is made during mould filling. The mould is evacuated to reduce chances of voids, porosity, pinholes etc. Casting liquid is sucked in the mould through bottom at slow pace for proper deaeration and reduced entrapped air. The filled mould is placed in ovens at elevated temperature for a week for curing and solidification. Curing is proceeded by two days of maturation period, when casting powder absorbs casting liquid at ambient temperature. Once solid homogenous mass of CDB is ensured after curing, mould is disassembled and central core is extracted (remotely). Ends of propellant grains are machined and if applicable, proper inhibition is applied at defined marked surface.

Manufacturing of CDB propellants needs safety and ballistic characterization. Heat sensitivity test through auto ignition and cook-off is carried out. Sensitivity testing for friction, impact and electrostatic discharge is also undertaken. The detonability test to assess exothermic chemical decomposition reaction coupled with shock wave is carried out. For quality control, moisture is responsible for improper gelatinization of NC. Strict control of moisture is mandatory for manufacturing CDB propellants. Since higher nitrogen content in NC prevents proper gelatinization in manufacturing of casting powder, NC with narrow band of nitrogen content is preferred for manufacturing CDB propellant. Lower nitrogen content of NC gives higher burning rate of propellant. It is not nitrogen content of NC, alone, but the alkalinity of NC is to be maintained low (< 0.2%) for proper incorporation of metal oxides and prevention of coagulation of NC.

Manufacture of Composite Propellants (CP)

Manufacturing of composite propellant is basically consisting of ingredient preparation, mixing of liquid ingredients, incorporation of solid ingredients, mixing of curing agent, vacuum casting, elevated temperature curing, decoring and demoulding, followed by characterization. The propellant manufacture revolves around special type of mixer. Generally two types of mixers are used for composite propellant manufacture (i) sigma blade mixer and (ii) Vertical Planetary mixer. A typical propellant formulation containing AP/Al/HTPB is described to illustrate manufacture of composite propellant. The complete cycle is shown in the Fig. 9.

Ingredient preparation for composite propellant manufacture is an important step. It is found that characterization of each of the ingredients for purity, viscosity, density, thermal properties, chemical properties, particle size, particle shape etc is needed for consistent end product. Usually, ammonium perchlorate is available in particle sizes of

200 m and 300 m from trade. Since burning rate is a strong function of particle size of oxidizer, obtained oxidizer from trade is ground to narrow particle size distribution by mechanical grinding. Ammonium perchlorate, in particle size range of 4 m to 300 m, is used in composite propellants. As particle size of ammonium perchlorate is lowered, tendency to agglomerate increases and suitable anti-caking agent must be incorporate, to store AP for longer duration. Otherwise, AP is to be ground just before use in manufacture of composite propellant. AP is hygroscopic in nature and it must be stored at elevated temperature to take care of moisture ingress. Mostly, AP is stored in tray-dryers at elevated temperatures. Aluminium powder is generally available in particle sizes of 15-25 m and since particle size is not very critical, only purity is checked for Al-Powder. For manufacturing propellant batch, the solid loading, NCO/OH ratio, viscosity, burning rate, etc dictates propellant composition, where percentage of each ingredient is fixed by weight. All ingredients must be weighed as per the schedule for pre-predicted results.

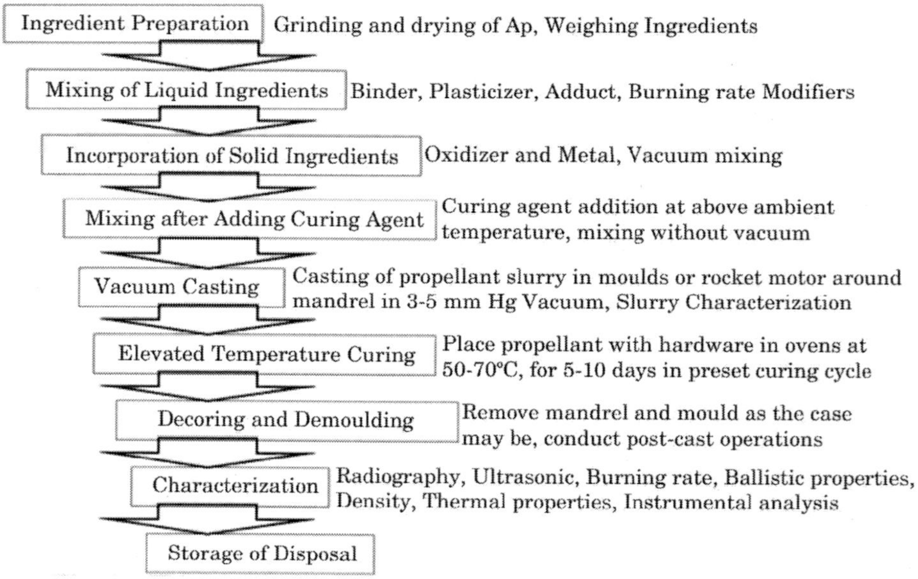

Fig. 9: Flow chart for manufacture of composite propellant by casting

Once all qualified raw materials, in properly weighed quantity are made available, mixing of ingredient is started. First of all binder, plasticizer and all liquid ingredients, except curing agent are added to mixer and agitated to get homogenous mix. The solid ingredients, which are to be incorporated in small amount, are also added during this mixing for their proper dispersion. Such ingredients include ballistic modifiers, mainly. Vacuum is also applied to a limited level for removing any entrapped air. After this, solid ingredients (AP and Al) are added in installments. Addition of solid ingredients needs proper care and absence of lumped mass (agglomerates, foreign materials) must be ensured before any addition. Vacuum is applied during solid ingredient incorporation. Application of vacuum removes entrapped gases and reduces viscosity of the mix, also. Propellant slurry temperature is raised to around 40°C by circulation of hot water around the mixer bowl to reduce viscosity of the propellant slurry further. At this temperature, curing agent is added to the propellant slurry and mixed for homogenous dispersion. No vacuum is applied during this operation, as it may lead to evaporation of curing agent. Curing agent addition, starts cross-linking reaction and viscosity starts building–up slowly. After 20-30 minutes of mixing with curing agent, propellant slurry is ready for casting operation. The propellant slurry sample is assessed for viscosity build-up at different time intervals to know the pot life of the slurry (time during which casting is to be completed).

Casting fixtures, which may include mould, rocket motor, mandrel, distributor, base plate, top cone etc. are assembled and placed in a vacuum casting chamber. Vacuum of the order of 3-5 mm Hg is applied. Propellant slurry is brought to the casting station and casting at the rate of around 10-20 kg/min is executed depending on flow area available and viscosity of the propellant slurry. Once casting is over, propellant is left for settling under vacuum for some time and then vacuum is released lowly. Propellant slurry along with casting fixtures is transferred to curing oven, where propellant is solidified at elevated temperature by curing reaction. The curing cycle depends on web size of the rocket motor, type of propellant slurry, length of rocket motor, type of rocket motor casing and mandrel, heat transfer pattern, thermal conductivity of the propellant slurry, etc. After curing cycle is complete, propellant is solidified and suitable cooling cycle is also devised to reduce generation of thermal stresses in the solidified propellant. After this, decoring and demolding operation is operation is carried out, as the case may be. End trimming, cleaning, finishing operations are also executed.

Parallel to each main casting, many auxiliary casting pieces are also kept and cured for characterization of propellant. Propellant is characterized for its burning rate, density, auto-ignition temperature, thermal conductivity, elemental analysis, etc. Radiography and ultrasonic testing are non-destructive methods employed to assess porosity, voids, cracks or unwanted mid-web anomalies. If propellant is cleared from all the characterization, it is sent for either static/dynamic/ flight trials or sent for storage. If propellant fails in any of the tests, it is sent for disposal, mostly by burning.

Although other varieties of propellants also exist, but these two classes of solid rocket propellants are mostly used in missiles and rocket applications. So, their manufacturing techniques and salient properties of raw materials as well as end products are illustrated in this chapter. They have practical relevance for rocketry. This, by no means curtails importance of other classes of propellants, but for the general understanding, the propellant manufacturing is explained, to use the features for rocket ballistics.

REVIEW QUESTIONS

1. How are ingredients of solid propellants for rockets classified as fuel and oxidizer?
2. Why double base propellants are also called (i) Homogeneous propellants (ii) Colloidal propellants (iii) Smokeless propellants?
3. What are general properties of double base propellants?
4. Compare general properties of double base propellants and composite propellants.
5. Write short notes on (i) Nitrocellulose (NC) (ii) Nitroglycerin (NG) (iii) Ammonium perchlorate (AP) (iv) Metallic Fuel (v) Hydroxy Terminated Poly Butadiene (HTPB) (vi) Burning rate modifier (vii) Opacifier (viii) Mechanical properties of propellants.
6. Why Double base propellants or composite propellants are fuel-rich in nature?
7. What are limitations in getting stoichiometric-balanced Double base propellants or Composite propellants?
8. What are minor ingredients (<10% by weight) in Double Base Propellants (DBP)?
9. Differentiate between Extruded Double Base (EDB) propellants and Cast Double Base (CDB) propellants.
10. Compare properties of Double Base propellants (DBP) and Composite propellants.
11. What are main ingredients of composite propellants (CP)?
12. What are heterogeneous propellants called composite propellants (CP)?
13. Why is Ammonium perchlorate (AP) used as universal oxidizer for composite propellants (CP)?

14. What are properties of Composite propellants (CP) with Ammonium nitrate (AN)?
15. What are limitations of using Ammonium perchlorate (AP) in composite propellants (CP)?
16. What are utilities of metallic fuel in composite propellants (CP)?
17. Why is Aluminium (Al) preferred as metallic fuel in composite propellants (CP)?
18. What are limitations of using Aluminium in Composite propellants (CP)?
19. Why is Hydroxy Terminated Poly Butadiene (HTPB), a work-horse binder (pre-polymer) for composite propellants (CP)?
20. What are additives (<5% by weight) to Composite propellants (CP)?
21. What are general properties of composite propellants (CP)?
22. What are limitations of using composite propellants (CP)?
23. Explain solventless process for extrusion of Double base propellants (DBP)?
24. What are limitations of producing Double base propellants (DBP) by extrusion?
25. Prepare flow-chart of extruded double base propellants (DBP)?
26. Explain manufacturing of double base propellants (DBP) by casting route?
27. Give flow-chart for the manufacture of composite propellants (CP) by casting?
28. What are case-preparation activities in composite propellants (CP)?
29. What are raw-material preparation activities for composite propellants (CP)?
30. What is difference in solidification process of double base propellants (DBP) and composite propellants (CP)?

3

Internal Ballistics: Domain and Derivations

CHAPTER SUMMARY

This chapter is heart of theoretical aspects of rocket ballistic. This starts with combustion gas properties and discusses ratio of specific heat, flame temperature, method of evolution of performance parameters, characteristic velocity and specific impulse. Nozzle theory is explained and important mathematical expressions are derived with adequate pictorial representations. Graphs for various ballistic parameters are given in this chapter and numerical values can be derived from tables given in Appendix – 2. The chapter discusses the following:

- Ratio of Specific heat of combustion gases
- Characteristics of flow through nozzle
- Shape and Area ratio of nozzle
- Variation of pressure, temperature, density and velocity in nozzle
- Calculation and expression for Thrust coefficient
- Thermo-chemical aspects of combustion
- Derivation of expression for characteristic velocity
- Specific impulse as overall performance parameter of rocket

INTRODUCTION

Combustion of propellant inside rocket motor chamber generates exhaust product of low mean molecular weight at high temperature, which is expelled at high velocity through the nozzle of rocket in rearward direction. The properties of combustion gases from propellants are important ballistic parameters. This includes ratio of specific heats, flame temperature and combustion gas constitution.

RATIO OF SPECIFIC HEAT

Specific heat of gaseous product of combustion is expressed under two conditions because of their compressibility. Specific heat is basically a measure of temperature rise for given heat input per unit mass of a given material. For constant heat input, given mass of material with high specific heat, result in low temperature rise. Reverse is true for material with low specific heat. Alternately, for a given rise in temperature, constant mass of a material with high specific heat needs more heat input. For solid and liquid, specific heat or heat capacity has single definition and numerical value. For example, water has specific heat of 4.1813 kJ/kg.K and steel has specific heat of 0.466 kJ/kg.K.

However, for gases, specific heat is defined under constant pressure and constant volume conditions and they are numerically different. Specific heat at constant pressure has higher numerical value, because at constant pressure, for compressible gases, the heat is needed for two simultaneous operations – heat gas at constant volume and second increasing volume of the gas to maintain pressure (expansion). Since higher heat is needed for constant pressure operation, high specific heat is realized for constant pressure condition. The difference in specific heat is equal to gas constant of the gas and their ratio is expressed by Greek letter Gamma (g) for ideal gases. It is also called *heat capacity ratio and adiabatic index*. It is basically *isentropic expansion factor*. Thermodynamically, it is also equal to ratio of enthalpy and internal energy of ideal gas.

The value of gamma is related to degree of freedom available in the molecule (f) and is defined as [1+ (2/f)]. Mono-atomic gases have 3 degree of freedom (all translational). They have 'γ' value of 1.67 (= 5/3). For diatomic gases, the value reduces to 1.333 (=4/3), because they have 5 degrees of freedom (3 translational + 2 rotational). At high temperatures, vibration degrees of freedoms are also included and the value of ratio of specific heat reduces further at high temperature. For more complex molecules, the value is further reduced. The lower limit of gamma is 1, when the gas behaves as incompressible material. General for atmospheric air at normal pressure and temperature, value of gamma is around 1.4. However, for the combustion gases, containing traces of fuel, the value of gamma is reduced. The reduction is dependent on molecular weight of combustion gases and their relative compressibility.

For the practical determination of ratio of specific heat of air, air at given pressure and temperature is expanded adiabatically, followed by isochoric compression to initial temperature. Initial pressure, pressure after adiabatic expansion and final pressure are denoted by P_1, P_0 and P_2. Initial temperature and temperature after adiabatic expansion are T_o and T_1 respectively, then expression for ratio of specific heat is given below.

$$\gamma = \ln (P_1/P_0) / \ln (P_1/P_2)$$

However, for such an experiment, the gas composition and condition should be invariant. Similar situation is difficult to get for combustion gases of rocket propellants. Hence, practical determination of ratio of specific heat is difficult.

As far as, solid rocket propellants are concerned, the value of gamma is dependent on composition, pressure and temperature. For a typical double base propellant at low temperature (less than 2000 K), gamma reduces with rise in temperature. Reduction of gamma may be from a level of 1.30 to around 1.205 at all pressures from 0.1 MPa to 60 MPa. The value of gamma increases thereafter with temperature. Rate of rise of gamma with temperature is higher at low pressures. *This indicates that value of gamma is retained at lower values for more range of temperatures at higher pressures. Generally, operating pressures in rocket motor varies from 3 MPa to 30 MPa and temperature of combustion gases exceeds 2000 K.* The value of gamma is a function of molecular weight or composition of combustion gases also. Since combustion gases of double base propellants contain carbon dioxide, carbon mono-oxide, water, nitrogen, hydrogen, oxygen, as major ingredient, the specific heat at constant pressure depends on their respective molecular weights and specific heats. As far as, molecular weights are concerned, the sequence of decreasing order of molecular weight is: - carbon dioxide (44), oxygen (32), carbon mono-oxide (28), nitrogen (28), water (18), hydrogen (2). However, ratio of specific heat of respective ingredients (at 20°C) has a different sequence: hydrogen (1.410), nitrogen (1.404), carbon mono-oxide (1.400), water (1.330), carbon dioxide (1.300). It is worth noting that despite having wide variation in molecular weights, the ratio of specific heat for nitrogen and hydrogen are comparable. This is because, both are diatomic gases. In fact for another diatomic gas, oxygen (32), also ratio of specific heat is 1.400. Ratio of specific heat of air is generally taken as 1.400. *The value of gamma reduces with increase in temperature. The value of ratio of specific heat also reduces with increase in operating chamber pressure of the rocket.* The rate of change with respect to pressure is high in the low pressure zones (below 3-5 MPa).

Value of gamma is taken as 1.21 for double base propellants, and 1.19 for AP/Al/HTPB propellants.

As far expressions for these parameters are concerned, following two relations are worth consideration, which leads to an expression of specific heat at constant pressure in terms of 'γ' and 'R'. In this case, 'R' is ratio of universal gas constant and mean molecular weight of combustion gases of propellants.

$$\text{Eqn 1: } C_p - C_v = R$$
$$\text{Eqn 2: } C_p/C_v = \gamma$$
$$\text{Eqn 3: } C_p = \gamma R/(\gamma - 1)$$

Flow through Rocket Nozzle

In a rocket operation, propellant is burnt in the combustion chamber to generate large amount of low molecular weight combustion products at elevated temperature. These combustion gases need an exit point to prevent regular monotonic rise in pressure due to accumulation of combustion gases in the rocket motor chamber. Rocket nozzle is the vent to restrict pressure rise.

Nozzle is a smooth passage of varying cross-section, where enthalpy of gas is converted into kinetic energy.

The above definition of nozzle does not states any restriction on the shape of the nozzle, but velocity of flow is continuously increasing during flow through the nozzle. Since first law of thermodynamic states that energy of an isolated system is fixed and it can only change form. There must be some form of energy, which is disappearing to give rise in the velocity (kinetic energy). Since flow through nozzle of a rocket has reduction of pressure and temperature both, these energies augment kinetic energy and increase flow velocity.

The pressure inside rocket chamber is basically dependent on equality of mass flow rate by two counteracting phenomena – (i) generation of hot gases by combustion of propellant, which leads to enhancement in pressure by injecting more mass in the empty volume and (ii) Discharge of combustion gases through nozzle. So, throat is responsible for generation and control of pressure, which leads to generation of thrust.

Thermodynamically, nozzle, located at the end of combustion chamber of a rocket, is responsible for exhaust of hot combustion gases. In a nozzle, enthalpy of gases is converted into velocity, in isentropic manner. While flow through a nozzle, pressure and temperature reduce and velocity of gases rises. Generally, it is assumed that incoming velocity at the nozzle inlet is zero or negligible (as it is very small as compared to exhaust velocity and calculation of exhaust velocity is not affected significantly by the inlet velocity of gases for the nozzle). *Nozzle can be defined as a flow passage for the combustion gases with varying cross-section, where gases are accelerated at the cost of enthalpy*. Flow through a nozzle is mathematically expressed under certain assumptions.

Assumptions for flow through nozzle in rockets have three components – propellant, combustion gases and nozzle. The assumptions are centered on these components. As far as propellants are concerned, it should be maintained at room temperature and temperature effects are incorporated in the developed mathematical formulation suitably. The burning rate of propellant is smooth in the operating pressure range. The combustion pattern of propellant is assumed steady and composition is assumed to be homogenous. Propellant burning surface is assumed to recede in the direction perpendicular to the initial surface. The propellant in rocket chamber burns to create combustion gases. During the conversion process, it is assumed that chemical equilibrium exists. This leads to continuous change in the chemical composition of the combustion gases. However, for ease of calculation, it is assumed that flow through nozzle is frozen flow and there is no change in the composition during flow through nozzle. This indirectly indicates that chemical and thermodynamic properties of combustion gases remain same during flow through the nozzle.

There are several assumptions made about the working fluid, which are hot combustion gases in case of rocket propulsion. The gases do not contain any condensed phase and are homogenous. There is no two-phase flow through nozzle and entire mass flows at same average cross-sectional speed. However, velocity changes in longitudinal direction. The properties of combustion gases do not change with temperature or pressure change. For mathematical ease, combustion gases are assumed to obey perfect gas laws. The combustion gases are assumed to have negligible velocity at the nozzle entrance.

The nozzle internal surface is assumed to be smooth and there is no surface heating due to friction. The formation of shock wave is excluded from normal calculations. The expansion process through a nozzle is assumed to be isentropic and walls of the nozzle are assumed to be insulated (no heat transfer). This ensures maximum conversion of heat (from combustion of propellant) into kinetic energy. The flow passage of nozzle is assumed to be free from any restrictions like boundary layers, eddies etc. The flow is always assumed to be under chocked condition and most of the time, the rocket nozzle has convergent-divergent shape. Flow through nozzle is assumed to be predominantly in axial direction. Under above mentioned assumptions, the flow through nozzle can be analyzed as simple quasi-one-dimensional flow. Applicability of conservation of mass, energy and momentum is valid.

For the derivation of various ballistic equations for performance prediction of rocket propellant, thermodynamic relations for perfect gases are utilized. Universal gas constant is a significant constant for perfect gases and this constant divided by molecular weight of gases is used for the derivation of ideal gas equation. The ratio of universal gas constant and mean molecular weight of combustion gases is denoted by 'R' and used in the equation. The ideal gas equation, which correlates pressure (p) with density (ρ) and temperature (T) of gases, is given as equation 4.

$$\text{Eqn 4: } P = \rho \times R \times T$$

Since flow through nozzle is assumed as frozen flow, molecular weight of combustion gas does not change during flow through the nozzle. This gives constant value of term 'R' in the Equation 4. Flow through the nozzle is also assumed to be isentropic and value of 'γ' (discussed in previous section) for the combustion gas is a constant. The isentropic expansion of combustion gas obeys Equation 5.

$$\text{Eqn 5: } P/\rho^{\gamma} = \text{constant}$$

Flow through nozzle is connected with velocity and sonic velocity (C) is assumed as reference for velocity calculation. The velocity of sound through a medium is dependent on temperature (T) and ratio of specific heat (γ). The expression is given as Equation 6. Also included in the expression is the term Mach Number (M), which is ratio of flow velocity (V) and sonic velocity (C).

$$\text{Eqn 6: } C = \sqrt{(\gamma \times R \times T)} \ \& \ M = V/C$$

Mass conservation equation for flow of gases through nozzle is expressed in terms of constant mass flow rate and is given by Equation 7. Mass flow rate (dm/dt) is derived as product of density (ρ) and volume flow rate. Volume flow rate is expressed as product of cross-sectional area (A) and flow velocity (V). Momentum conservation equation is given by Equation 8 and it contains contribution of pressure and velocity mainly. The energy equation is given by Equation 9, which is nothing but mathematical expression for the conversion of enthalpy of gas into kinetic energy.

$$\text{Eqn 7: } (dm/dt) = \rho \times A \times V = \text{constant}$$

$$\text{Eqn 8: } P + \rho V^2/2 = \text{constant}$$

$$\text{Eqn 9: } h + V^2/2 = C_pT + V^2/2 = \text{constant}$$

If nozzle, as flow passage has zero pressure difference across entrance and exit, there will not be any flow and the kinetic energy of outgoing gases will be zero. If pressure is reduced at the nozzle outlet, some velocity of flow is gained. As pressure difference is increased by reducing outlet pressure, flow velocity continues to rise. This happens because upstream gases are signaled by reduction in outlet pressure (increase in pressure difference) to increase the mass flow rate. The transfer of such signal occurs at the sonic velocity. As flow velocity increases continuously, at certain point of time, it exceeds sonic velocity. When sonic velocity is achieved by the flow velocity (M = 1), further rise in mass flow rate (flow velocity) cannot be achieved by reduction in outlet pressure. The upstream is unable to sense further pressure reduction and increment in mass flow rate is not observed. So, maximum mass flow rate is achieved, when flow velocity attains sonic velocity within the nozzle. This is depicted in Fig. 10. The flow accelerates to supersonic speeds (M > 1) downstream, but mass flow rate remain constant thereafter. In supersonic region, velocity rises, flow area also rises, but mass flow rate remain same. This is because of corresponding high reduction in density (refer Fig. 12) of the combustion gases in this section, which balances rise due to velocity and area and gives constant mass flow rate. However, it is clear that flow through nozzle may be sub-sonic (M < 1), sonic or supersonic (M > 1), during movement through the nozzle.

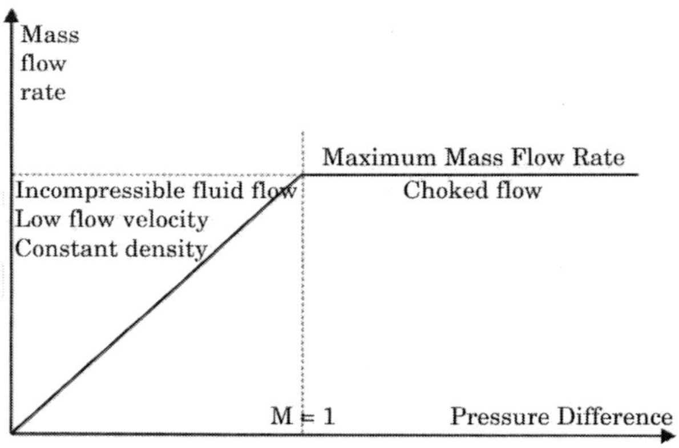

Fig. 10: Variation of mass flow rate with pressure difference

For determining nature of shape of nozzle for isentropic flow, differential form of mass conservation and momentum conservation equation along with perfect gas equation is used.

Mass conservation equation (Eqn7) gives $d\rho/\rho + dA/A + dV/V = 0$

Momentum conservation equation (Eqn8) gives $dP/\rho + VdV = 0$

Isentropic gas equation (Eqn5) gives $dP/P - \gamma d\rho/\rho = 0$

$dA/A = - d\rho/\rho - dV/V$ (from derivative form of Eqn 7)
$= - (1/\gamma) \times dP/P - dV/V$ (from derivative form of Eqn 5)
$= - (1/\gamma) \times dP/P + dP/\rho V^2$ (from derivative form of Eqn 8)
$= (dP/\gamma P) \times [(\gamma P/V^2) - 1]$ (simplification)
$= (dP/\gamma P) \times [(\gamma RT/V^2) - 1]$ (using Eqn 4)
$= (dP/\gamma P) \times [(C^2/V^2) - 1]$ (using Eqn 6)
$= (dP/\gamma P) \times [(1/M^2) - 1]$ (using Eqn 6)
$= (dP/\rho V^2) \times [1 - (\rho V^2/\gamma P)]$ (alternate simplification)
$= (dP/\rho V^2) \times [1 - (V^2/\gamma RT)]$ (using Eqn 4)
$= (dP/\rho V^2) \times [1 - (V^2/C^2)]$ (using Eqn 6)
$= (dP/\rho V^2) \times [1 - M^2]$ (using Eqn 6)
Eqn 10: $dA/A = (dP/\gamma P) \times [(1/M^2) - 1] = (dP/\rho V^2) \times [1 - M^2]$

From equation 10, it is clear that during flow through a nozzle, pressure will reduce, indicating negative value of dP. If middle term is considered, then value of 'γ' and 'P' are always positive. This indicates that the term under first bracket is always negative. Same is situation for the extreme right expression, where density and square of velocity is always positive. The variation of area of nozzle is a function of Mach number (M). For subsonic flow, Mach number is less than 1 and second bracket of the extreme right expression of Equation 10 is positive. This indicates that area variation is negative and area must reduce to gain velocity. From central expression of Equation 10, also, if $M < 1$, $1/M^2$ is more than 1 and the second bracket is positive. So, for subsonic flow, nozzle area must reduce and the shape of nozzle for subsonic flow is converging.

When flow is supersonic, Mach number is more than 1. This gives second bracket of extreme right expression of Equation 10, as negative. As first bracket is always negative for flow through a nozzle, the variation of area becomes positive. So for supersonic flow through nozzle, area must increase and cross-section is diverging. Actually, nozzle for incompressible fluid flow is always converging. This is because rise in velocity is compensated by reduction of area and density of fluid is assumed invariant. For compressible fluid flow, density is invariant in subsonic flow. However, for supersonic nozzle, the density reduction due to reduction in pressure (and temperature) surpasses area reduction requirements due to velocity rise, to maintain same mass flow rate. Ultimately more flow area is demanded by fluid in supersonic region and the cross-sectional area becomes diverging. The shape of the nozzle is shown in Fig. 11.

Nozzle has both converging and diverging section with least diameter of each are equal and joined together. The figure of nozzle has certain definition of subscripts. Nozzle entrance is denoted by 'i', while nozzle exit plane is depicted by subscript 'e'. Nozzle receives combustion gases from rocket motor chamber (combustion chamber) at stagnant condition ($V \simeq 0$). The rocket chamber condition is denoted by subscript 'c'. The nozzle discharges gases in open atmosphere and this condition is denoted by subscript 'a'. The least cross-section is called throat and it lies at the summit of convergent and divergent portion of the nozzle. This is denoted by subscript 't'. Any trial arbitrary cross-section at the nozzle length is denoted by subscript 'x'.

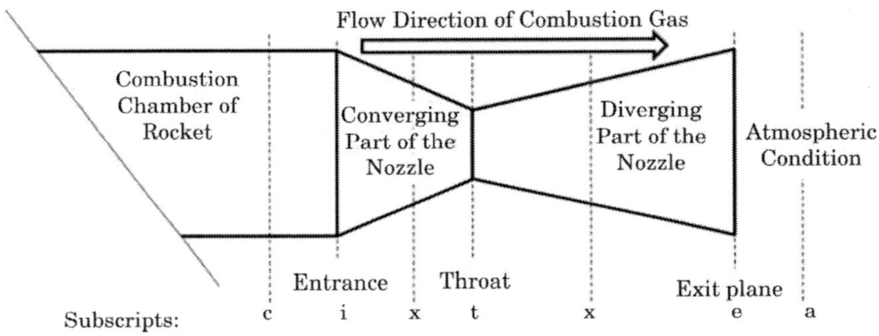

Fig. 11: Description of nozzle parameters

From energy equation between entrance and any point at nozzle length,

$$C_p\, T_i + V_i^2/2 = C_p\, T_x + V_x^2/2$$
$$V_x^2/2 = C_p\,(T_i - T_x) + V_i^2/2$$
$$= [\gamma R/(\gamma - 1)] \times (T_i - T_x) + V_i^2/2 \text{ (Using Eqn 3)}$$
$$V_x = \sqrt{[2\gamma R/(\gamma - 1)] \times (T_i - T_x) + V_i^2}$$

So, outlet velocity from the nozzle is given by equation 11.

$$V_e = \sqrt{[2\gamma R/(\gamma - 1)] \times (T_i - T_e) + V_i^2}$$
$$\text{Eqn 11} \quad = \sqrt{[2\gamma R\, T_i/(\gamma - 1)] \times (1 - T_e/T_i) + V_i^2}$$
$$= \sqrt{[2\gamma R\, T_e/(\gamma - 1)] \times (T_i/T_e - 1) + V_i^2}$$

For a typical trial case, the value of velocity of combustion gas at the nozzle exit plane is calculated. Following input parameters are considered:

$$\gamma = 1.19,\ MW = 29,\ T_i = 3000\ K,\ V_i = 50\ m/s$$

The value of exit plane velocity using Equation 11 is tabulated for exit plane temperature varying from 500 K to 2000 K. Same calculation is repeated after neglecting entrance velocity of the combustion gas (V_i = 0 m/s). The result is shown in Table 5.

From the table, it is clear that higher temperature at exit means temperature difference across nozzle is less and less energy is converted into velocity. Same is depicted by second column of the Table 5. The value of Mach number at exit is also computed and depicted as third column of the Table 5. With assumption of zero inlet velocity, the value of velocity at exit is depicted in fourth column. Although value of exit velocity by neglecting inlet velocity (column 4) is less than that with proper consideration of inlet velocity conditions (column 2), percentage change in exit velocity is very less. This is clear from the comparison of second and fourth column of the table. So, the equation of velocity can be further simplified by assuming stagnation condition at nozzle entrance. This means V_i = 0 and nozzle inlet condition may resemble stagnation pressure and temperature. For any cross-section at the nozzle length, the expression for velocity can be further simplified to arrive at temperature ratio.

Table 5: Computation of exit plane velocity

T_x (K)	V_e (m/s)	$M = V_e / \sqrt{\gamma R T_x}$	V_e (m/s) $V_i = 0$
500	2996.788	7.255772	2996.371
600	2936.257	6.489798	2935.832
700	2874.453	5.881916	2874.018
800	2811.289	5.381127	2810.845
900	2746.674	4.956768	2746.219
1000	2680.502	4.589113	2680.035
1100	2612.654	4.264795	2612.175
1200	2542.996	3.974365	2542.505
1300	2471.376	3.710905	2470.871
1400	2397.618	3.469194	2397.096
1500	2321.517	3.245181	2320.979
1600	2242.836	3.035639	2242.273
1700	2161.292	2.83793	2160.714
1800	2076.549	2.649833	2075.946
1900	1988.196	2.469421	1987.567
2000	1895.731	2.294956	1895.071

$V_x = \sqrt{[2\gamma R\, T_x/(\gamma - 1)] \times (T_i/T_x - 1)}$ (assume stagnation at inlet, $V_i = 0$)

$= \sqrt{[2C_x^2/(\gamma - 1)] \times (T_i/T_x - 1)}$ (using Eqn 6)

$T_i/T_x = 1 + (\gamma - 1) \times (V_x^2/C_x^2)/2$ (simplification)

Eqn 12: $T_i/T_x = [1 + (\gamma - 1) \times M_x^2 / 2]$ (using Eqn 6)

It is indicated that at the throat, sonic velocity is realized and value of Mach number is 1 there. Applying this assumption, exit plane temperature can be calculated from inlet temperature for different Mach numbers. Percentage reduction in temperature over inlet temperature at different Mach number is depicted in Fig. 12. The value of g is taken as 1.19 for the calculation and generation of the curve. Using Equation 4 and 5, percentage reduction in pressure and density is also calculated and plotted in same figure.

From the figure, it is clear that reduction in temperature during flow through a nozzle is at the slowest pace. At Mach number unity, representing throat section, temperature reduces only by 91.32% of the nozzle entrance temperature. Contrary to this, pressure reduces to around 56.64% of the stagnation pressure (chamber pressure). Similarly, density reduces by around 62.02%. The pace for reduction of pressure and density is faster in sub-sonic (M < 1) region. The pace of reduction slows down at higher Mach number (M > 2). At Mach number 4, temperature is only 40% of the stagnation temperature. For such super-sonic condition, pressure and density reduces to less than 1% of the nozzle entrance values.

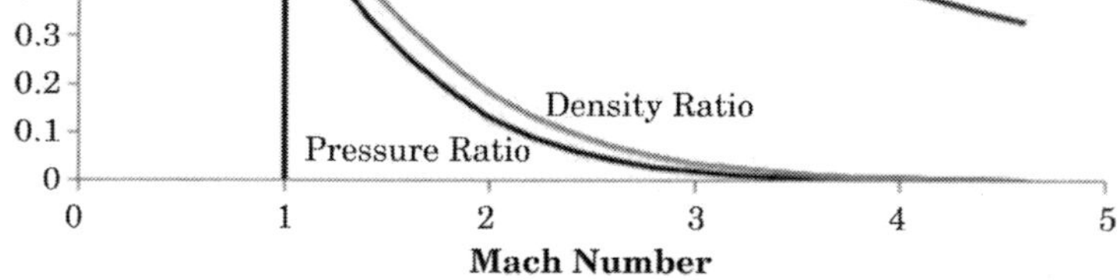

Fig. 12: Variation of thermodynamic parameter during flow through nozzle

Another parameter that affect, these thermodynamic parameters is ratio of specific heat. For another value of g (= 1.25), the curves are superimposed on already plotted Fig. 12. It is clear from Fig. 13, that there is insignificant variation in pressure and density reduction. However, at larger value of g, the temperature reduction is on lower side. At throat, temperature is around 88.88% of the nozzle entrance temperature.

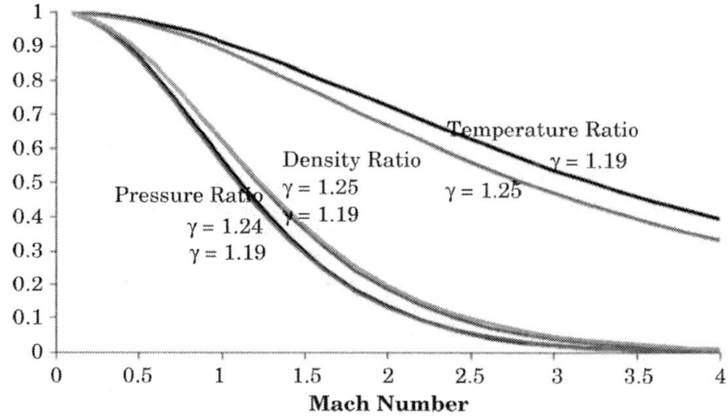

Fig. 13: Thermodynamic variations with different g

By putting value of M equal to unity in Equation 12, the temperature ratio between nozzle entrance and throat conditions can be calculated. The same is expressed as Equation 13.

Eqn 13: $T_t/T_i = 2 / (\gamma + 1)$

$$V_t = \sqrt{[2\gamma R\, T_i /(\gamma - 1)] \times (1 - T_t/T_i)} \quad \text{(Using Eqn 11)}$$
$$= \sqrt{[2\gamma R\, T_i /(\gamma - 1)]} \times [1 - 2/(\gamma + 1)] \quad \text{(Using Eqn 13)}$$
$$= \sqrt{[2\gamma R\, T_i /(\gamma + 1)]} \quad \text{(simplification)}$$

Eqn 14: $V_t = \sqrt{[2\gamma R\, T_i /(\gamma + 1)]} = \sqrt{[\gamma R\, T_t]}$

For the input condition given earlier, Temperature at throat section is 2739.7 K. This means that temperature reduction (from 3000 K) in convergent section is insignificant, as compared to that in divergent section. Same is true for pressure and density variation through the nozzle also. The value of velocity at throat is calculated as 966.81 m/s. This clearly indicates that flow accelerates beyond throat and attains a higher velocity after throat section. For different exit plane temperatures, velocity at throat section is same. So, all sections, which are at downstream to the throat are starting with same throat condition in terms of temperature, pressure, density and flow velocity.

Nozzle throat is responsible for rocket chamber pressure. As dimension of nozzle is pre-decided, during operation of rocket, rocket chamber pressure adjusts itself to realize sonic velocity at throat. So, throat governs pressure developed inside rocket motor chamber.

Area Ratio of the Nozzle

The mass flow rate through the nozzle is dependent on throat condition and it seems section of nozzle after the throat is not significant from mass flow rate point of view. Throat area is generally considered as standard reference for nozzle calculations. Area ratio can be expressed as ratio of area at any cross-section and at throat section. The calculation starts with conservation of mass flow rate (Equation 7). Velocity at the throat is sonic given by Equation 6.

$$\text{Mass flow rate, } dm/dt = \rho_t A_t V_t = \rho_x A_x V_x$$

$$A_x/A_t = \rho_t V_t/\rho_x V_x = (\rho_t/\rho_x)\,(1/M_x)\,\sqrt{(T_t/T_x)} \text{ (using Eqn 6)}$$

$$= (T_t/T_x)^{(\gamma+1)/2(\gamma-1)}\,(1/M_x) \text{ (using Eqn 4 and 5)}$$

$$= [(T_t/T_i)\,(T_i/T_x)]^{(\gamma+1)/2(\gamma-1)}\,(1/M_x) \text{ (simplification)}$$

$$= [\{2 + (\gamma-1)\,M_x^2\}/(\gamma + 1)]^{(\gamma+1)/2(\gamma-1)}\,(1/M_x)$$

$$\text{Eqn 15: } A_x/A_t = (1/M_x) \times [\{2 + (\gamma-1)M_x^2\}/(\gamma + 1)]^{(\gamma+1)/2(\gamma-1)}$$

For $\gamma = 1.19$, the value of area ratio of nozzle is plotted against Mach number realized in the nozzle (Fig. 14). The plot also contains numerical values of Mach number and corresponding area ratio. For example, at Mach number of 0.10, area ratio is 5.96, at M = 0.20, area ratio is 3.03, at M = 0.30, area ratio is 2.08, at M = 2.00, area ratio is 1.98, at M = 2.40, area ratio is 3.06, at M = 3.60, area ratio is 16.80 etc. At nozzle throat represented by unity Mach number, nozzle area ratio is also 1, representing reference condition. For sub-sonic (M < 1) region, the area continues to reduce. In other words, if we move from throat section to lower Mach numbers, the area increases. For super-sonic (M > 1), as flow velocity increases (Mach number increases), area also rises. *In fact, for both sub-sonic and super-sonic region, area increases continuously compared to throat section*. So, for single area ratio, two flow velocities are possible – one in sub-sonic region and another in super-sonic region. For area ratio of 2, two corresponding Mach numbers are 0.30 and 2.00.

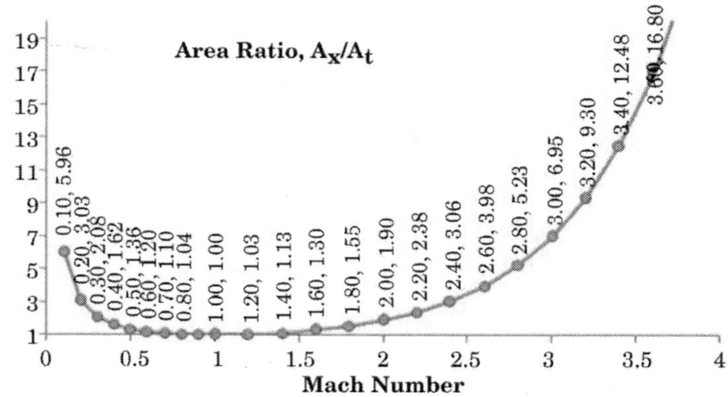

Fig. 14: Variation of area ratio of nozzle with mach number

The above diagram is also helpful in understanding the shape of the nozzle. The convergent length of the nozzle is generally made small. However, the divergent section is generally made with small divergent angle of around 10-15°. If larger angles are used, chances of flow separation and deterioration of performance is observed. However, a higher area ratio leads to longer nozzle, resulting in enhancement in inert weight of the rocket.

The nozzle area ratio is a function of ratio of specific heat also. The calculations are repeated for $\gamma = 1.25$ and superimposed curve is shown as Fig. 15. The numerical values of Mach number and area ratio for certain discrete points are also, depicted for $\gamma = 1.25$. It is clear from the curve that converging section is not affected significantly by value of ratio of specific heat. There is no change in sub-sonic flow pattern by change of γ. However, in the super-sonic region, the area ratio requirement for getting a given Mach number reduces. In fact, in sub-sonic region also, numerical reduction is observed in cross-sectional area requirements, but the difference is very small. For example, at M = 0.20, the value are area ratio changes from 3.03 to 3.01. This indicates that *if ratio of specific heat is enhanced, the divergent with lower cross-sectional area can give comparable Mach number.*

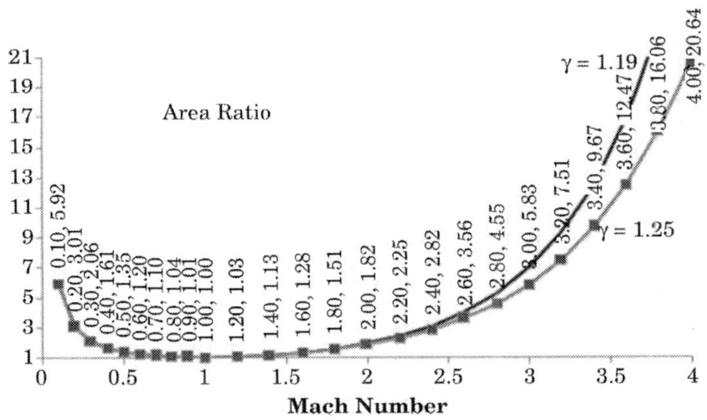

Fig. 15: Variation of area ratio for different γ

However, since rocket operation is associated with the generation of thrust, the exit plane velocity is also important. This exit plane velocity contributed to momentum thrust directly. In addition to this, exit plane area of nozzle directly influences the pressure thrust of the rocket. In fact addition of divergent section to the nozzle, leads to thrust augmentation. The area ratio of nozzle refers to enlargement of exit area with respect to throat area. The value of area ratio (A_e/A_t) can be calculated by taking mass flow rate conservation across throat and exit plane.

$$\text{Mass flow rate, } dm/dt = \rho_t A_t V_t = \rho_e A_e V_e$$

$$A_t/A_e = \rho_e V_e / \rho_t V_t = (P_e / P_t)^{1/\gamma} (V_e / V_t) \text{ (using Eqn 5)}$$

$$= (P_e / P_i)^{1/\gamma} (P_i / P_t)^{1/\gamma} (V_e / V_t) \text{ (simplification)}$$

$$= (P_e / P_i)^{1/\gamma} [(\gamma+1)/2]^{1/(\gamma-1)} [(\gamma+1)/(\gamma-1)] x [1-(P_e/P_i)^{(\gamma-1)/\gamma}]$$

$$\text{Eqn 16: } A_e/A_t = 1/(P_e / P_i)^{1/\gamma} [(\gamma+1)/2]^{1/(\gamma-1)} \sqrt{[(\gamma+1)/(\gamma-1)] \times [1-(P_e/P_i)^{(\gamma-1)/\gamma}]}$$

For a given pressure ratio, area ratio can be calculated using Equation 16, if value of g is known. The fact depicted in Fig. 15 is reproduced in Fig. 16 also, as higher value of g, need lower area ratio. However, in Fig. 16, it is variation with respect to pressure, which is depicted. In Fig. 16, all the parameters are non-dimensional. For design purpose, Equation 16 as well as Fig. 16 is adequate, because once pressure is fixed, nozzle dimensions can be arrived at, using these two features. For $\gamma = 1.19$, area ratio is plotted against pressure ratio and best fit curve by regression analysis is obtained as $A_e/A_t = 0.356643 (P_i/P_e)^{0.767096}$. The regression coefficient is 0.999 and the best fit curve is very reliable, also. Similar equations can be derived for different values of g. This simplifies the calculation of area ratio for given pressure ratios. For design purpose, this approach may be helpful. So, a general law is derived by regression analysis as $A_e/A_t = (0.142029g + 0.187662) (P_i/P_e)^{(1.464684 - 0.585579g)}$.

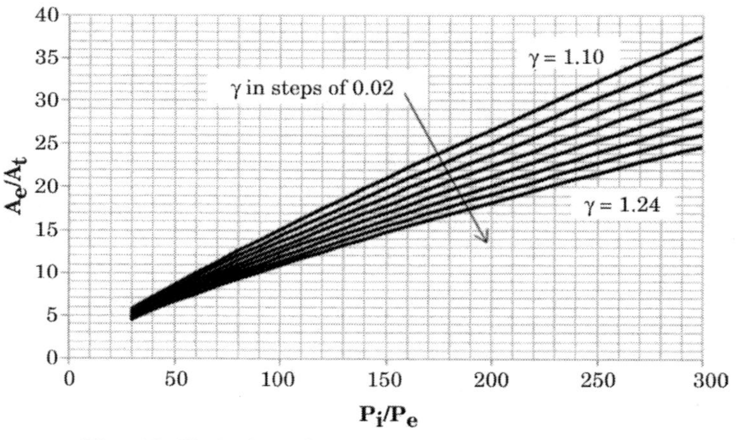

Fig. 16: Variation of area ratio with pressure ratio

However, in real life, nozzle geometry is known beforehand and it is invariant. For a given geometry, if pressure ratio is to be calculated, it becomes a trial and error type iterative procedure. So, regression analysis is carried out for pressure ratio (P_i/P_e) varying from 30 to 300 and g varying from 1.10 to 1.25. The equation valid for the domain is given by Eqn 17 and it has regression coefficient of 0.999. The equation can be helpful for performance prediction.

$$\text{Eqn 17: } P_i/P_e = (1.868083\gamma+1.611678) \times (A_e/A_t)^{(0.971091\gamma + 0.148019)}$$

Thrust Coefficient

Rocket is a reaction device and thrust is produced as internal reaction of the high velocity outgoing jet of combustion gases. Higher the velocity of the outgoing gases, higher will be the thrust produced by the rocket. Accelerating combustion gases to high velocities is main aim of providing nozzle in rockets. Alternately, thrust in a rocket can be expressed in terms of reference pressure and cross-sectional area. For rockets, nozzle inlet pressure (or rocket chamber pressure, P_i or P_c) is considered as reference pressure. Similarly, throat area (A_t) is considered as reference area. If free body diagram of rocket internal surface is generated, pressure acts at the entire inner surface. All pressures in lateral direction are balanced adequately. However, pressure acting at the throat area of rocket remains unbalanced and generates forward thrust. Product of these two reference parameters (pressure and throat area) is not numerically equal to thrust generated by the rocket. The thrust value, actually realized, may be from 0.8 times to 2.0 times this product. The factor representing thrust augmentation over the reference value is called thrust coefficient and is represented by C_f. It is a non-dimensional number. Thrust generated in a rocket is given by the following expression.

$$\text{Eqn 17: Thrust, } F = C_f\, P_i\, A_t$$

In general, rocket generates thrust by two actions – first is *momentum change* and second is *pressure imbalance*. For a rocket operation, mass flow rate of hot combustion gases through nozzle at certain velocity gives momentum thrust. Product of mass flow rate of gases and their exhaust velocity is the numerical measure of momentum thrust. At the exit plane, outgoing pressure of gases is never equal to outside atmospheric pressure. This results in pressure thrust, which is equal to pressure difference multiplied by exit area. Total thrust generated in a rocket is summation of both the thrusts. When pressure at exit plane is equal to outside atmospheric pressure, there is no pressure thrust and the condition is called optimum expansion. However, in course of flight of a rocket, the condition of optimum expansion is seldom satisfied. This is because; the pressure of outgoing gases from a nozzle is never constant throughout propellant combustion. Since burning area of propellant changes with time, the pressure inside rocket motor chamber changes and the pressure at the exit plane are altered. Similarly, during rocket flight, it travels through various layers of atmosphere, where density, temperature and pressure are different. This leads to optimum expansion in rocket flight for very small duration. For all sub-optimal conditions, rocket thrust has contributions from both types of thrusts. However, for optimum expansion, pressure thrust is zero and only momentum thrust is prevalent.

Momentum thrust is given the following expression:

$$\begin{aligned}
\text{Momentum thrust} &= (dm/dt) \times V_e \\
&= \rho_t A_t V_t V_e \text{ (using Eqn 7)} \\
&= P_i A_t (\rho_t/P_i) V_t V_e \text{ (simplification)} \\
&= P_i A_t (\rho_t/\rho_i)(\rho_i/P_i) V_t V_e \text{ (simplification)} \\
&= P_i A_t [2/(\gamma+1)]^{1/(\gamma-1)} (\rho_i/P_i) V_t V_e \text{ (using Eqn 13)}
\end{aligned}$$

$$\begin{aligned}
\text{Momentum thrust} &= P_i A_t [2/(\gamma+1)]^{1/(\gamma-1)} (1/RT_i) V_t V_e \text{ (using Eqn 4)} \\
&= P_i A_t [2/(\gamma+1)]^{1/(\gamma-1)} (1/RT_i) \sqrt{[2\gamma R\, T_i/(\gamma+1)]}\ \sqrt{[2\gamma R\, T_i/(\gamma-1)] \times (1 - T_e/T_i)} \\
&\qquad \text{(using Eqn 11 and Eqn 14)} \\
&= P_i A_t [2/(\gamma+1)]^{(\gamma+1)/2(\gamma-1)} \sqrt{[2\gamma^2/(\gamma-1)] \times (1 - T_e/T_i)} \text{ (simplification)} \\
&= P_i A_t [2/(\gamma+1)]^{(\gamma+1)/2(\gamma-1)} \sqrt{[2\gamma^2/(\gamma-1)] \times [1 - (P_e/P_i)^{(\gamma-1)/\gamma}]} \text{ (using Eqn 4 and 5)}
\end{aligned}$$

On similar line, pressure thrust is also obtained, as pressure difference at the nozzle exit plane exists. If there is no pressure difference at the nozzle exit plane, there is no pressure thrust development. The nozzle is said to be operating under optimum expansion condition. The outgoing gas from the nozzle at pressure of P_e is discharged in open atmosphere at pressure of P_a gives a pressure thrust.

$$\text{Pressure thrust} = (P_e - P_a) A_e = P_i A_t [(P_e/P_i) - (P_a/P_i)] (A_e/A_t)$$

Thrust coefficient (C_f) is defined as thrust augmentation over product of chamber pressure (or nozzle inlet pressure, P_i) and nozzle throat area (A_t). In fact, product of chamber pressure and throat area has dimension of thrust and thrust coefficient is a nozzle characteristic defining enhancement in thrust due to nozzle-action.

$$\begin{aligned}
\text{Thrust coefficient} &= (\text{Total thrust})/P_i A_t \\
&= (\text{Momentum thrust} + \text{Pressure thrust})/P_i A_t
\end{aligned}$$

$$\text{Eqn 18: } C_f = \sqrt{[2/(\gamma+1)]^{(\gamma+1)/(\gamma-1)} [2\gamma^2/(\gamma-1)] \times [1-(P_e/P_i)^{(\gamma-1)/\gamma}]} + [(P_e/P_i)-(P_a/P_i)](A_e/A_t)$$

Thrust coefficient is performance parameter of the nozzle and is independent of the combustion phenomena. It is a function of γ, A_e/A_t and P_i/P_e. However, thrust coefficient is independent of nozzle entrance pressure (P_i or chamber pressure, P_c). If outside atmosphere is vacuum, the value of $P_a = 0$. The thrust coefficient is called vacuum thrust coefficient and the value of thrust thus obtained is called vacuum thrust. The vacuum thrust is ideal thrust obtained in a rocket. The variation of vacuum thrust coefficient with pressure ratio is shown as Fig. 17.

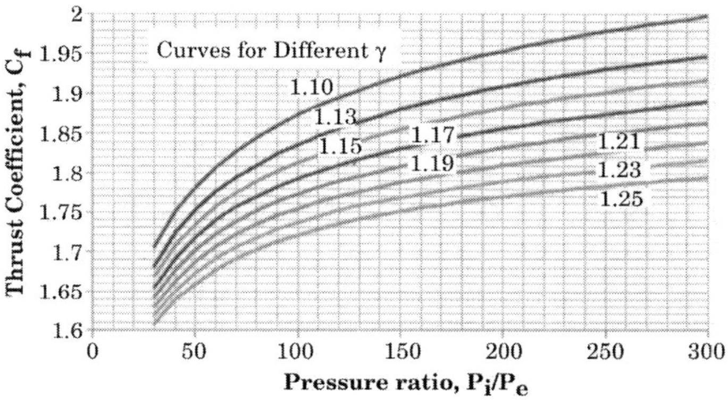

Fig. 17: Variation of vacuum thrust coefficient

Under firing in vacuum, the rockets are supposed to deliver higher thrust. One argument is reduction in skin friction drag in vacuum, which leads to enhancement in thrust. But major factor comes from enhancement in pressure thrust. This clearly indicates that, optimum expansion (exhaust pressure equal to outside vacuum pressure) is not possible for the calculation of vacuum thrust coefficient. In the Fig. 17, the x-axis of the plot represents pressure ratio (inlet pressure/exit pressure). High value on x-axis represents higher nozzle inlet (or rocket motor chamber pressure) for a given outlet pressure. But most of the time chamber pressure or nozzle inlet pressure is fixed and higher pressure ration means lower value of exit pressure. For a given ratio of specific heat, thrust coefficient increases with increase in pressure ratio. The rate of increase in thrust coefficient is higher at lower pressure ratio, which levels off at higher values of pressure ratios. This indicates that the value of vacuum thrust coefficient for infinite value of pressure ratio represents nozzle exit pressure as almost zero pressure (same as vacuum). This maximum value of thrust coefficient for infinite pressure ratio (vacuum pressure at exit plane of the nozzle) is a finite value and not an infinite parameter. This indirectly indicates that propellants possess finite energy and infinite thrust cannot be obtained, even if ideal expansion to vacuum condition is assumed. This maximum value depends on ratio of specific heat only and the values are tabulated as Table 6. Higher value of ratio of specific heat gives lower value of maximum vacuum thrust coefficient. From the curve (Fig. 17), it is also clear that for constant pressure ratio, lower value of ratio of specific heat results in higher thrust coefficient (or higher augmentation of thrust).

For higher thrust coefficient, lower value of ratio of specific heat and higher value of chamber pressure (pressure ratio of nozzle inlet to outlet pressure) is desirable.

Table 6: Maximum vacuum thrust coefficient

g	*Vacuum* C_f	*g*	*Vacuum* C_f	*g*	*Vacuum* C_f
1.10	2.947271	1.17	2.384247	1.24	2.109342
1.11	2.832236	1.18	2.334116	1.25	2.080984
1.12	2.732796	1.19	2.288418	1.26	2.054527
1.13	2.645851	1.20	2.246578	1.27	2.029787
1.14	2.569093	1.21	2.208119	1.28	2.006605
1.15	2.500767	1.22	2.172643	1.29	1.984840
1.16	2.439510	1.23	2.139813	1.30	1.964367

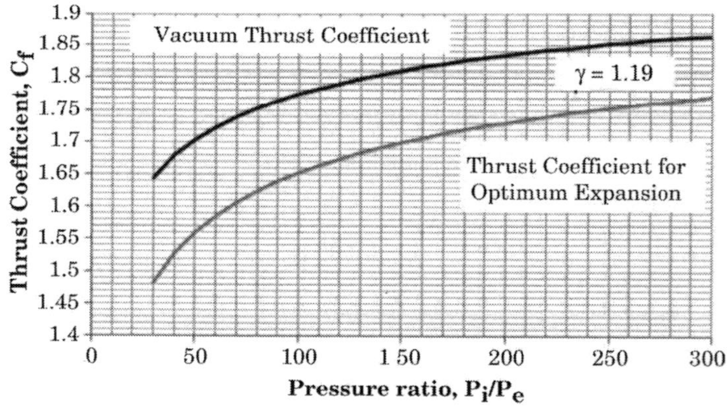

Fig. 18: Comparison of thrust coefficients in vacuum and under optimum expansion

Although vacuum thrust coefficient indicates upper limit of the thrust coefficient, but for tactical rockets or rockets operating in atmospheric condition, the external atmospheric pressure cannot be taken zero and pressure thrust will have certain finite value. However, for suitably designed nozzle, the exit plane pressure can be equalized to outside atmospheric pressure and pressure thrust can be avoided. Although this situation occurs for a very small duration during rocket operation, but this gives a situation of optimum expansion. Under optimum expansion, total thrust of the rocket is entirely contributed by momentum of outgoing exhaust gases. Thrust coefficient under optimum expansion will be definitely lower than vacuum thrust coefficient. A comparison of thrust coefficient under both the conditions is shown in Fig. 18 for ratio of specific heat equal to 1.19. It is clear that for constant ratio of specific heat (= 1.19), lower pressure ratio gives higher difference between both the thrust coefficients. Both the thrust coefficients give same value for infinite pressure ratio (as given in Table 6). Similar curves can be generated for different values of ratio of specific heat. However, the trend will have same nature of variation.

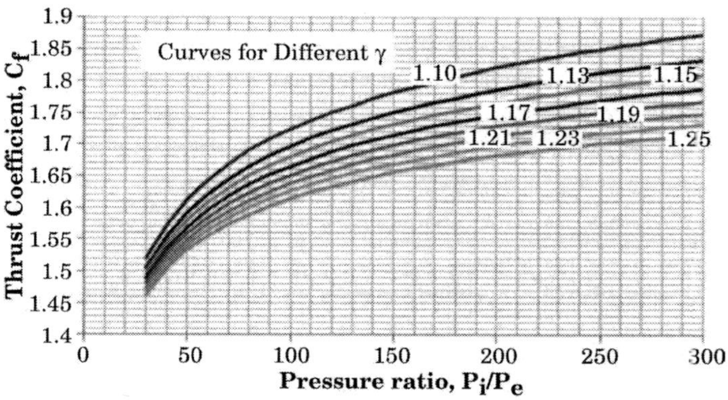

Fig. 19: Variation of thrust coefficient under optimum expansion

The value of thrust coefficient under optimum condition (area ratio adjusted to give exit plane pressure equal to outside pressure) is plotted against pressure ratio in Fig. 19. It is clear again that higher pressure ratio and lower value of ratio of specific heat results in higher value of thrust coefficient. The trend is same as depicted in Fig. 17 and only numerical values are different. In actual operation, nozzle efficiency is taken as 95%, which can be raised to 98% for well designed systems. So, thrust value is lower than value calculated using thrust coefficient for optimum expansion through the nozzle. So thrust coefficient is said to represent performance of the rocket nozzle.

Characteristic Velocity

As obvious from the previous section that thrust is given by the following expression.

$$\text{Eqn 17: Thrust, } F = C_f\, P_i\, A_t$$

On similar line mass flow rate can also be expressed in terms of nozzle inlet pressure (or rocket chamber pressure, $P_i = P_c$) and throat area (A_t).

Mass flow rate, $dm/dt = \text{constant} \times P_i\, A_t$

The constant term in the expression is called discharge coefficient and is represented as C_d in fluid mechanics. Russian literature use discharge coefficient for rocket operation, also. However, for rocket operations, inverse of this term is used for calculation and also for expressing propellant combustion efficiency. Since dimensionally inverse of discharge coefficient has same dimension as that for velocity, this term is called characteristic velocity (C^*). This term is defined as inverse of discharge coefficient ($1/C_d$). So, $C^* = P_i\, A_t/(dm/dt) = A_t\, \grave{o}P\, dt/\, m_p$.

Since nozzles in rocket operation always operate in choked condition, the mass flow rate is governed by throat conditions. The expression for characteristic velocity can be derived from the mass flow expression.

Mass flow rate, $dm/dt = r_t A_t V_t$ (using Eqn 7)

$= P_i A_t\, \rho_t V_t/P_i = P_i A_t\, (P_t/P_i)\, (\rho_t\, /P_t)\, V_t$
$= P_i A_t\, [2/(\gamma + 1)]^{\gamma/(\gamma-1)}\, (1/RT_t)\, V_t$ (using Eqn 13)
$= P_i A_t\, [2/(\gamma + 1)]^{\gamma/(\gamma-1)}\, (1/RT_t)\, \sqrt{[\gamma R\, T_t]}$ (using Eqn 14)
$= P_i A_t\, [2/(\gamma + 1)]^{\gamma/(\gamma-1)}\, (\gamma/\sqrt{[\gamma R\, T_t\,]})$ (simplification)
$= P_i A_t\, [2/(\gamma + 1)]^{\gamma/(\gamma-1)}\, (\gamma/\sqrt{[\gamma R\, T_i\,]})\, \sqrt{(T_i/T_t)}$ (simplification)
$= P_i A_t\, [2/(\gamma + 1)]^{\gamma/(\gamma-1)}\, (\gamma/\sqrt{[\gamma R\, T_i\,]})\, \sqrt{[(\gamma + 1)/2]}$ (using Eqn 13)
$= P_i A_t\, [2/(\gamma + 1)]^{(\gamma+1)/2(\gamma-1)}\, (\gamma/\sqrt{[\gamma R\, T_i\,]})$
$= P_i A_t/C^*$ (by definition)

Eqn 19: $C^* = \sqrt{[gR\, T_i]\, /\, [2/(\gamma + 1)]^{(\gamma+1)/(\gamma-1)}/\gamma} = A_t \int P\, dt/\, m_p$

The equation shown above is thermodynamic definition of C^*. The value of C^* is independent of nozzle parameters like throat diameter, exit diameter, nozzle area ratio etc. The derived thermodynamic expression indicates dependence of C^* on combustion gas properties. The controlling parameters are γ, T_i and molecular weight of the combustion gases. The dependence is expressed, one by one for each of the parameter. In general, for non-aluminized composite propellants, these values are 1.21, 2816 K and 25.3 kg/mol, respectively. For aluminized composite propellants, these values are around 1.17, 3371 K and 29.3 kg/mol, respectively. Nominal values of these parameters are taken as 1.19, 2500 K and 27 kg/mol. In general, temperature of combustion gases at the nozzle inlet can have a value in the range of 1500 K to 3500 K for various types of solid rocket propellants. Similarly, molecular weight of the combustion products in nozzle may acquire a value varying from 20 kg/mol to 30 kg/mol. It may be noted that higher molecular weight results in complex molecules and lower value of γ.

Additionally, at higher temperatures, the other vibration modes (in addition to translation and bending modes), are also activated resulting in lower value of g. However, for the assessment of trend, molecular weight, flame temperature and g are independently varied. First of all, variation of ratio of specific heat is considered and values are changed from 1.10 to 1.25. Value of temperature and molecular weight are combined as their ratio (T_i/MW). Upper limit of this parameter is obtained as 175 K-mol/kg with higher temperature (3500 K) and lower molecular weight (20 kg/mol) of combustion products. Similarly lower value of the parameter is 50 K-mol-kg for lower temperature (1500 K) and higher molecular weight (30 kg/mol). So this parameter can have an operational range of 50 K-mol/kg to 175 K-mol/kg. The value of variation of C* with these parameters is plotted as Fig. 20.

One peculiar observation is lower C* for the higher value of γ. So, *for getting higher value of C*, lower value of ratio of specific heats is desirable*. This indirectly indicates that higher molecular weight complex molecules can give higher value of C*, if other parameters are unchanged. However, the dependence of C* on is so less that in the working range, the effect of γ on C* is relegated to secondary status due to other more prominent terms. Major observation of Fig. 20 is that as ratio of temperature and molecular weight of the combustion gases increase, the value of C* increases. This increase is substantial, as compared to the increase due to lower γ. For example, for γ = 1.19, when T_i/MW = 50 K-mol/kg, C* = 997 m/s, which increases to 1854 m/s for T_i/MW = 175 K-mol/kg.

Another way to look at the graph is to estimate temperature and molecular weight combination for a given C*. If C* = 1550 for γ = 1.19, then value of T_i/MW = 120 K-mol/kg. This indicates different combination

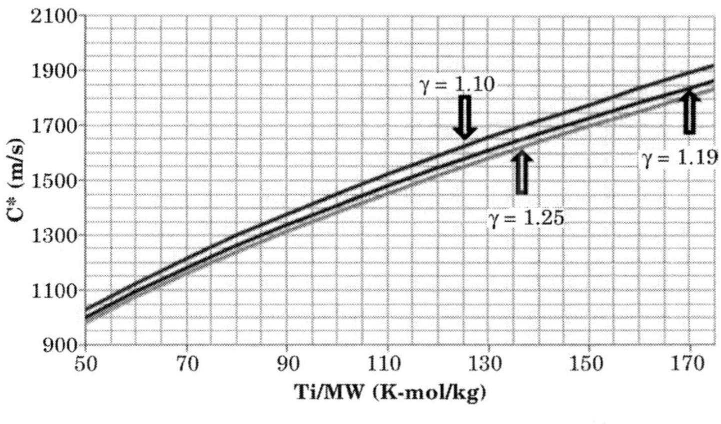

Fig. 20: Variation of C* against combustion gas parameters

of molecular weight and temperature. If temperature of combustion gases is 2400 K, molecular weight is 20 kg/mol. It also indicates that since molecular weight below 20 kg/mol is practically not obtained in case of solid propellants, the flame temperature of the combustion gases at nozzle inlet is never less than 2400 K. Similarly, to satisfy the upper limit of temperature as 3500 K, the molecular weight of combustion products will never exceed 29.16 kg/mol. This type of range prediction is possible with the above figure. Since all the values of γ cannot be shown on the Fig. 20, γ is taken as x-axis and the curve is re-plotted as Fig. 21.

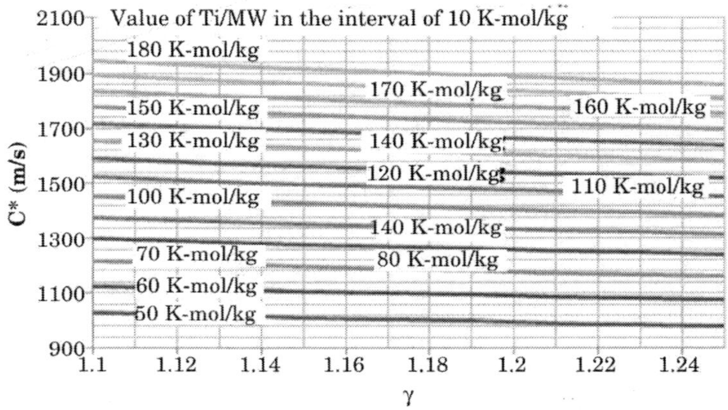

Fig. 21: Variation of C* with g for different combustion gas parameters

From above curve, it is clear, that increase in g reduces value of characteristic velocity. So, for higher value of C*, lower g and higher T_i/MW of combustion gases is needed. Suppose, a propellant composition exhibit C* equal to 1550 m/s at g = 1.19, from Fig. 21, it can be seen that value of T_i/MW $\simeq$ 120 K-mol/kg. If mean molecular weight of combustion gases is 25 kg/mol, the temperature of combustion gases is 3000 K. If under certain circumstances, after addition of certain exotic ingredients, realized C* is 1600 m/s. This increase in C* may be due to reduction in value of g. Keeping other parameters same, value of g for C* = 1600 m/s is 1.08, which is difficult to realize in the propellant combustion gases. In general, g does not change significantly. At g = 1.19, the value of T_i/MW is 130 K-mol/kg for C* = 1600 m/s. If temperature of gas is unaltered, the molecular weight of combustion gases will be 23.08 kg/mol. On the other hand, if molecular weight of combustion gases is kept constant (= 25 kg/mol), the value of temperature is 3250 K. Both these variations are acceptable from prevailing propellant compositions.

Sine C* is purely a parameter, indicating efficiency of propellant combustion, estimation of C* is possible from thermodynamic calculations. Since reaction during solid propellant combustion takes place through a number of pathways and many competing elementary reactions will be simultaneously generating and consuming species, the calculation for combustion for simple molecules is also very involving. Generally, software is developed for chemical kinetics calculations, where large numbers of chemical species, reactions, kinetics are considered. In fact, high temperature adds further complications. The application of heat balance requires product species to be complete known. The constitution of product gases depend on temperature realized. So, the complete process forms a loop and the process of calculation of C* is iterative in nature. For actual propellant formulations, dedicated software is available in open literature.

To show the iterative steps for hand calculation of thermo-chemical characteristics, propellants containing only carbon, hydrogen, nitrogen and oxygen are considered. Let the equivalent propellant formulation be $C_aH_bN_dO_e$ and a, b, c and d are known. The chemical reaction for this equivalent propellant may be given below, along with element balance.

$C_aH_bN_dO_e \rightarrow f\,CO_2 + g\,CO + j\,H_2 + k\,H_2O + m\,N_2$

Where, a = f+g, b = 2j+2k, d = 2m, e = 2f + g + k, by element balance.

Or, k = b/2 – j, f = e +j – a – b/2, g = 2a+b/2-e-j, by readjustment.

Since, only four elements are involved, water gas equation can be utilized, which is well characterized by known variation of equilibrium constants with temperature of reaction.

$$CO_2 + H_2 \rightarrow CO + H_2O\text{: water gas equation}$$

The expression of equilibrium constant (K) is reproduced below from open literature (Smith Byron R.J. *et al.*, A Review of the Water-Gas Shift Reaction Kinetics, International Journal of Chemical Reactor Engineering, Vol 8, Review R4, ISSN 1542-6580, 2010).

$$\text{Ln}\,(1/K) = (5693.5/T) + [1.077 \times \ln(T)] + (0.000544 \times T) - (0.0000001125 \times T^2) - (49170/T^2) - 13.148.$$

$$\text{Where, } K = [CO][H_2O] / [CO_2][H_2] = g \times k / (f \times j)$$

$$= (2a+0.5b-e-j)(0.5b-j)/(e+j-a-0.5b)j$$

$$\text{Or, } Kj(e+j-a-0.5b) = (2a+0.5b-e-j)(0.5b-j)$$

$$\text{Or, } j^2\,(K-1)+j(Ke-Ka-0.5Kb+2a+0.5b-e+0.5b)-0.5b(2a+0.5b-e)=0$$

$$\text{Or, } j^2\,(K-1)+j(Ke-Ka-0.5Kb+2a+b-e)-0.5b(2a+0.5b-e)=0$$

The above equation is a quadratic equation in 'j' (coefficient of H_2), whose solution can be obtained readily. Once 'j', is known, values of 'f', 'g', 'k' and 'm' can be obtained. So product of combustion can be made available. The salient properties of product of combustion are given from various thermodynamic tables in Table 7. The value of rationalized specific heat (C_p/R) is expressed as expression given below and the constants (A, B, C) are also compiled in Table 7.

$$C_p/R = A + 10^{-3}\ B \times T + 10^5\ C/T^2$$

Table 7: Properties of combustion products

Species	*Mol Wt*	*Heat of formation (kJ/mol)*	*A*	*B*	*C*
CO_2	44	–393.52	5.457	1.045	–1.157
CO	28	–110.53	3.376	0.557	–0.031
H_2	2	0	3.249	0.422	0.083
H_2O	18	–241.83	3.470	1.450	0.121
N_2	28	0	3.280	0.593	0.040

After that heat balance of the equation is worked out. The heat of formation of reactants should be equal to sum of cumulative heat of formation of products and heat needed to raise the temperature of combustion products to the flame temperature. Temperature is adjusted to give a matching result.

For example, suppose Nitrocellulose (NC) is considered as a test case for the calculation. The molecular formula is $C_6H_7N_2O_{10}$ and heat of formation is -694 kJ/mol. For this case, a = 6, b = 7, d = 2 and e = 10. As a start point, assume temperature is 2000 K. The equilibrium constant for the water-gas equation for this condition is obtained from the above mentioned equation as 4.434303. This gives quadratic equation in 'j'.

$$3.434303 \times j^2 + 11.21715 \times j - 19.25 = 0$$

The value of 'j' is obtained as 1.243046.

$$f = e + j - a - b/2 = 1.743046$$

$$g = 2a + b/2 - e - j = 4.256954$$

$$k = b/2 - j = 2.256954$$

$$m = d/2 = 1$$

$$C_6H_7N_2O_{10} \rightarrow 1.743046\ CO_2 + 4.256954\ CO + 1.243046\ H_2 + 2.256954\ H_2O + 1\ N_2$$

Specific heat is evaluated for each of the combustion product.

For CO_2, $C_{p@2000K}$ = 8.3143 x [5.457 + 0.001045 x 2000 – 115700/ (2000^2)] = 62.508 kJ/mol/K

For CO, $C_{p@2000K}$ = 8.3143 x [3.376 + 0.000557 x 2000 – 3100/ (2000^2)] = 37.325 kJ/mol/K

For H_2, $C_{p@2000K}$ = 8.3143 x [3.249 + 0.000422 x 2000 + 8300/ (2000^2)] = 34.048 kJ/mol/K

For H_2O, $C_{p@2000K}$ = 8.3143 x [3.47 + 0.00145 x 2000 + 12100/ (2000^2)] = 52.987 kJ/mol/K

For N_2, $C_{p@2000K}$ = 8.3143 x [3.28 + 0.000593 x 2000 + 4000/ (2000^2)] = 37.14 kJ/mol/K

Heat available = heat of formation of NC = -694 kJ

Heat needed by combustion product = heat of formation + heat needed to raise temperature to 2000 K = [1.743046 x (-393.52) + 4.256954 x (-110.53) + 2.256954 x (-241.83)] + (2000-298) x [1.743046 x 62.508 + 4.256954 x 37.325 + 1.243046 x 34.048 + 2.256954 x 52.987 + 1 x 37.14] = -907.58 kJ. Since heat needed is less than heat available, the assumed temperature is on lower side and temperature must be raised to higher values to get the balance of heat.

Hence, new temperature is assumed as 2500 K and entire process is repeated again. The governing reaction is expressed below. There is no change in the heat available, but heat needed changes and after calculation, heat needed is found as -590.63 kJ/mol.

$$C_6H_7N_2O_{10} \rightarrow 1.598469\ CO_2 + 4.401531 CO + 1.098469\ H_2 + 2.401531\ H_2O + 1\ N_2$$

Although the iteration can continue with any assumed temperature in between 2000 K and 2500 K, but from these two values linear interpolation is done to obtain, better estimate of combustion temperature.

$$T\ (K) = 2000 + (2500\text{-}2000)^*\ (\text{-}694\text{+}907.58)/(\text{-}590.63\text{+}907.58) = 2336.93\ K$$

After more iteration, the value of final temperature, satisfying mass (element or species) and heat balance, is obtained as 2342.74 K for NC monopropellant. The combustion product combination and specific heat values at specified temperatures are shown in Table 8. Heat needed is obtained as -694 kJ.

From moles and specific heat of the combustion products, specific heat and molecular weight of the combustion product is obtained by weighed average.

Table 8: Final combustion product

Species	*Moles*	*Cp (kJ/mol/K)*
CO_2	1.636382	65.551
CO	4.363618	38.914
H_2	1.136382	35.246
H_2O	2.363618	57.112
N_2	1	38.828

Molecular weight = (1.636382 x 44 + 4.363618 x 28 + 1.136382 x 2 + 2.36382 x 18 + 1 x 28) / (1.636382 + 4.363618 + 1.136382 + 2.36382 + 1) = 25.4285

Specific heat = (1.636382 x 65.551 + 4.363618 x 38.914 + 1.136382 x 35.246 + 2.36382 x 57.114 + 1 x 38.828)/(1.636382 + 4.363618 + 1.136382 + 2.36382 + 1) = 46.756 kJ/mol/K

Ratio of specific heat, g = $C_p/(C_p-R)$ = 46.756/(46.756 – 8.3143) = 1.2168

Characteristic velocity, C* = 1336.378 m/s

Similar calculation is repeated for nitroglycerin (NG). The molecular formula and heat of formation are $C_3H_5N_3O_9$ and -386 kJ/mol, respectively. After iteration, the value of combustion temperature is obtained as 3470 K. this is very high as compared to combustion temperature obtained for NC (earlier). The molecular weight of combustion product is obtained as 32.42. The value of specific heat is obtained as 69.326 kJ/mol/K. Both these values are much higher than that obtained for NC. However, value of ratio of specific heat is very low as 1.13. The value of characteristic velocity is obtained as 1480.58 m/s. But during this calculation, moles of some of the product species are negative.

The process is repeated for RDX, which is an explosive, but now utilized to enhance energy of the conventional propellants. The molecular formula and heat of formation are $C_3H_6N_6O_6$ and +59.94 kJ/mol, respectively. The combustion temperature, molecular weight of combustion products, specific heat of combustion products are obtained as 2898.29 K, 24.667, 49.164 kJ/mol/K, after iteration. This results in value of ratio of specific heat as 1.2035, which is similar to that obtained for NC. This may be because, both exhibited near equal values of specific heats of combustion products. The value of characteristic velocity is 1515.58 m/s, which is much higher than what is obtained for NC. In fact compared to NC, RDX gives higher combustion temperature and lower mean molecular weight of combustion products, resulting in enhancement in performance. This is surprising to note that although NG gives higher temperature of combustion products, but the same is not reflected in characteristic velocity. This is because molecular weight of combustion products of NG offsets the benefits of high combustion temperature. So, RDX gives higher characteristic velocity.

Although this method of calculation is iterative, but a very good estimate of characteristic velocity of single molecules as well as propellant compositions containing only 4 elements (C, H, N and O) can be obtained by this method. But this hand calculation strategy is by no means a method to substitute the software based calculations, where many governing equations and equilibrium constants are used and solved simultaneously.

Specific Impulse

Specific impulse is combined rocket performance parameter. It takes into account effects from propellant combustion, combustion gas parameters and nozzle parameters. *The specific impulse is product of characteristic velocity and thrust coefficient for a rocket, rationalized by gravitation acceleration ($9.81\ m/s^2$) to give a unit of time (seconds).* In fact, this rationalization is valid in space also, where acceleration due to gravity is zero. So, rather than expressing it as a function of weight, mass is generally considered with proper rationalization to adjust the unit to seconds. It is represented by I_{sp} and is used to compare performance of different rockets. The value of C_f (Eqn 16) and C* (Eqn 19) can be substituted to get expression of specific impulse in terms of propellant and rocket parameters. Various expressions for calculation of specific impulse of rocket are given as Equation 20.

$$\text{Eqn 20}: I_{sp} = \sqrt{[2gR\,T_i/(g-1)]x[1-(P_e/P_i)^{(g-1)/g}]}\,/g = V_e/g$$
$$= \int F\,dt/m_p = F/(dm/dt) = C^* \times C_f/g$$

Specific impulse is a function of pressure ratio (P_e/P_i), ratio of specific heat (g), combustion gas temperature (T_i) and Mean molecular weight of combustion products. Higher pressure ratio and combustion gas temperature result in higher specific impulse. Low mean molecular weight of combustion products gives higher specific impulse. High value of ratio of specific heat gives lower specific impulse. So, lower value of specific heat ratio is desirable. However, this is always associated with corresponding rise in mean molecular weight of exhaust products. So, effect of lowering value of ratio of specific heat is nullified by rise in mean molecular weight of combustion products.

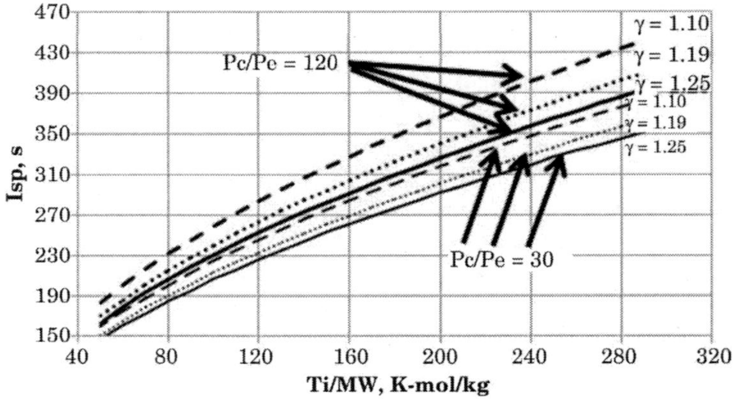

Fig. 22: Variation of specific impulse of rocket

Variation of specific impulse with various control parameters namely pressure ratio, ratio of specific heat, combustion temperature and molecular weight of combustion products is shown in Fig. 22. It is clear that higher chamber pressure gives higher value of specific impulse. Enhancement in thrust coefficient due to increase in chamber pressure is responsible for higher specific impulse at high pressure ratio. Higher combustion temperature and lower molecular weight of combustion products, which is combined as a single parameter T_i/MW, result in higher specific impulse. In fact, this property is derived from variation of characteristic velocity. Dependence of specific impulse on ratio of specific heat of combustion products is clear and lower value of g is desirable for higher specific impulse. Both characteristic velocity and thrust coefficient contribute to realize higher specific impulse at lower value of g.

From rocket operation point of view, specific impulse is directly connected to thrust produced and operational time of the rocket. Area under thrust time profile for a rocket operation is called total impulse and it is nominal definition of impulse used in physics basic curriculum. However, *specific impulse is defined as total impulse divided by mass of the propellant. This can alternately be expressed as thrust produced per unit mass flow rate*. This indicates that for higher value of specific impulse, same thrust has to be produced with smaller mass flow rate. *An alternate definition of specific impulse may be time taken to produce unit thrust by unit mass flow rate of propellant*. The more time unit mass of propellant takes to produce unit thrust, higher will be the specific impulse.

So for high value of specific impulse (i) Flame temperature of combustion gas should be high (ii) Pressure ratio (Exit to chamber) should be low (iii) Molecular weight of combustion product should be low (iv) Ratio of specific heats of combustion gases should be low.

REVIEW QUESTIONS

1. Why gases exhibit two numerically different specific heats?
2. What is significance of ratio of specific heats?
3. How ratio of specific heats changes for double base propellants with variation of temperature and pressure?
4. What are characteristics of nozzle in a rocket motor?
5. What are main assumptions in the analysis of flow through nozzle of rocket?
6. Derive expression for area variation with flow velocity (Mach number) inside rocket nozzle.
7. Derive expression for exit velocity from rocket nozzle.
8. Express velocity at the throat of nozzle in terms of inlet (stagnation) temperature.
9. Why flow through nozzle accelerates in the divergent section?
10. What is utility of divergent section of the nozzle?
11. What is effect of ratio of specific heats on the nozzle area ratio?
12. How area ratio of the nozzle changes with pressure ratio?
13. What is thrust coefficient?
14. Why is thrust coefficient always more than 1?
15. Derive expression for thrust coefficient.
16. Why is thrust coefficient for vacuum and for optimum expansion different?
17. How is thrust coefficient a nozzle performance parameter?
18. What is characteristic velocity?
19. Derive expression for characteristic velocity.
20. What are factors affecting characteristic velocity?
21. Compare the value of characteristic velocity for RDX and NG.
22. What is specific impulse?
23. What are factors affecting specific impulse in a rocket?
24. How specific impulse is overall performance parameter of rocket motor?
25. How is specific impulse related to gas velocity?
26. Combustion gases at 10 bar pressure is adiabatically expanded to 2 bar and brought back to initial temperature by isochoric

compression. For an ideal gas behavior, if final pressure is 2.5 bar, find the value of ratio of specific heat of the combustion gas. [Ans: 1.16].

27. Find sonic velocity in a gas of molecular weight 28, maintained at 2800K. Assume ratio of specific heats = 1.3. [Ans: 1039.6 m/s].
28. Find exit plane velocity of flow through nozzle for the combustion gas properties given in problem 27 and exit plane temperature is 1200 K. Assume inlet velocity = 100 m/s. [Ans: 1438.32 m/s].
29. Solve question 28 with ratio of specific heat = 1.2. [Ans: 2398.82 m/s].
30. For inlet temperature of 3200 K and ratio of specific heat of 1.2, find temperature at (i) Throat (ii) point on nozzle where Mach number is 3 (iii) Exit plane, where Mach number is 6. [Ans: (i) 2909 K (ii) 1684 K (iii) 695.65 K].
31. For chamber pressure of 70 bar and ratio of specific heat of 1.19, find pressure at (i) Throat (ii) point on nozzle where Mach number is 3 (iii) Exit plane, where Mach number is 6. [Ans: (i) 39.65 bar (ii) 1.46 bar (iii) 0.0343 bar].
32. Combustion gas of propellant has molecular weight of 28, ratio of specific heat of 1.2 and flame temperature of 3000 K. Find value of (i) Temperature (ii) Flow velocity at the Throat. [Ans: (i) 2727.3 K (ii) 985.8 m/s].
33. Solve question 32 with molecular weight of 25 for the combustion gas. Other parameters remain same. [Ans: (i) 2727.3 K (ii) 1043.3 m/s].
34. Solve question 32 with flame temperature of 2800 K for the combustion gas. Other parameters remain same. [Ans: (i) 2545.4 K (ii) 952.4 m/s].
35. Solve question 32 with ratio of specific heats of 1.16 for the combustion gas. Other parameters remain same. [Ans: (i) 2777.8 K (ii) 978.2 m/s].
36. During flow through nozzle combustible gases in frozen condition expands from chamber pressure of 70 bar to 1 bar, then find the value of area ratio. Assume ratio of specific heats = 1.2. [Ans: 9.0623].
37. Solve question 36 with chamber pressure of 100 bar. Other parameters are invariants. [Ans: 11.871].
38. Solve question 36 with ratio of specific heat of 1.15. Other parameters are invariants. [Ans:10.0566].
39. Find pressure ratio (Exit plane pressure to chamber pressure) for flow through rocket nozzle with area ratio of 10. Assume ratio of

specific heats equal to 1.2. [Ans: 0.012537 by iteration with Eqn 16 or 0.0126 using empirical Eqn 17].

40. For optimum expansion of combustion gas with ratio of specific heat 1.2 and pressure ratio (Exit to chamber) of 1/70, find the thrust coefficient. [Ans: 1.6].
41. Combustion gases (ratio of specific heats = 1.2) in a rocket is expanded through nozzle of area ratio 10 from chamber pressure of 70 bar in vacuum. Find (i) pressure ratio (Exit to Chamber) (ii) Exit plane pressure (iii) Thrust coefficient. [Ans: (i) 0.0126 (ii) 0.8829 bar (iii) 1.7423].
42. Solve question 41 for expansion in atmosphere at 1 bar. [Ans: (i) 0.0126 (ii) 0.8829 bar (iii) 1.5994].
43. Solve question 41 for nozzle area ratio of 6. [Ans: (i) 0.02467 (ii) 1.7269 bar (iii) 1.6725].
44. Solve question 41 for nozzle area ratio of 15. [Ans: (i) 0.007406 (ii) 0.5184 bar (iii) 1.79].
45. Solve question 41 for rocket operating at chamber pressure of 100 bar. [Ans: (i) 0.0126 (ii) 1.2613 bar (iii) 1.7423].
46. Find characteristic velocity for the combustion gas, whose properties are described in question 27. [Ans: 1776.47 m/s].
47. Solve question 46 with ratio of specific heats equal to 1.2. [Ans: 1687.2 m/s].
48. Solve question 46 with mean molecular weight of combustion product as 30. [Ans: 1716.24 m/s].
49. Solve question 46 with flame temperature of combustion product as 2500 K. [Ans: 1678.6 m/s].
50. Find specific impulse for a rocket with following combustion gas properties: Ratio of specific heat = 1.19, Molecular weight of combustion products = 30, Flame temperature = 3000 K, Exit to Chamber pressure = 0.011. [Ans: 227.7 s].

4

Staging of Rockets: Essence and Examples

"The Earth is the cradle of the mind, but we cannot live forever in a cradle".

KONSTANTIN EDUARDOVICH TSIOLKOVSKY

CHAPTER SUMMARY

This chapter is dedicated to Russian school teacher Tsiolkovosky, who has given rocket equation to correlate velocity (gain) of rocket with the velocity of outgoing gases. The unique equation has been derived and utilized for explaining the staging of rockets to gain higher velocity with same propellant mass. The expression for range is also derived. After reading this chapter, following understanding is imparted:

- Concepts and ideas in Rocketry by Tsiolkovosky
- Derivation of rocket equation of Tsiolkovosky
- Utility of rocket equation for staging of rockets
- Parametric study for unequal stages
- Expression for range
- Analysis of PSLV velocity gain
- Numerical examples and parametric studies

Definition of terms/abbreviations used

F = Force acting on a body
M = Mass of body
V = Gain in rocket velocity at ABP
ABP = All burnt point of rocket
t = time elapsed
M_i = Initial mass of rocket

= mass of fuel, hardware and payload

M_f = mass of rocket at ABP

= mass of hardware and payload

M_r = mass ratio of rocket = M_i/M_f

w = mass flow rate of exhaust gases

v_e = exhaust velocity of outgoing combustion gases

Values of variables taken in the chapter, unless otherwise specified

v_e = 2500 m/s

M_i/M_f = 5

Payload mass fraction = ratio of payload mass to initial mass = 2%

INTRODUCTION

Rocket performance is measured in terms of range and payload transportation capacity. For higher values of these operating parameters, rocket should gain highest velocity at the end of powered-boost phase. Within various constraints of space, hardware, fuel and payload, achieving highest velocity at complete consumption of the propellants needs various innovative skills. Staging is one such operational needed to enhance ABP velocity of a rocket. The basics mathematical model of staging was theoretically established in the beginning of last century by *Russian School teacher Konstantin Eduardovich Tsiolkovsky*.

Tsiolkovsky's Equation

Konstantin Eduardovich Tsiolkovsky (1857-1935) is generally recognized as "*Father of Astronautics*". Though, rocketry was not known at that time and reaching outer space was one of the challenges before mankind at that time, Tsiolkovosky theorized several events, concerning space travel. In childhood, he suffered from scarlet fever and became hard of hearing. This limited his formal education through any primary school and he was forced to complete his early education at home, without visiting any school. He was self-taught in mathematics, physics and astronomy. In 1903, he published his theoretical study of rocket fuels and rocket motor efficiency. He wrote about utilization of solar energy for satellites, ether or space suits, use of plants to provide food and fresh oxygen on long journey in space. He conceived centrifugal shower for taking bath in weightless condition, which would be prevailing in spaceship. Without any practical model or demonstrated spaceflight at that time, it was remarkable that half a century before realization of actual spacewalk through sputnik, he conceived and established various principles related to manned mission and space flights correctly. He stated "liquid fuels and multiple stages would probably be the most efficient propulsion techniques to explore outer space".

Tsiolkovsky used *Newton's second law of motion* effectively to put forth his concept mathematically. The law states that rate of change of linear momentum is proportional to applied force and change takes place in the direction in which force acts. It is also known as principle of conservation of linear momentum Linear momentum of a system is given by product of its mass and velocity. Total mass of a rocket is sum of fuel mass and inert hardware mass. It does not remain constant, because fuel is consumed continuously leading to reduction in total mass. A rocket has highest mass in the beginning, which becomes minimum at ABP (all-burnt point as explained earlier). A constant rate of reduction of fuel mass can be assumed as 'w'. Inert hardware mass can also be divided into two parts one is payload and other is combustion chamber mass, which is needed for proper confinement of the fuel combustion products.

As far as velocity components of linear momentum is concerned, there are two terms, first is velocity of the rocket in forward direction and second is exhaust velocity of combustion gases from the nozzle in rearward direction. Exhaust velocity of combustion gases id a function of type of fuel, rocket chamber parameters, free volume available, combustion efficiency, and nozzle parameters. It can be assumed constant for all practical purposes for ease of calculation.

$$\text{Force, } F = d(MV)/dt = V\,dM/dt + M\,dV/dt$$

Where, M and V are mass and velocity of rocket respectively. Derivative is taken with respect to independent variable time and is denoted by dt. Let M_i is initial mass of the rocket, which becomes M_f at ABP and $M_f = M_i - w \times t_{ABP}$.

Since there is no force acting on the system, left hand side (F) becomes zero. Regarding first part of right hand side, since only mass of fuel changes, velocity of exhaust gases, generated by combustion of fuel are to be considered only. Let exhaust velocity of combustion gases is v_e and mass flow rate of exhaust gases is w, then equation above gives $v_e \times w + (M_i - wt)\,dV/dt = 0$. Rearrangement of terms gives $dt/(wt - M_i) = dV/(v_e w)$.

Integrating this equation from ground position at zero velocity to all burnt point, we get *Tsiolkovsky equation* as Velocity of rocket at ABP, $V_{ABP} = v_e \ln(M_i/M_f) = v_e \ln(M_r)$, where M_r is called mass ratio and is equal to ratio of initial mass to final mass of the rocket. Value of M_r is always more than 1 and velocity at all burn point is independent of mass flow rate (w).This exhaust velocity of combustion gases can be correlated to specific impulse (Isp) and is equal to v_e/g. So the final form of Tsiolkovosky equation can be written as

$$V_{ABP} = g \times Isp \times \ln (M_i/M_f) = g \times Isp \times \ln (M_r),$$

Velocity of rocket at ABP depends on exhaust velocity and mass ratio of the rocket only. For higher ABP higher value of exhaust velocity of combustion gases and higher mass ratio are desirable. Exhaust velocity is a function of chemical energy available with propellant and conversion efficiency of this energy into kinetic energy of exhaust gases. Higher mass ratio demands highest fuel content and bare minimum hardware mass.

For example, exhaust velocity of exhaust gases from a rocket is 2500 m/s and it is operating at a mass ratio of 5 then velocity of rocket at all burnt point will be given by 2500 x ln(5) = 2500 x 1.609 = 4023.6 m/s. If exhaust velocity of combustion gases is doubled, rocket velocity at ABP will be doubles due to linear relation depicted above. However variation of rocket velocity at ABP is slower than variation in mass ratio. If mass ratio is made 10, then rocket velocity at ABP will be 5756.5 m/s, which is less than twice the earlier depicted rocket velocity at ABP. The value of rocket velocity at ABP is tabulated below (Table 9) for mass ratio varying from 2 to 10.

Table 9: Variation of rocket velocity at ABP with different mass ratio

Mass ratio (initial/final)	***Rocket velocity (m/s)***
2	1732.868
3	2746.531
4	3465.736
5	4023.595
6	4470.399
7	4879.399
8	5198.604
9	5493.061
10	5756.463

It is clear from the table that if mass ratio is doubled, rocket velocity at ABP increases by 1732.868 m/s. This value is product of exhaust velocity of combustion gases (at present, v_e = 2500 m/s) and ln(2). Similarly for any tripling of mass ratio in the present example, rocket velocity at ABP increases by 2746.531 m/s (2500 x ln 3). *It is surprising to note that velocity of rocket at ABP is independent of rate of consumption of fuel and even if complete fuel is* consumed instantaneously, rocket will gain above mentioned velocity *at ABP but distance traveled will be zero. So, higher range demands combustion of propellant at a slower pace.* It is not velocity at ABP that is alone governing factor but range has to be considered in conjunction.

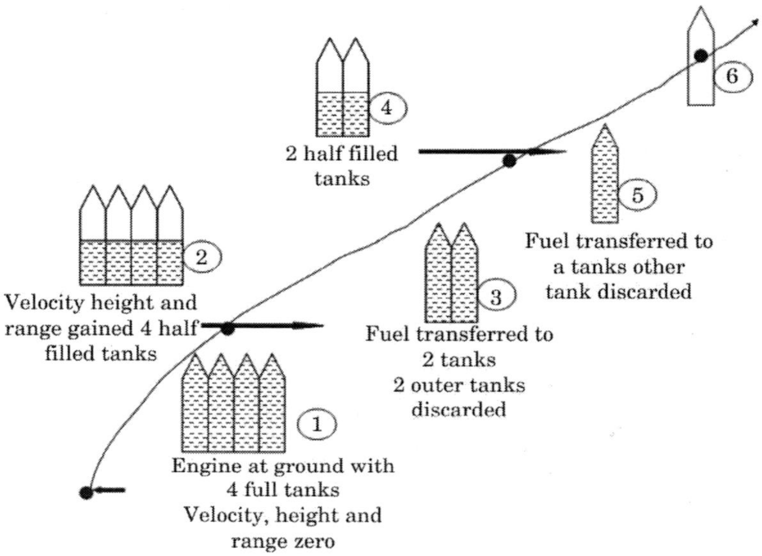

Fig. 23: Operation of a multi-stage rocket as conceived by Tsiolkovsky.

The concept of staging as explained by Tsiolkovsky is explained in Fig. 23. He conceived a liquid fueled engine with 4 fuel tanks (step 1). He assumed uniform rate of consumption of fuel from all the tanks. He explained that when half the fuel from each of the tank is consumed (step 2), two external tanks can be discarded after filling their fuel in two internal tanks (step 3). Now engine will have only two tanks and it will have some initial velocity already gained during flight with 4 tanks. With two tanks engine will move further and gain velocity and range. When the tanks are half empty (step 4), fuel from one tank will be transferred to second and first tank will be discarded. Now engine will have only one full tank (step 5). Engine continues to move till complete fuel is consumed (step 6). Tsiolkovsky explained that this system would give higher range compared to a system where complete fuel is stored in a single tank.

Mathematical Treatment of Staging

Tsiolkovsky equation is ideal for rockets moving in gravity-free frictionless environment. Rocket velocity at ABP can be obtained directly. Let us assume staging of the rocket, explained earlier in such a way that propellant and hardware mass are equally distributed amongst various stages. Let us take a mass ratio of 5 and exhaust velocity of combustion gases as 2500 m/s. Rocket is supposed to deliver a payload of 2% of its initial mass. Rocket velocity at ABP is depicted above as 4023.595 m/s. Rocket mass is made of 80% fuel mass and hardware weight excluding payload is 18% of total rocket mass.

If this rocket is split into two stages, both the stages will have 40% fuel mass each and hardware mass will be 9% of total mass of rocket each. For first stage mass ratio is 1.6666 and velocity gain in first stage will be 1277.064 m/s. For the second stage, mass ratio is 4.636364 and gain in rocket velocity is 3834.826 m/s. Finally, at the end of two stages, total gain in velocity is 5111.89 m/s (1277.064 m/s + 3834.826 m/s), which is higher than gain in single stage rocket velocity (4023.595 m/s). Similarly with increasing number of equal stages, gain in rocket velocity at ABP is tabulated below as Table 10.

Table 10: Variation of rocket velocity at ABP due to staging

No of stages	*Initial wt*	*Propel-lant*	*Hard-ware*	*Final wt*	*Mas ratio*	*Stage Vel*	*Final vel*
1	100	80	18	20	5	4023.595	4023.595
2	100	40	9	60	1.666667	1277.064	-
2	51	40	9	11	4.636364	3834.826	-
3	100	26.66667	6	73.33333	1.363636	775.3873	-
3	67.33333	26.66667	6	40.66667	1.655738	1260.617	-
3	34.66667	26.66667	6	8	4.333333	3665.843	5701.847
4	100	20	4.5	80	1.25	557.8589	-
4	75.5	20	4.5	55.5	1.36036	769.3741	-
4	51	20	4.5	31	1.645161	1244.596	-
4	26.5	20	4.5	6.5	4.076923	3513.356	6085.185

A more realistic estimate can be obtained, if gain in velocity at ABP is plotted against number of stages (Fig. 24). It is obvious that adding stages increases gain in rocket velocity but rate of gain of rocket velocity by adding a stage slows down, once number of stages exceeds 4. Complexities of staging surpass gain in rocket velocity and practical systems are generally conceived with maximum 4 stages. Amongst Indian satellite launch vehicles, PSLV (Polar Satellite Launch Vehicle) is a 4-stage vehicle. However, GSLV (Geosynchronous Satellite Launch Vehicle) is conceived in a 3-stage configuration.

Calculation for velocity gain in PSLV (Polar Satellite Launch Vehicle) is carried out to get the practical velocity rise. The data is taken from internet from the reliable site, but authenticated input is not used. The input may be taken as a test case without any resemblance to PSLV, also. PSLV has 4 stages and 6 strap-on boosters, which operates along

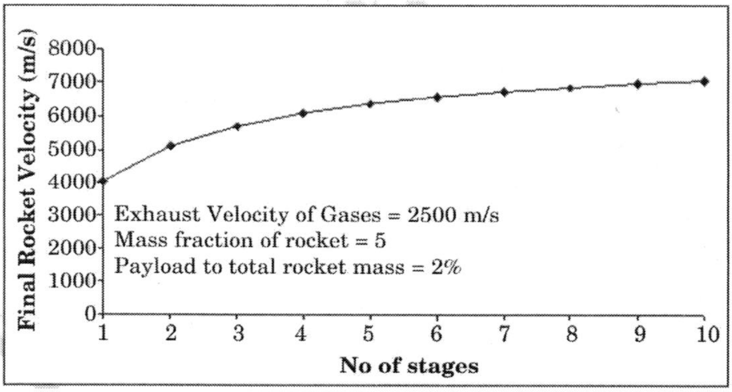

Fig. 24: Gain in rocket velocity at ABP with increasing number of stages

with first stage. The velocity gain is reported as 10 km/s Geo-Transfer Orbit (GTO) after fourth stage burnout. The weight of stage, propellant weight of each stage and specific impulse of each stage is considered for calculation. For stage 1, stage weight is 168 tons, propellant weight is 138 tons and specific impulse is 237 seconds. For each strap-on booster, stage weight is 11 tons, propellant weight is 9 tons and specific impulse is 229 seconds. For stage 2, stage weight is 45.8 tons, propellant weight is 40.8 tons and specific impulse is 296 seconds. For stage 3, stage weight is 8.4 tons, propellant weight is 7.3 tons and specific impulse is 294 seconds. For stage 4, stage weight is 3 tons, propellant weight is 2.1 tons and specific impulse is 308 seconds. The payload is 1.1 tons. Since burn time of strap on booster (45 s) is less than that of stage 1 (98 s), for calculation, strap-on boosters are taken as zero stage, operating before stage 1. Total weight of the PSLV at launch is 292.3 tons (66 + 168 + 45.8 + 8.4 + 3 + 1.1). After strap-on burn out, weight of launch vehicle is (292.3 tons – 54 tons =) 238.3 tons. Velocity gain after strap-on burn out is 229 x 9.81 x ln (292.3/238.3) m/s = 458.846 m/s. Initial weight, final weight and velocity gain in stage 1 are (292.3 tons – 66 tons =) 226.3 tons, (226.3 tons – 138 tons =) 88.3 tons and (237 x 9.81 x ln [226.3/88.3] =) 2188.079 m/s. Initial weight, final weight and velocity gain in stage 2 are (226.3 tons – 168 tons =) 58.3 tons, (58.3 tons – 40.5 tons =) 17.8 tons and (296 x 9.81 x ln [58.3/17.8] =) 3445.031 m/s. Initial weight, final weight and velocity gain in stage 3 are (58.3 tons – 45.8 tons =) 12.5 tons, (12.5 tons – 7.3 tons =) 5.2 tons and (294 x 9.81 x ln [12.5/5.2] =) 2529.593 m/s. Initial weight, final weight and velocity gain in stage 4 are (12.5 tons – 8.4 tons =) 4.1 tons, (4.1 tons – 2.1 tons =) 2.0 tons and (308 x 9.81 x ln [4.1/2] =) 2168.938 m/s. Total velocity gain after burnout of fourth stage of PSLV is (458.846 + 2188.079 + 3445.031 + 2529.593 + 2168.938 =) 10790.5 m/s $\simeq$ 10.8 km/s. This value approximately matches with the reported values on internet site of ISRO.

Velocity after strap-on burnout = 0.4588 km/s (53 s)

Velocity after first stage burnout = 2.6469 km/s (98 s)

Velocity after second stage burnout = 6.0919 km/s (245 s)

Velocity after third stage burnout = 8.6215 km/s (354 s)

Velocity after fourth stage burnout = 10.7905 km/s (870.6 s)

The relative worth of adding a stage can be depicted in terms of percentage gain in rocket velocity at ABP. The curve plotted as Fig. 25 depicts the same. It is clear that for the practical example considered, single stage to 2-stage rocket registers a gain of around 27% in absolute terms. However, when rocket is converted to 3-stage system, it shows an absolute gain of 41.71% in rocket velocity over single stage rocket. This can be calculated that converting a 2-stage rocket into 3-stage rocket will result in relative gain of only 11.71%. Addition of further stages reduces percentage gain in rocket velocity to less than 10% and finally adding more stages beyond 4 is practically uneconomical and non-efficient. World over maximum 2 to 3 stages are used and now concept of single-stage-to-orbit (SSTO) is being implemented for advanced systems.

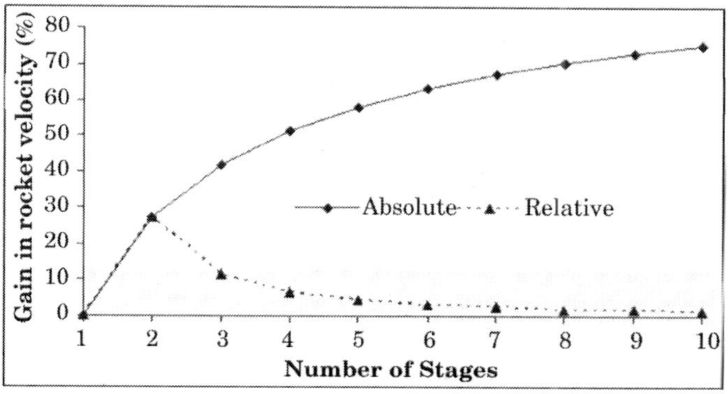

Fig. 25: Percentage gain in rocket velocity at ABP number of stages

The most popular space vehicle ARIANE has three stages till ARIANE-4 version, which was reduced to 2 only, in ARIANE-5. The largest operational rocket at present is Space Shuttle, which is implementing SSTO concept, but owing to use of both solid and liquid fuel for propulsion, this is referred as 2-stage propulsion. NASA announced on 30-June-2006 the names of the next generation of launch vehicles that will return humans to the moon and later take them to Mars and other destinations. The crew launch vehicle will be called Ares I, and the cargo launch vehicle will be known as Ares V. This vehicle is configured with 2 stages, similar but advanced compared to Space-Shuttle launch system.

Equal distribution of rocket mass amongst various stages of a multi-stage rocket is not always advantageous. For 2% payload mass fraction and 5 as mass ratio of rocket, equal distribution of mass over two rocket stages results in rocket velocity of 1732.868 m/s at constant exhaust velocity of 2500 m/s.

To assess the effect of mass distribution in 2-stage rocket lower and upper stages are conceived with different mass distributions. Lower stage mass fraction is changed from 0% to 100%. 0% indicates only upper stage, while 100% indicates only lower stage. It is clear that for both the extreme cases, rocket behaves as single stage rocket and final rocket velocity will be 4023.595 m/s only. However, plot for variation of rocket velocity (Fig. 26) indicates that higher fraction of mass of lower stage is always advantageous. For the test case, maximum gain in rocket velocity is around 90% of lower stage mass fraction with payload mass fraction of 2% and maximum gain in velocity at that moment is 6015.16 m/s.

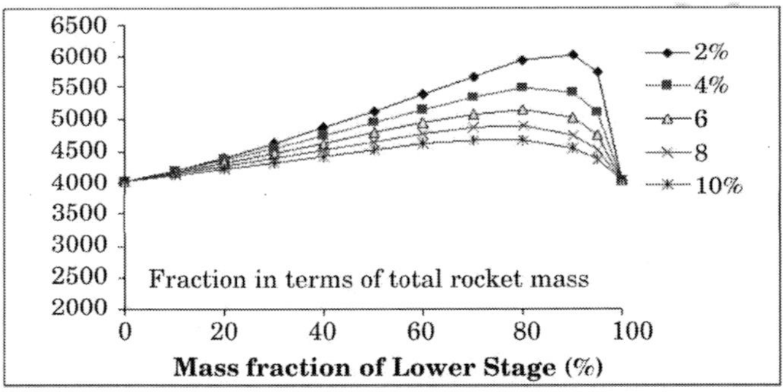

Fig. 26: Mass distribution in 2-stage rocket

Curves are also generated with higher payload mass fractions. It is clear from the graph that, when payload fraction increases, the maximum gain in velocity demands lower mass fraction of lower stages. However, mass fraction of lower stage never goes less than 70%, even if payload mass is increased to 10% of total mass of rocket at launch.

In nutshell, using Tsiolkovsky equation, derived from Newton's second law of motion, flight of a rocket can be simulated mathematically in gravity-free frictionless environment. It is obvious that gain in velocity of a rocket depends on only two parameters, namely velocity of exhaust gases and mass fraction of rocket. It is seen that rate of mass flow through rocket does not affect rocket velocity but it affects range. Converting a single-stage rocket into multi-stage rocket always results in gain in rocket velocity at ABP. However, practical conditions limit maximum number of stages in a rocket to 3. For a 2-stage rocket, lower stage mass fraction should have 70-80% of total rocket mass for highest gain in rocket velocity at ABP. If payload is made heavier, then a 2-stage rocket demonstrates highest gain in velocity at lower value of lower stage mass fraction.

Parametric Studies for 2-stage Rocket

The velocity gain in staging can be tailored by changing mass fraction of the stages effectively. Any rocket, conceived as a number of stages can gain more velocity by adding more and more stages, but the advantages of adding stages is offset by complexity of adding a stage. Adding a stage after third stage is less advantageous (in percentage) than adding a stage after first or second stage. Staging is basically an operation to split the rocket into smaller rockets, working in sequence, so that better structural and ballistic performance can be achieved. Another method to enhance velocity gain is accomplished by making unequal stages in 2 stage rockets. Making lower (initial, first) stage heavier than upper stages gives better velocity gains. In addition to this, there also exist the concept of parallel stages called strap-on booster or zero stage, which operates with first stage (to make lower stage heavier and more thrust producing) to give higher velocity gain. These boosters have lower burning time than first stage and they are ejected on burnout. Delta Rocket is conceived with 3,6 and 9 boosters. Titan III has 2 and 4 boosters. Space shuttle has 2 solid booster placed on either side of the main liquid fuel tank. Soyuz has 4 (kerosin + liquid oxygen based) boosters. These boosters result in less gravitational losses, as boosters are consumed faster but they result in bulky rocket and enhancement of drag.

Rather than taking numerical example, it is better, if algebraic derivation of staging is made and performance prediction is done. For ease of calculation, only two stages are considered initially. Assume that

payload fraction of initial total weight is 'p',
propellant to hardware ratio is 'r' and
ratio of weights of first and second stages is 'q'.

Assume, initial weight of rocket = 100
Weight of the payload = p
Weight of launching part = (100-p)

Single stage configuration:
Initial weight = 100

Weight of propellant = (100-p) x r / (r+1)

Weight of hardware = (100-p)/(r+1)

Final weight = p+[(100-p)/(r+1)] = 100–[(100-p)xr/(r+1)] = (pr+100) /(r+1)

Velocity gain = g x Isp x ln [(pr+100)/{(r+1)(100-p)}]

The highest value of velocity gain can be obtained directly for those cases which makes denominator of the fraction under logarithm infinity. p = 100 and r = -1 are trivial cases representing extreme conditions of the velocity gain, but both the cases are impractical.

2-stage configuration:

Stage 1:

Initial weight = 100

Stage weight = (100-p) x q / (q+1)

Weight of propellant = (100-p) x q x r / [(q+1) x (r+1)]

Weight of hardware = (100-p) x q / [(q+1) x (r+1)]

Final weight = [100(q+r+1)+pqr] / [(q+1)(r+1)]

Velocity gain = g x Isp x ln [100(q+1)(r+1)/{100(q+r+1)+pqr}]

Stage 2:

Initial weight = 100 – [(100-p) x q / (q+1)] = (100+pq)/(q+1)

Stage weight = (100-p) / (q+1)

Weight of propellant = (100-p) x r / [(q+1) x (r+1)]

Weight of hardware = (100-p) / [(q+1) x (r+1)]

Final weight = [100+p(q+r)+pqr] / [(q+1)(r+1)]

Velocity gain = g x Isp x ln [(100+pq)(r+1)/{100+p(q+r)+pqr}]

If same propellant is used in both the stages, specific impulse (Isp) is same for both. Overall velocity gain is given by velocity gain of each stage.

Overall velocity gain = gxIspxln

$$\frac{[100(100+pq)(q+1)(r+1)^2]}{[\{(100+pr)q+100(r+1)\}\{(r+1)pq+(100+pr)\}]}$$

= g x Isp x ln (H)

For getting optimum values, 'H' is to be integrated with respect to q and dH/dq should be made zero. This will lead to a value of 'q', which result in highest value of velocity gain.

Differentiation:

$$(100+p+2pq)x[\{(100+pr)q+100(r+1)\}\{(r+1)pq+(100+pr)\}] - \{100+(100+p)q+pq^2\}[(100+pr)\{(r+1)pq+(100+pr)\}+p(r+1)\{(100+pr)q+100(r+1)\}] = 0$$

The above expression is a cubic equation in 'q' and coefficient of each power of 'q' can be separately evaluated. Terms with different powers of 'q' are to be collected separately for evaluation of this expression.

$$\text{Cube of q: } q^3 \rightarrow 2p(100+pr)(pr+p)-p\{(100+pr)(pr+p)+p(r+1)(100+pr)\} = (100+pr)(2p^2r+2p^2-p^2r-p^2-p^2r-p^2) = 0$$

$$\text{Square of q: } q^2 \rightarrow (100+p)(100+pr)(pr+p) + 2p(100+pr)^2 + 2p(100r+100)(pr+p) - p(100+pr)^2 - p(pr+p)(100r+100) - (100+p)(100+pr)(pr+p) - (100+p)(pr+p)(100+pr) = (100+pr)\{(100+p)p(r+1)+2p(100+pr) - p(100+pr) - (100+p)p(r+1)-(100+p)p(r+1)\} + 100p^2(r+1)^2 = (100+pr)\{p(100+pr) - (100+p)p(r+1)\} + 100p^2(r+1)^2 = 100p^2(r+1)^2 - (100+pr)(100r+p)p = 100p^2r^2 + 200p^2r + 100p^2 - 10000pr - 100p^2 - 100p^2r^2 - p^3r = -pr(p-100)^2$$

$$\text{Linear Term of q: } q \rightarrow (100+p)(100+pr)^2 + (100+p)100p(r+1)^2 + 200p(r+1)(100+pr) - 100(100+pr)p(r+1)-100p(r+1)(100+pr)-(100+pr)(100+pr)^2 - (100+p)100p(r+1)^2 = 0$$

$$\text{Constant Term} \rightarrow (100+p)100(r+1)(100+pr) - 100(100+pr)^2 - 100^2p(r+1)^2 = (100+pr)\{(100^2+100p)(r+1) - 100(100+pr)\} - 100^2p(r+1)^2 = 100(100+pr)(100r+p) - 100^2p(r+1)^2 = 100r(p^2 - 200p + 100^2) = 100r(p-100)^2$$

So the expression of $dH/dq = 0$ becomes $100r(p-100)^2-pr(p-100)^2q^2 = 0$

$$\rightarrow q = 10/\sqrt{p}$$

The above derived expression can be used to get rid of term 'p', from the expression for overall velocity gain. The maximum overall velocity gain will depend only on 'q' and 'r'. The value of maximum velocity gain is given by the following expression:

$$\text{Maximum overall velocity gain} = g \times Isp \times \ln [\{q^2(r+1)^2\} / (q+r)^2]$$

This expression clearly indicates that for maximum velocity gain, product pq^2 should be constant (100). If 'q' is high, 'p' should be less and vice-versa. This means, if weight fraction of first stage (q) is increased, then payload weight fraction (p) should be decreased for maximum velocity gain. Generally, a compromise should be attained between payload weight fraction and first stage weight fraction. A numerical example is taken to prove the point, where specific impulse is taken as 240 seconds.

Payload mass fraction (p) is taken as 1, then for maximum velocity gain in 2-stae rocket, the first stage to second stage weight ratio (q) should be 10 (from the derived expression). If initial weight of the rocket is 100 units, payload weight is 1 unit. Weight of launch weight less payload weight is 99 units. For stage weight ratio of (q =) 10, weight of first stage is 90 units and that of second stage is 9 units. For a typical case, propellant to hardware weight ratio is assumed to be (r =) 5, then for the first stage, propellant and hardware weights are 75 units and 15 units, respectively. Velocity gain in first stage is 240 x 9.81 x ln (100/25) m/s = 3263.891 m/s. For the second stage, propellant and hardware weights are 7.5 units and 1.5 units, respectively. Velocity gain in second stage is 240 x 9.81 x ln (10/2.5) m/s = 3263.891 m/s. Total velocity gain = 6527.78 m/s. Alternately, maximum overall velocity gain = 240 x 9.81 x ln (100 x 36/225) m/s = 6527.78 m/s.

If stage weight ratio is taken other than 10, the velocity gain will be less. Assume stage weight ratio of (q =) 8 with payload weight fraction of (p =) 1. Weight of first stage is 88 units and that of second stage is 11 units. Propellant to hardware weight ratio is assumed to be (r =) 5, then for the first stage, propellant and hardware weights are 73.333 units and 14.667 units, respectively. Velocity gain in first stage is 240 x 9.81 x ln (100/26.667) m/s = 3111.913 m/s. For the second stage, propellant and hardware weights are 9.167 units and 1.833 units, respectively. Velocity gain in second stage is 240 x 9.81 x ln (12/2.833) m/s = 3398.465 m/s. Total velocity gain = 6510.38 m/s. It is clear that although total weight of propellant is the same in both the cases (82.5 units), but the velocity gain is less in second case. This reduction in velocity gain is because there are different velocity gains in both the stages due to non-optimal value of 'q'. The condition depicted above leads to equal velocity gains in both the stages of the rocket, thus making it an optimal condition. The efficacy of derived expression is established by the numerical example discussed.

A very interesting observation is negligible role of propellant to hardware weight fraction (r) in these maximization efforts of velocity gain. Although mass fraction plays a very important role in velocity gain calculation, but in a 2-stage rocket, propellant to hardware weight fraction does not affect maximum velocity gain. Keeping all conditions same (p = 1, q = 10), only propellant to hardware weight ratio is changed to (r =) 6. Definitely, now more mass of propellant will be loaded in each stage with higher propellant to hardware weight ratio. Although total velocity gain will rise, but the maximum velocity gain for this situation will be obtained, when q = 10. Higher value of 'r' means higher propellant weight and for similar other parameters, higher velocity gain is expected. However, the maximum of velocity gain is found independent of value of 'r'.

If payload weight fraction is (p =) 4, the stage weight ratio for maximum velocity gain is (q =) 5. The propellant to hardware weight ratio is (r =) 5. Maximum velocity gain 240 x 9.81 x ln (25 x 36/100) m/s = 5173.15 m/s. If first stage is made lighter by taking q = 3, then velocity gain in two stages is 240 x 9.81 x ln (100 x 28/40 x 8) m/s = 5106.82 m/s, which is less than earlier calculated maximum velocity gain (5173.15 m/s). Contrary to this, if first stage is made heavier by taking q = 7, then velocity gain in two stages is 240 x 9.81 x ln (100 x 16/30 x 6) m/s = 5143.9 m/s, which is less than earlier calculated maximum velocity gain (5173.15 m/s). It is clear that by keeping (r) propellant to hardware weight ratio constant, total combined weight of propellant in both the stages is maintained same, but due to changing relative weight of first stage, maximum overall velocity gain varies. High payload needs lighter first stage (smaller q) for maximum overall velocity gain, if other fractions are kept constant.

Displacement in Vertical Launch

Although Tsiolkovosky equation is mostly used in rocketry for the calculation of velocity of rocket at all burn point, but the calculation can be extended to get displacement till all burn point. The section title refers to height, which is displacement. To get rid of angle of launch, vertical displacement is analyzed. The starting equations include instantaneous mass of rocket and velocity at all burn point as given below. Integrating instantaneous velocity with respect to time can be used to get height gained by the rocket.

$$V = dy/dt = v_e \ln (M_i/M) = v_e \ln [M_i/(M_i - w\,xt)]$$

$$\text{Height, } H_{ABP} = \int dy = \int v_e \ln [M_i/(M_i - w\,xt)]\, dt = v_e \int \ln [M_i/(M_i - w\,xt)]\, dt$$

$$= v_e [M_i - M_f - M_f \ln (M_i/M_f)]/w = v_e\, t_{ABP} - M_f V_{ABP}/w$$

Since vertical rise of a rocket is always resisted by gravitational acceleration of the Earth, one more term is to be added in the developed equation. The equation for velocity and height at all burn point gets modified as given below.

$$V_{ABP} = v_e \times \ln (M_i/M_f) - g \times t_{ABP} = g \times Isp \times \ln (M_i/M_f) - g \times t_{ABP}$$

$$H_{ABP} = g \times Isp \times t_{ABP} - M_f V_{ABP}/w - \tfrac{1}{2} g\, t_{ABP}^2$$

The relations given above can be interpreted in several ways. One of the important observation is value of distance travelled is a function of mass flow rate (velocity is independent of mass flow rate). Higher mass flow rate results in lower value of distance travelled. *So, to get larger displacements (height), lower mass flow rate is preferred*. During definition of specific impulse (chapter 3), it is expressed that if same thrust is delivered by a slower mass flow rate, the specific impulse is on higher side for a given rocket. A similar interpretation is derived here for the calculation of displacements, also. Lower mass flow rate indicates higher time to consume the propellant or higher burning time of the rocket.

Higher exhaust velocity of combustion gases result in higher displacement of rockets. Higher difference between initial weight and final weight of the rocket leads to higher displacement. This indirectly refers to higher value of mass ratio. This can be alternately interpreted as lower value of final weight of the rocket, higher relative weight of the propellant or lower weight fraction of hardware and payload. Most of these observations can be derived in absence of the derived equations also, but the equations provide a conclusive means to express the results and relations through numerical examples.

Numerical values are considered for different variables for understanding the efficacy of the derived equations. Rocket hardware of 85 kg is used to propel 50 kg payload in 5 seconds burning time using 100 kg propellant of specific impulse 240 seconds.

Input parameters

Initial weight of the rocket, M_i = 85 + 50 +100 kg = 235 kg

Final weight of the rocket, M_f = 85 + 50 kg = 135 kg

Specific impulse, Isp = 240 seconds

Time at all burn point (ABP), t_{ABP} = 5 seconds

Calculations

Mass ratio, M_r = 235/135 = 1.74

Exhaust velocity of combustion gases, v_e = 240 x 9.81 m/s = 2354.4 m/s

Weight flow rate of combustion gases, w = 100/5 kg/s = 20 kg/s.

Gravity free flight

Velocity at ABP, V_{ABP} = 2354.4 x ln(1.74) m/s = 1305.1 m/s
Height at ABP, H_{ABP} = 2354.4x5 – 135x1305.1/20 m = 2962.6 m ≃ 3 km

Gravity-assisted flight

Velocity at ABP, V_{ABP} = 1305.1 – 9.81 x 5 = 1256.1 m/s
Height at ABP, H_{ABP} = 2962.6 – ½x9.8x5^2 m = 2839.9 m ≃ 2.8 km

For the example cited above, the exhaust velocity of combustion gas is higher than final velocity of the rocket. This is because, rocket has very low mass ratio. *So long as mass ratio is less than (e or) 2.7183, velocity of rocket at all burn point will be less than exhaust velocity of the combustion gases from the rocket.* Another important observation is negligible effect of gravitational acceleration over height as well as velocity at all burn point. Since rocket velocity is very high, the gravity is reducing rocket velocity by less than 5%. Same is the case of the calculated height of the rocket for gravity free and gravity assisted flights.

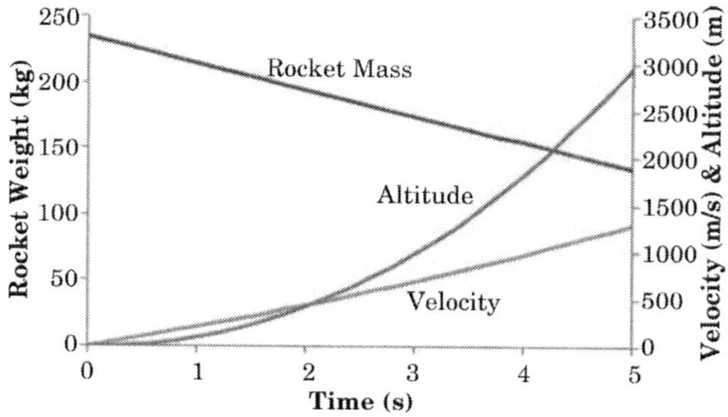

Fig. 27: Variation of different rocket parameters with time

Rocket mass varies linearly with time due to propellant consumption and ejection as combustion gas. In 5 seconds, it changes from 235 kg to 135 kg. This is depicted in Fig. 27. The velocity of rocket gradually increases in non-linear way during 5 seconds. The velocity gain is sluggish in the beginning, which increases at later part of the burning. This indicates that rocket is accelerating. As specific impulse (240 s) and weight flow rate (20 kg/s) is constant, the rocket produces constant thrust (4800 kg) during its operation (5 s). As mass of rocket reduces, the acceleration increases with time and velocity increases with time. Initial acceleration is (4800 x 9.81 / 235 =) 200.4 m/s^2, while acceleration at 5 s is (4800 x 9.81 / 135 =) 348.8 m/s^2. The greater change is observed in gain in altitude. Till around 3 s, rocket has gained, only one-third of the total altitude gained (~ 3km). In last 2 s, rocket gains two-thirds of total altitude gained. The variation of velocity and altitude gains is depicted in Fig. 5.

The above example is for a single stage rocket. In this case single burning time, single specific impulse (exhaust velocity of combustion gases) and single mass flow rate of propellant is considered. However, in actual multi-stage rocket, the situation becomes more complex. The calculation is repeated for PSLV, which is a multi-stage rocket. As PSLV operation is quite complex, where zero stage has 6 strap-on boosters, each having burning time of 45 seconds. 4 Boosters are ignited at ground for giving lift-off thrust and they are not separated immediately after their burnout at 45 seconds. They will coast with the main vehicle till 68 seconds, before jettisoning. The other 2-boosters are ignited mid-air, 25 seconds after lift-off. Their burnout occurs at 70 seconds, after lift-off, but they are separated at 90 seconds after lift-off. First stage of the PSLV has burning time of 98 seconds and it ignites on ground. It is separated 113 seconds after lift-off at an altitude of around 68.5 kms. The second stage is ignited thereafter and continues to work till 266 seconds, and it has a burning time of around 147 seconds. Launch vehicle is at an altitude of 248 kms, at the shutdown of second stage and ignition of third stage. Third stage is shutdown at 389 seconds (burning time of around 109 seconds), but a coasting period is available before shutdown. At the shutdown of third stage, launch vehicle is at an altitude of around 425 kms. At the end of fourth stage burnout and coasting, the launch vehicle attains an altitude of around 800 kms. The fourth stage has burning time of 516.6 seconds. Since burnout time and jettisoning time is not matching, direct application of derived equations is not possible, but approximate calculations can be made.

For calculation sake, it is assumed that all stages are jettisoned immediately after burnout and there is no carriage of weight of previous stage in coasting mode. Another assumption is about simultaneous ignition of all the 6-strap-on boosters on ground. All strap-on boosters are jettisoned at 45 seconds. Since, first stage has a burning time 98 seconds and it is ignited on ground, it is assumed that it work for additional 53 seconds, only, after burnout and jettisoning of strap-on boosters. This results in lowering of burning time (from 98 seconds to 53 seconds) for first stage and enhancement in mass flow rate. All stages are assumed to eject after their respective burnout. The payload of 1.1 tons remains with the launch vehicle till end along with 0.9 tons of fourth stage hardware. For all the calculations, gravitational acceleration is neglected, as this is negligible at higher altitudes and is insignificant at lower altitudes due to high velocity of rockets. It is also assumed that all stages has constant mass flow rate given by total propellant mass divided by burning time of the stage. The calculations are made for an ideal situation with minimum consideration for the practical staging constraints. The transfer of thrust generation between different stages is very smooth and there is no acceleration or reduction in velocity during such transitions.

Stage zero: Strap-on boosters

Mass flow rate = 1.2 tons/s

Average thrust generated = 1.2 x 229 kg = 274.8 tons

Jettisoning time from lift-off = 45 seconds

Altitude gained = 9.973 kms

Cumulative altitude = 9.973 kms

Stage 1:

Mass flow rate = 2.604 tons/s

Average thrust generated = 2.604 x 237 kg = 617.15 tons

Jettisoning time from lift-off = 98 seconds

Altitude gained = 49.020 kms

Cumulative altitude = 58.993 kms

(Reported altitude at 90 seconds from lift-off = 42.6 kms)

(Reported Time of jettisoning = 113 seconds: altitude = 68.5 kms)

Stage 2:

Mass flow rate = 0.2755 tons/s

Average thrust generated = 0.2755 x 296 kg = 81.55 tons

Jettisoning time from lift-off = 245 seconds

Altitude gained = 204.278 kms

Cumulative altitude = 263.271 kms

(Reported altitude at 266 seconds from lift-off = 248 kms)

Stage 3:

Mass flow rate = 0.06697 tons/s

Average thrust generated = 0.06697 x 294 kg = 19.69 tons

Jettisoning time from lift-off = 345 seconds

Altitude gained = 117.964 kms

Cumulative altitude = 381.235 kms

(Reported altitude at 389 seconds from lift-off = 425 kms)

Stage 4:

Mass flow rate = 0.004065 tons/s

Average thrust generated = 0.004065 x 308 kg = 1.252 tons

Jettisoning time from lift-off = 870.6 seconds

Altitude gained = 493.779 kms

Cumulative altitude = 875.014 kms

(Reported altitude at fourth stage burnout after coast $\simeq$ 800 kms)

Variation of total weight of the multi-stage rocket analyzed is shown as Fig. 28. Since calculation is highly idealized, variation in mass flow rate in each stage is clear on the graph by change of slope of the weight variation curve. As assumed, the stages are acting one-after the other and strap-on acts as zero stage. Although, first stage starts in the beginning and end at 98 seconds, the entire propellant of first stage is consumed only in 53 seconds (after 45 seconds : burnout of strap-on boosters). This gives a very high weight flow rate for stage-I, as depicted in the figure. Each change of stage is depicted in the figure by kinks or change in slope. Velocity variation also has similar kinks. Rate of gain in velocity is height for stage-I, but due to higher burning time, total gain in velocity in stage-II is the highest. Velocity in fourth stage rise slowly, but it is nearing escape velocity from the Earth.

In fact, variation of altitude with time is also similar as variation of velocity. Only difference is that variation of altitude is of higher order and at faster pace than variation of velocity. Similar trend is depicted

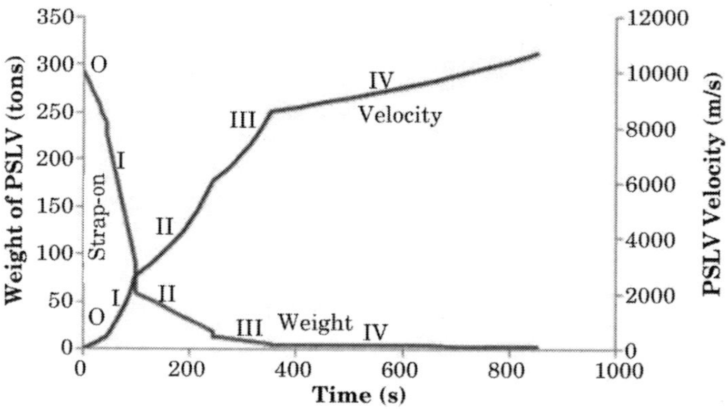

Fig. 28: Variation of kinetic parameters for multi-stage rocket

in Fig. 27, earlier also. For more clarity variation of altitude against velocity is plotted as Fig. 29. The variation of velocity as well as altitude is depicted for each of the stages. It is clear that velocity gain is the highest in stage-II, but variation of altitude is the highest for stage-IV. It is because of cumulative effect of higher velocity, lower mass and higher burning time of the stage.

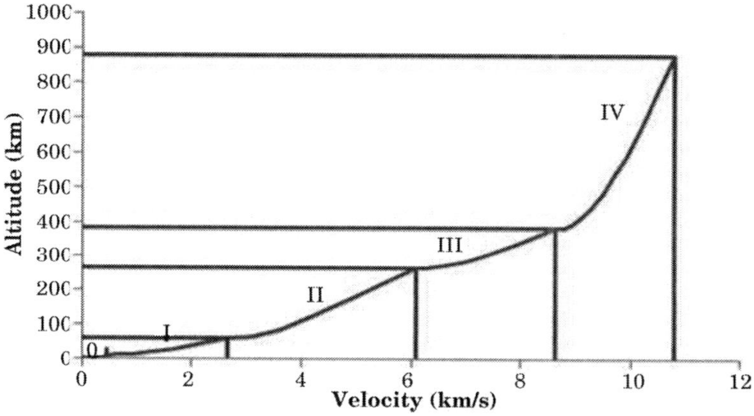

Fig. 29: Variation of altitude with velocity

Due to several idealizations, exact matching of the calculated velocity and altitude with actual values cannot be ensured, but the calculation strategy for multi-stage rocket is depicted in this chapter adequately. If data (mass variation, burning time, specific impulse) is available, velocity and altitude can be calculated for multi-stage rockets.

REVIEW QUESTIONS

1. What is thrust – time variation needed for a rocket with consumption of propellants?
2. What is staging of rockets? What are its advantages?
3. Explain Tsiolkovosky's concept of staging for liquid propellant based rockets.
4. Derive Tsiolkovosky's rocket equation.
5. What is significance of Tsiolkovosky's rocket equation?
6. What are effects of specific impulse on rocket velocity at All Burn Point?
7. Whether Tsiolkovosky's rocket equation can give velocity of rocket at maximum range?
8. What is role of mass ratio on the velocity of rocket at all burn point?

9. Whether velocity of rocket at all burn point is affected by mass flow rate of combustion gas or fuel?
10. What is difference in rocket velocity, if fuel is consumed instantly and when it is consumed over a prolonged period of time?
11. Comment on gain in velocity by splitting single stage into 2-identical stages with same mass ratio and payload.
12. Why relative gain in velocity of rocket at all burn point over immediately lower number of stages reduces with increase in number of stages?
13. What are disadvantages of staging?
14. Why is concept of Single Stage to Orbit (SSTO) effective and advocated in modern era?
15. Enlist merits and demerits of staging.
16. Derive expression for range of rocket at all burn point.
17. Comment on statement – "Lower mass flow rate gives higher range of rocket at all burn point".
18. What are limitations for increasing mass fraction of a rocket?
19. If a rocket with specific impulse of 245 s shreds 80% of its weight in the powered phase of travel, find (i) mass fraction (ii) exhaust velocity (iii) Rocket velocity at all burn point. [Ans: (i) 5 (ii) 2403.45 m/s (iii) 3868.2 m/s].
20. If mass of a rocket with specific impulse of 245 s reduces to 10% of initial mass during powered phase, find (i) mass fraction (ii) exhaust velocity (iii) Rocket velocity at all burn point. [Ans: (i) 10 (ii) 2403.45 m/s (iii) 5534.15 m/s].
21. A rocket has two identical stages, each with mass ratio of 5. Find velocity at all burn point, if payload mass fraction is 2% of initial weight and exhaust velocity of combustion product is 2500 m/s. What will be velocity at all burn point for single stage rocket? If three identical stages are used, then find velocity of rocket at all burn point [Ans: (i) 5111.89 m/s (ii) 4023.59 m/s (iii) 5701.847 m/s].
22. For the two stage rocket described in Q 21, if lower stage is twice as heavy as upper stage, find the value of velocity of rocket at all burn point. [Ans: 5571.193 m/s].
23. For the two stage rocket described in Q 21, if lower stage is 5 times as heavy as upper stage, find the value of velocity of rocket at all burn point. [Ans: 5994.738 m/s].
24. Derive expression for height gained by rocket in vertical flight.

5

Performance Prediction of Rockets

CHAPTER SUMMARY

This chapter is the basis for next section of the book. It covers various concepts, terms and definitions, related to rocketry and internal ballistics. The main parameters are associated with the geometry of the propellants. Final expressions for getting pressure time profile for a rocket is derived and used through illustrations. The chapter discusses the following:

- The concept of burning rate and associated laws
- The coefficients of burning rate law as per different laws
- The geometrical concept of web, burning area, port area, loading
- The internal ballistic equation using global analysis
- Example and parametric study for internal ballistic calculations for the rockets

BURNING RATE

Burning rate of solid rocket propellant is one of the important characteristics, which are significant for internal ballistic prediction of rockets. Although combustible materials are plenty in nature but their characterization in terms of burning rate is rarely observed. Burning rate of wood, leaves, fruits, organic substance etc are never reported. Similarly for explosives, also burning rate is not mentioned. Propellant burning is characterized by a term called deflagration, which is subsonic combustion. The phenomena of deflagration is present in solid as well as gases. Premixed gases deflagrate due to combined effect of convection, diffusion and reaction. The rate of propagation in premixed gases is directly proportional to square root of the reaction rate and has a value ranging from 0.2 to 10 m/s. the thickness of flame zone may be of the order of 0.2 to 5 mm. Smaller thickness of flame zone gives higher propagation speed due to better feedback of heat to unburnt premixed gases, which surpasses heat loss due to convection and radiation. This raises temperature of unburnt gases to auto-ignition temperature, quickly. Deflagration in solid propellants, specially double base propellants (DBP) occur due to exothermic heat release in gaseous phase, which gives sufficient heat feedback for preheating solid phases for decomposition or evaporation. Composite propellants (CP) deflagrate by combined effects of gas phase and solid phase heat and mass interactions. Deflagration imparts self-sustenance to the combustion of solid rocket propellants. Explosives detonate or explode and undergo a volume combustion rather than surface combustion in propellants. Other natural combustible materials simply burn without any self-sustenance and self-containment.

So, for the solid propellants, which burn faster than natural combustible materials but slower than explosives, burning rate is specified. It is defined as rate at which propellant surface moves during combustion. Basically, burning rate has the dimension of velocity and is measured perpendicular to the burning surface. In the definition of burning rate of propellants, one important law was stated in 1839.

Piobert's Law → burning of propellant takes place in parallel layers

This indirectly states steady uniform combustion of propellants. Propellant burning is not irregular as one observes during combustion of a sheet of paper. A sheet of paper ignited uniformly at an edge does not burn in parallel layer and extinguishing paper at any instance in-between, gives irregular shape of paper on the burning edge. If end face of a cylinder is ignited and lateral surface is restricted from combustion, the propellant will burn in parallel layers and burning area remains constant for such configuration. Similarly, if a tubular configuration of propellant is initiated at its inner lateral surface with inhibition on outer lateral and end faces, the inner diameter of the propellant configuration will remain cylindrical during combustion. The diameter of inner surface will increase but shape of propellant remains tubular during combustion.

There are many factors, which affect propellant burning rate. All these factors can be classified under two broad heads – (i) propellant composition and (ii) Operating condition. Propellant composition is one of the major factors affecting burning rate. For the double base propellant, processing parameters like incorporation, number of passes for extrusion, maturation time, additives, etc affect propellant burning rate. For composite propellants, increasing binder percentage at the cost of oxidizer quantity gives lower burning rate. In fact, reduction of particle size of oxidizers, in solid propellant formulation, increases burning rate of composite propellants. Another compositional parameter is type of oxidizer and binder used for manufacture of propellant. One of the important ingredients in composite propellant is burning rate modifier, which increases or decreases burning rate of propellants. If addition of a substance in propellant composition reduces its burning rate, they are called burning rate retardants. Lithium fluoride, oxamide, nitroguanidine etc act as burning rate retardants for composite propellants. However, Lithium fluoride is reported as a burn rate enhancer for DBP. Ingredient based burning rate variations are considered as inherent property of a given propellant and it is rarely written as burning rate law.

Burning rate laws are relation to express dependence of burning rate of solid rocket propellants on environmental conditions like pressure, temperature and flow velocity. The most popular burning rate law is Vieille's law. Sometimes, it is also called Saint Robert's law.

$$\text{Vieille's Law} \rightarrow r = aP^n,\ a = \text{coefficient},\ n = \text{index}$$

This law is conventionally a curve fit for data of burning rate of pressure at different pressures, derived from strand burner or ballistic evaluation motor (BEM) tests. The value of burning rate index (n) is less than 1 for stable burning of propellant inside rocket motor. If 'n' is more than 1, any increase in pressure leads to higher burning rate, which increases pressure and finally rocket leads to burst under excessive pressure due to synergistic rise in burning rate and pressure. This condition on 'n' is derived later as stability criteria of rocket operation. On the other side with 'n' less than 1, if pressure reduces, burning rate rises at lower rate, which reduces pressure further till stable operating pressure is achieved in the rocket chamber. The value of 'n' for DBP is generally between 0.4 and 0.9 and for CP, it is from 0.2 to 0.5. The value of burning rate index depends on pressure range and in different pressure ranges, different values of 'n' is recorded in different pressure ranges.

However, for the normal operating pressure range of rocket (2-20 MPa), burning rate law as per Vieille' law is plotted in Fig. 30. Two different burning rate indexes (0.3 and 0.6) are considered. Higher burning rate index gives higher burning rate at a given pressure and rate of rise of burning rate is also on higher side. The rate of rise of burning rate with pressure is given by the expression $dr/dP = nr/P$ and numerically P has higher numerical value, thus leading to *lower rate of rise of burning rate with pressure at higher pressures*. However, at lower values of pressure, rate of rise of pressure is non-linear and faster. The variation has quick rise at lower pressure (2-5 MPa), but above 5 MPa, the variation is more or less linear. To capture, this linearity of burning rate at higher pressures (above 5 MPa), linear burning rate in the form of Muraour's law is conceived.

Variation of burning rate with temperature is covered by including a term called temperature sensitivity coefficient of burning rate and in the Vieille's law, the burning rate coefficient is altered in the following form to take into account effect of temperature on burning rate.

$$a = a_0 \times \exp\{\pi_{rp}(T-T_0)\},\ \pi_{rp} = \text{Temperature sensitivity coefficient of burning rate}$$

In the above expression, temperature sensitivity coefficient is a positive number. Subscript '0' represent standard or reference condition. If $T > T_0$, the term in exponent is positive and value of burning rate coefficient (and also burning rate) increases for higher values of temperature 'T'. If $T < T_0$, the term in exponent is negative and value of burning rate coefficient (and also burning rate) is lower than the same for standard reference condition. Burning rate index 'n' is generally assumed to be independent of temperature variation. The value of π_{rp} for DBP is 0.3 to 0.8 %/K and for CP, it is 0.2 to 0.4 %/K.

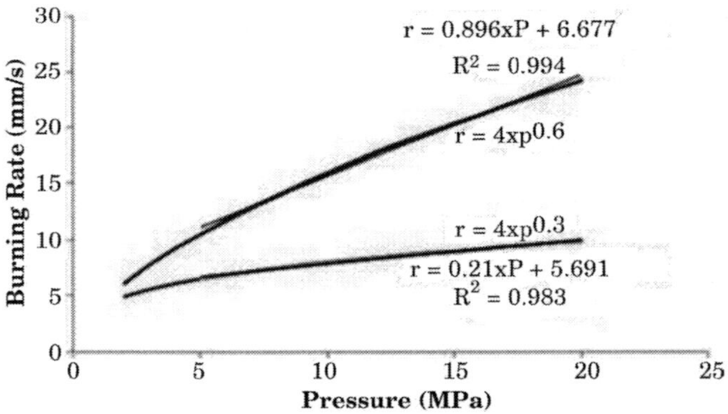

Fig. 30: Vieille's and muraour's law of burning rate

Another contemporary burning rate law for the solid rocket propellant is Muraour's law. This is a linear rate of burning in true sense. Vieille's law indicates zero burning-rate at zero pressure, but propellant should burn at zero pressure also, as it carries own oxygen. On investigation, it is observed that as per Vieille's law, in the normal operating range of pressure (2 MPa to 20 MPa), the burning rate has reducing slope with rise in pressure. However, in actual practice, at higher pressures, higher rise in burning rates for same pressure rise is observed. This is contrary to mathematical formulation of Vieille's law. In fact, in the normal operating pressure ranges of the rocket motor, even variation of burning rate with pressure as per Vieille's law resembles more or less that of a straight line. Compared to this Muraour's law gives a straight line variation of burning rate with pressure for all the pressure ranges.

Muraour's Law → r = H+S xP, H = Zero pressure burning rate, S = rate of change of burning rate with Pressure.

It is stated that *Muraour's law* is valid for pressures above 20 MPa. For homogenous propellants, the value of parameter H is stated to be 10 mm/s. The parameter S is expressed as a function of flame temperature. However, from Fig. 1, it is clear that Muraour's law can represent burning rate variation with pressure with equal ease as Vieille's law in the normal operating pressure range of rockets. However, at lower pressures (less than 2 MPa) the value using Vieille's law has larger rate of change of burning rate with pressure than normal linear variation.

Both the above mentioned law can be expressed in generic form $r = a+bP^n$. For the Vieille's law, a = 0 and 'b' is represented by 'a' in the equation above. The value of 'n' is less than 1 (derived later as stability criteria). If value of 'n' is equal to 0, the propellant is said to be exhibiting Plateau effects. This indicates that burning rate is not dependent on pressure. If 'n' is less than 0 (negative), the propellant burning is said to be exhibiting MESA effect and pressure increment reduces burning rate. Since Vieille's law is most popular burning rate law, most of the derivations are also based on this law, only. For Muraour's law, n=1, a=H, b=S.

One of the burning rate laws is *Summerfield's Law*. This law is derived in phenomenological way and has a different form of burning rate variation with pressure.

Summerfield's law → $p/r = a+bp^{2/3}$, a = chemical reaction time parameter, b = Diffusion time parameter.

The chemical kinetics affect burning rate parameters at lower pressures and parameter 'a' becomes important at low pressure. In this phase, burning is exhibited through premixed flame. This law is said to be valid for solid rocket propellant from 0.1 to 10 MPa. There are several other burning rate laws valid up to 10 MPa.

Penner's Law α $(1/r)^2 = (a/P)^2 + [b/P^{1/3}]^2$

Von Elbe's Law α $1/r = a/P + C$

Rastogi's law α $(r/P)^2 = a/P - b$

The effect of flow velocity on burning rate is observed only when the flow surpasses certain minimum mass flow rate criteria. *This is called erosive burning*. High flow rate gives higher burning rate of propellant, but the universally accepted relation is still missing for the same. Creation of turbulent boundary layer at the surface of propellant enhances heat flux from gas phase to solid phase and linear burning rate of propellant increases. Two competing phenomena is reported to occur at the surface of propellant. Lower boundary layer thickness (more turbulence) is preferred if convective heat transfer is increased and it gives enhanced heat flux. High burning rate gives more input to boundary layer, thus increasing its thickness. At low mass flux, both effects balance each other and burning rate is unaltered. When threshold mass flux is attained, convective heat transfer becomes dominant. This lower boundary layer thickness and burning rate increases.

From burning rate point of view, both higher and lower values of burning rates are sought after in the field of rocketry. Most of the booster rocket motors require burning rates in the range of 20 mm/s to 200 mm/s. The sustainer rockets require burning rate of 5 mm/s to 20 mm/s. For auxiliary propulsion requirements like spin, separation, correction, stabilization motors, burning rate requirements may be of the order of similar to booster propellants. For prolonged burning time in space applications, burning rate requirements may be as low as 1 mm/s. For gas generator applications, burning rate requirements may be low. However, these burning rates are nominal and the real burning rate requirements may be determined by considering burning area of the propellants also. For gas generator applications, if burning area is less, higher burning rates may be required for a gas generator propellant. If burning area of booster propellant is high, burning rate requirement as low as 4 mm/s is possible for booster propellant.

For enhancement or lowering of burning rates, there are two means available – *first is chemical means and second is physical means.* In chemical means, selection of oxidizer, adding burning rate catalyst (both for enhancement or reduction of burning rate), change solid loading, etc are possible. Amongst physical means, throat area variation for control of pressure, control on particle size of oxidizer or addition of metallic additives is preferred. The addition of metallic additives is highly specialized and very limited research has taken place in this area. It is observed that for a base composition with burning rate of 12 mm/s, addition of thin metallic wire enhances burning rates. The burning rate became 29 mm/s, 46 mm/s, 59 mm/s and 67 mm/s, for the inclusion of wires of aluminium, tungsten, copper and silver, respectively. It is practically observed that addition of other forms also enhances burning rate. Hollow spheres (50-100%), Hollow tubes (10-200%), threads (50-400%) and foils (50-450%) enhances burning rates, when included in propellant formulations.

ROCKETS WITH SOLID PROPELLANT ROCKET

Solid propellants has certain fixed geometry in the beginning and burning in parallel layers result in same shape of the propellant with changed dimensions. If initial propellant configuration is tubular, this configuration is retained during entire course of burning of propellant or rocket operation. However inner diameter, outer diameter and/or length may vary with time. For solid rocket propellants, several geometrical features of propellant in rocket motor casing can be derived. This section gives a fairly good idea of these features for different propellant geometries.

***Web*:** Web is the minimum distance to be burned, when propellant pieces become unconnected or consumed. As per Piobert's law, propellant burns in layers in the direction perpendicular to the burning surface. As the entire propellant's uninhibited burning surface recede, the distance to be burned can be calculated. Web for a solid sustainer (cylinder, burning at one end) is equal to length of the propellant. A long tubular propellant grain, burning at inner and outer surface has web equal to one-fourth the difference between outer and inner diameters. If burning of the propellant for the long tubular propellant configuration, is restricted to only at the inner diameter, then propellant web is only half the difference between outside and inside diameter. *Web is indicative of burning time, indirectly, if burning rate of the propellant is known.* It is clear that solid sustainer gives long burning time due to higher available web.

The concept of web in gun propellant is slightly different from what is used in rocket propellants. As concept of inhibition is in infancy for gun propellants, the web for gun propellant charges is equal to distance between two receding surfaces, which meet at the end of burning. For a simple tubular configuration, it is equal to half of the difference between outer and inner diameter. For rocket propellant with similar configuration, web is equal to minimum distance that is burnt for complete consumption of propellant and for simple uninhibited long tubular propellant grain web is equal to one-fourth the difference between outer diameter and inner diameter. *So, the web in gun propellant is twice that in rocket propellant.* For cylindrical configuration, web is equal to diameter (if length is longer than diameter, which is generally true) in gun propellants. There is no concept of solid sustainer (Chapter 6) in gun propellant, as gun propellants are always uninhibited.

Web in rocket propellant configuration needs special attention, because for complicated geometries like multi-perforated (Chapter 8), star, fin, fin-o-cyl or other complex configurations definition of web is not correlated to complete consumption of the propellant. In such cases, web is defined as maximum distance that is burned, when unconnected pieces of propellants, called slivers, are formed in course of combustion. Web in a star port configuration (Chapter 9) is not defined till complete consumption of propellant but it is up to that point when star flat (top) surface reached outer diameter of the propellant grain. Accordingly, web burning time is separately specified for such configuration. It is also assumed that after consumption of web, sliver formation takes place and propellants remain in unconnected pieces. These pieces burn in highly regressive manner and this situation leads to formation of tail-off in the burning profile (monotonically reducing pressure or thrust level in course of burning). As this portion of burning profile does not impart useful energy for the motion of rockets or missiles, it is generally web burning time, which gains importance in rocket ballistics. So, *web may be defined as point of beginning of sliver formation or point of complete consumption of propellant.*

Burning Area of the Propellant (A_b): Burning area of the propellant is purely a geometrical parameter, derived from the mathematical calculations. In course of burning, propellant burns in layer and burning surface changes with time. So *burning area of the propellant is to be calculated at each time step in course of burning.* However, to make this parameter purely geometrical, burning area is calculated at certain intervals of web burnt. The nature of propellant burning is decided by the variation of burning area of the propellant with web burnt. If burning surface area increases with web burnt, the burning is called progressive burning. If it reduces in course of burning, burning profile is called regressive burning. If burning surface area does not change with web burnt, the burning is called neutral burning. A solid sustainer behaves as neutral burning propellant configuration, because burning surface area remains constant ($=\pi D^2/4$) with web burnt. For a tubular propellant configuration (DxdxL) burning at all the exposed surfaces, burning profile is regressive. The burning surface area at any web burnt (derived in Chapter 7) is given by π (D+d)(D+2L-d-8w)/2.

Ratio of the burning surface area of rocket propellant to throat area of the rocket motor is one of the important characteristic for rocket operation. This ratio is called Klemmung (A_b/A_t). High value of this parameter gives higher pressure inside rocket chamber, if all other parameters are kept constant.

Port Area (A_p): Port area is feature of solid propellants with central longitudinal perforations. This includes tubular, star, finocyl or similar configurations, where central recess is available for flow of gases. End burning solid sustainer (cigarette-mode of burning) propellant grain or spherical propellant grain does not exhibit this feature of port area. Port area may be constant throughout the length of the rocket motor or it may be variable. High port area means bigger central perforation and smaller propellant loading in the rocket motor casing. *This condition is favorable for better structural integrity of the propellants, but is unfavorable from energy content and performance. Contrary to this, lower port area gives higher propellant loading in a given volume, better performance but poor structural margin of safety for the propellants.*

The above mentioned definition of port area is valid for case-bonded propellant configurations, where only central perforation for flow of combustion gases is available. In case of cartridge loaded propellant configuration, a secondary port is created between propellant grain and rocket motor casing. As central port and annular port communicate at head end and nozzle end only, the longitudinal pressure distribution in both the port depend on propellant burning area and flow cross-section. If pressure equalization is not proper between these two ports, propellant is subjected to radial compressive loads. Any differential pressure may be catastrophic. To overcome this, radial holes are drilled at various longitudinal positions for pressure equalization between both the ports. Additionally, propellant burning area to port area in both the ports (central core and external annulus) is also made equal for proper pressure management.

Port area is seldom uniform throughout the length of the propellant grain. Extruded propellant grains have uniform port area along length, but propellant made by casting have draft taper for extraction of mandrel (core). Generally port area is larger on nozzle end side in most of the rockets. However, reverse taper is also observed in some specific situations. If taper at port is of higher angle, then combustion gases accelerates in port section also. This situation may lead to internal secondary nozzle formation. Port area is a purely geometrical parameter, but it affects rocket motor performance to a much larger extent. There are two types of ratios, involving port area, which gives insight into propellant combustion behavior inside rocket motor chamber.

Port area at nozzle end side, in conjunction with throat area is responsible for assessment of erosive burning of the propellant. High ratio of throat area (A_t) to port area (A_p) leads to erosive burning. A limit of 0.35 is practically observed for this parameter, above which erosive burning is observed.

Another ratio is burning area (A_b) to port area (A_p). For a typical cylindrical propellant configuration, at central perforation, A_b/A_p is equal to 4L/d. This is indirect indication of accumulation of combustion gases in the port to generate abnormally high pressure inside rocket chamber. It is observed that in high performance rocket motors, where port area is smaller due to high volumetric loading and length to diameter ratio (L/d) of the rocket motor is high, propellant gases are accumulated in head end section, which may be catastrophic in specific situations.

***Propellant loading*:** Fraction of area or volume of the rocket motor filled with propellant is indicated by propellant loading. Case-bonded propellant configuration has higher propellant loading as compared to cartridge-loaded propellant configuration. Two types of propellant loading parameters are generally specified for solid propellants inside rocket chamber – *cross-sectional loading and volumetric loading*. For a propellant configuration with uniform area of cross-section along the length, both these loadings are equal. However, cross-sectional loading varies along length for other configurations. As name suggests, fraction of any cross-section filled with propellant is called cross-sectional loading. *High cross-sectional loading leads to higher web, lower port area, more chances of erosive burning and high stress-concentration at propellant port*. Contrary to this, fraction of volume of rocket motor (chamber and throat till throat) filled with propellant is indicative of volumetric loading of the propellant. Volumetric loading is lower than cross-sectional loading because of addition of convergent and other empty spaces inside rocket motor chamber, where cross-sectional loading of the propellant is zero.

***Solid loading*:** is a term used in an entirely different context in case of solid rocket propellants. It is indicative of propellant composition and is amount of fraction of solid ingredients by weight in the propellant formulation. This term is generally used for composite propellants, where oxidizer and metal powders are loaded in binder matrix. Of course, after curing, entire mass becomes solid. *This solid loading has nothing to do with propellant configuration and is quite different from propellant loading fractions*.

Solid sustainer type of propellant configuration has 100% cross-sectional loading, while volumetric loading may be of the order of around 95-98% because of empty space at the nozzle end side. Tubular propellants can be configured with cross-sectional loading of up to 90%, which may give a volumetric loading of around 86-88%. Star shaped configuration can give propellant loading fraction of 80-85%. These configurations are discussed in detail in section II of the book.

***Neutrality*:** Although Neutrality represents variation of pressure or thrust during a rocket firing, but it is affected directly by variation of burning area with time for a rocket motor. *Neutrality is roughly defined as ratio of average pressure to maximum pressure and is expressed in percentage.* In this definition, maximum value is deterministic but the calculation of average value need area under pressure time profile, which is to be divided by burning time. The definition of burning time is also not very consistent. So, overall, calculation of neutrality is not exact.

If neutrality is expressed as average burning area to maximum burning area obtained for a given geometry, then the value is different from neutrality obtained from pressure time profile. However, calculation of average area is relatively simple and it is equal to volume of propellant divided by maximum web available for complete consumption of propellant. For a progressively burning tubular propellant grain (DxdxL), maximum burning area is πDL. Average burning area is π(D+d)L/2. Neutrality is given by 50[1+(d/D)]. The value of neutrality is independent of length because length variation in course of burning is not taking place for progressively burning tubular propellant grain. *If tubular grain has OD of 115 mm and ID of 60 mm, then neutrality is 76%.* For regressively burning tubular configuration, there is no inhibition at any surface and maximum burning area occurs in the beginning. The initial burning area consists of surface area at inner diameter, outer diameter and both the end surfaces. The maximum burning area is (D+d)(D+2L-d)/2. Average burning area (volume/ maximum web) is given by π(D+d)L. Neutrality on the basis of burning area, alone is given by 2L/(D+2L-d). The value of neutrality depends on length of propellant grain also. *If a tubular propellant grain with 115 mm OD, 60 mm ID and 200 mm Length is considered, then neutrality is calculated as 87.9%.* Definitely, variation in case of regressive burning tubular propellant grain is more neutral, indicating less variation of pressure for same geometry.

For conventional star burning propellant grain (number of star points = N), if star half angle is given by a, then condition for the best neutrality is derived as Cota+a=π/2+π/N. The star half angle for different number of star points is calculated and is given below.

N	3	4	5	6	7	8	9	10
a	24°32'.8	28°12'.9	31°07'.7	33°31'.7	35°33'.4	37°18'.4	38°50'.4	40°12'

Internal Ballistic Equations

Internal ballistic calculations deal with phenomena taking place inside rocket motor chamber and are in simple form derived from mass balance equation. Although for advanced calculations, energy and momentum conservation equations are also utilized, but for most of the cases, simple mass balance calculations are sufficient to calculate pressure time and thrust time profiles inside rocket motor chamber. Internal ballistic calculation for performance prediction of solid rocket propellants is carried out using simple mass balance equation. Rate of mass accumulation inside rocket motor chamber is equal to difference of rate of mass generation inside rocket motor chamber by combustion of propellant and mass discharge through nozzle. Mathematically, this expression is written in equation form.

$$\frac{d}{dt}\left[\rho_g V\right] = \dot{m}_g - \dot{m}_d$$

In the above expression, r_g represents density of combustion gas and V is empty volume of the rocket motor and is given by inside volume of the rocket motor minus remaining solid propellant volume. Right hand side of the expression is difference of rate of mass generation and rate of mass discharge. For steady state situation, left hand side of the above expression is zero and rate of mass generation is equal to rate of mass discharge. However, steady state condition may not be present during ignition transience and tail-off portion of the rocket firing. But for most of the central portion of the firing, steady state condition prevails and it is used for calculation of stable pressure inside rocket motor chamber.

Rate of mass generation represents rate at which solid rocket propellant is getting converted into gaseous form by combustion. It is given by density of propellant multiplied by rate of volume consumption of propellant. The rate of volume consumption is given by product of burning area of propellant and the burning rate of the propellant. The expression for the mass generation rate due to consumption of propellant is given below.

$$\dot{m}_g = \rho_p A_b r$$

Rate of mass discharge is a function of chamber pressure and throat diameter. Higher the value of chamber pressure, higher will be rate of mass discharged. Same way higher is throat diameter, higher will be mass discharge. However, for exact knowledge of mass discharge, a factor is to be used called discharge coefficient, as used in fluid mechanics. However, in rocketry, inverse of discharge coefficient, called characteristic velocity is used and expression of rate of mass discharged through nozzle of the rocket motor is given below.

$$\dot{m}_d = \frac{P_c\ A_t}{C^*}$$

For steady state condition, rate of mass discharge is equal to rate of mass generation and using different burning rate laws, expression for pressure inside rocket chamber can be derived. In all the derivations, following parameters appear as a bunch and can be expressed by symbol χ.

$$\chi = \left[\frac{\rho_p\ A_b\ C^*}{A_t}\right]$$

Using Vieille's law, the expression for pressure is derived below.

$$\dot{m}_g = \rho_p\ A_b\ r = \rho_p\ A_b\ aP_c^n = \dot{m}_d = \frac{P_c\ A_t}{C^*}$$

$$P_c = \left[\frac{a\ \rho_p\ A_b\ C^*}{A_t}\right]^{\frac{1}{1-n}} = \left[a\ \chi\right]^{\frac{1}{1-n}}$$

Using Muraour's law, the expression for pressure is derived below.

$$\dot{m}_g = \rho_p\ A_b\ r = \rho_p\ A_b\ (H + S \times P_c) = \dot{m}_d = \frac{P_c\ A_t}{C^*}$$

$$P_c = \frac{H}{\left[\frac{A_t}{\rho_p A_b\ C^*} - S\right]} = \frac{H}{\frac{1}{\chi} - S}$$

Using Summerfield's law, the expression for burning rate becomes $r = P_c / \chi$, because in this case, $\chi = a + bP_c^{2/3}$. The expression for the pressure inside rocket motor chamber is given below.

$$P_c = \left[\frac{\chi - a}{b}\right]^{3/2}$$

From the derived expressions, it is clear that in addition to the burning rate pressure relationship, the components of parameters *c* are important for estimation of pressure inside rocket motor chamber. The first term is density of the propellant, which is more or less fixed before the rocket firing and seldom, has very wide variations. The value of density of propellants may vary from 1.60 g/cc to 1.83 g/cc. The parameters C* is also very less sensitive to type of propellants and values in the range from 1200-1600 m/s is obtained. The throat diameter is known before rocket firing and in absence of any erosion, the throat diameter after firing also remains constant. Only in case of nozzleless boosters, the throat diameters changes significantly during the firing. This parameter is also more or less invariant in normal rocket firing. In case of throat erosion or deposition on slag at nozzle throat, proper correction factors must be considered. *So, overall burning surface area of the rocket propellant becomes governing parameter for getting pressure variation in the rocket operations. This burning surface area variation is so important for solid propellants, that in this book, one full section is devoted to evolution of burning surface area of solid propellants for different configuration. From these derivations, pressure inside rocket motor chamber can be predicted comfortably.*

In addition to derivation of pressure based on steady state condition, there are stability criteria for the rocket operation. Any variation in pressure due to any eventuality must not result in abnormal pressure rise, else rocket motor may burst. The stability criteria are given by the following expression and for each of the burning rate law, expression for stable operation of rocket motor is also derived.

Stability criteria and their derivations for different laws

$$\frac{d\dot{m}_d}{dP} > \frac{d\dot{m}_g}{dP}$$

Vieille's law	$n < 1$
Muraour's law	$cS < 1$
Summerfield's law	

$$\frac{1}{r} > \frac{a}{P_c} + \frac{b}{3P_c^{1/3}}$$

Propellant Design Examples

As all derivations for the performance prediction are completed, in this portion, an attempt is made to understand design of solid propellants for rockets from the system requirements, through some examples.

Problem statement is taken as design rocket configuration with a neutral burning tubular propellant grain (End Inhibited) to give total impulse of 80000 N-s for 2 seconds in a 200 mm outer diameter and 800 mm length of propellant. Throat diameter of rocket motor is 60 mm. The operating pressure is considered as 10 MPa and available propellant with density of 1750 kg/m^3 and burning rate of 14 mm/s at 7 MPa is to be used. Take C* of the propellant as 1540 m/s.

Given

Total impulse, $\int Fdt$ = 80000 N-s
Burning time, T = 2 seconds
Outer diameter of propellant, D = 200 mm
Length of the propellant, L = 800 mm
Throat diameter, D_t = 60 mm
Operating Pressure, P_c = 10 MPa
Density of the propellant, r_p = 1750 kg/m^3
Burning rate of the propellant, r = 14 mm/s at 7 MPa
Characteristic velocity, C* = 1540 m/s

Calculation

Average thrust, $F = \int Fdt/T$ = 40000 N.
Throat Area, $A_t = \pi D_t^2/4$ = 2827.43 mm^2.
Thrust coefficient, $C_f = F/P_cA_t$ = 1.4147.
Mass discharge through nozzle, $\dot{m}_d = P_cA_t/C^*$ = 10 x 2827.43/ 1540 = 18.36 kg/s
Mass of propellant required, m_p = 18.36 x 2 kg = 36.72 kg.
Specific impulse, $Isp = \int Fdt/\, W_p$ = 222 sec.

As mass of propellant is known volume can be calculated using density.

Volume of propellant required, V_p = 20982857 mm^3.

As cross-section is constant, uniform cross-sectional area can be calculated using length of propellant.

Cross-sectional area of propellant = 26228.57 mm^2.
Cross-sectional area of a simple tube can be given by $\pi(D^2\text{-}d^2)/4$.
Inner diameter of the propellant, d = 81.27 mm.
Propellant web available = (D-d)/4 = 29.68 mm.
Burning rate at operating Pressure = 29.68/2 = 14.84 mm/s.

Now values of burning rates are available at two pressures:- (i) at 7 MPa, it is 14 mm/s and (ii) at 10 MPa, it is 14.84 mm/s. From these two burning rate index as per Vieille's law can be calculated.

Burning rate index, $n = \ln(r_1/r_2)/\ln(p_1/p_2)$ = 0.16337.

So, Stability criteria ($n<1$) is satisfied.

Burning area of the propellant, $A_b = \pi(D+d)L = 706908.6$ mm^2.

Mass generation rate, $= \dot{m}_d = \rho_p A_b r = 18.36$ kg/s.

As rate of mass generation is equal to rate of mass discharge through the nozzle, steady state condition of rocket motor operation is satisfied.

For the parametric study, the input variables are changed one by one and their effects on calculated parameters are verified. For this parametric study, only one parameter is changed at a time.

Suppose, keeping all other parameters same, only burning time of rocket operating is changed to 2.3 seconds. As burning time is increased for same total impulse, average thrust will reduce (to 34782.6 N) and accordingly thrust coefficient will also reduce (to 1.2302). Reduction in thrust coefficient reduces specific impulse (to 193 s). As operating pressure, throat diameter and characteristic velocity are not changed the mass flow rate of propellant remains same. However, enhancement in burning time results in increased mass of propellant (18.36 x 2.3 = 42.23 kg) for same mass flow rate. As length and outer diameter of propellant is unchanged, the inner diameter is to be reduced to accommodate more propellant. As burning time is enhanced, web required is more (40.01 mm) and for a given outer diameter, the value of inner diameter must be reduced (to 39.94 mm). The value of burning rate at the operating pressure will increase (to 17.39 mm/s) and burning rate index will also increase (to 0.609). So *net effect of increasing burning time is reduced (i) average thrust (ii) thrust coefficient (iii) Inner diameter and enhanced (i) mass of propellant (ii) web thickness (iii) burning rate index.*

Suppose throat diameter is increased to 63 mm and no other parameter is changed. Now larger throat diameter requires a higher mass flow rate to maintain same operating pressure and the value goes to 20.242 kg/s. So, propellant required will increase (to 40.484 kg). Accordingly inner diameter will reduce (to 56.41 mm) and propellant web will increase (to 35.898 mm). As burning time is not changed, the value of burning rate at the operating pressure will increase (to 17.949 mm/s) and burning rate index will also increase (to 0.6966). In fact, throat enhancement results in lowering of thrust coefficient (1.283) and specific impulse (201.437) also.

Reduction in characteristic velocity has the same effect as increase in throat diameter. This result in *increased (i) mass flow rate (ii) weight of propellant (iii) web thickness of propellant (iv) burning rate at the operating pressure (v) burning rate index, and reduced (i) specific impulse, (ii) inner diameter of the propellant.* If specific burning rate at defined pressure is changed, keeping other parameters constant, no calculated parameter is changed, except burning rate index. The summary of parametric study is expressed in tabular form below as Table 11.

Table 11: Parametric study on end inhibited tubular propellant configuration

Parameters	*Units*	*Reference*	*BT (s) [2-2.3]*	*BR (mm/s) [14-12]*	*C* (m/s) [1540-1400]*	*Dt (mm) [60-63]*
Thrust	N	40000	34782.6	40000	40000	40000
Throat area	mm^2	2827.433	2827.43	2827.433	2827.433	3117.245
Thrust coeff.		1.414711	1.23018	1.414711	1.414711	1.283184
Mass flow rate	Kg/s	18.35996	18.3599	18.35996	20.19595	20.24185
Weight of propellant	Kg	36.71991	42.2279	36.71991	40.39191	40.48371
Specific impulse	s	222.085	193.117	222.085	201.8955	201.4377
Inner diameter	mm	81.27007	39.9443	81.27007	57.14286	56.40761
Web	mm	29.68248	40.0139	29.68248	35.71429	35.8981
BR at operating pressure	mm/s	14.84124	17.3973	14.84124	17.85714	17.94905
Burning rate index	-	0.163602	0.60912	0.59579	0.682263	0.693656

REVIEW QUESTIONS

1. What is meaning of burning rate of rocket propellants?
2. What is Piobert law of burning of propellants? What are consequences of this law?
3. What is Vieille's law of solid propellant burning?
4. What is Muraour's law of burning of solid propellants?
5. Enlist various burning rate laws.
6. What is significance of two terms in (i) Vieille's law (ii) Muraour's law?
7. What is temperature sensitivity coefficient of burning rate of solid propellants?
8. What is meaning of plateau and MESA effects of burning?
9. What are different burning rate laws for rocket propellants?
10. Which ballistic parameter is affected by web of the solid propellants?
11. How is web defined from rocket propellants? How it is different from that in gun propellants?

12. What is burning area of solid propellants?
13. What are consequences of higher burning area of solid propellants?
14. What is port area of solid propellant?
15. Whether cartridge loaded tubular propellants has two ports? What should be relation between both the ports?
16. What is volumetric loading of solid propellants in rocket chamber?
17. How is volumetric loading different from cross-sectional loading of solid propellants? Explain with examples.
18. What is neutrality of propellant burning? How is it defined?
19. What is meaning of tail-off of solid rocket propellants?
20. Derive expression for pressure inside rocket chamber using global analysis.
21. What is basis for internal ballistic calculation using global analysis?
22. How is stability of burning defined for different burning rate laws?
23. What is effect of higher burning rate index on pressure generated inside rocket chamber?
24. What is effect of increasing throat diameter of rocket?
25. What is significance of characteristic velocity in the internal ballistic calculation of rockets?
26. If all parameters are same, except characteristic velocity, then what is meaning of getting of equal pressure for two different throat diameters?
27. Whether burning area versus web burn profile is similar to pressure-time profile?
28. What is Klemmung?
29. What is significance of port to throat area of rocket propellant?
30. Explain effect of varying (i) burning time (ii) burning rate (iii) characteristic velocity (iv) throat diameter on internal ballistic performance parameters of a neutral burning solid cylindrical shaped rocket propellant with known average thrust and burning time.

Section II:

Predictive Rocket Propellant Ballistics

This section gives derivation for burning surface area evolved for different solid propellant geometries during rocket operation (propellant combustion or consumption). Solid propellant for rockets burns in layers and analytical technique is implemented to get burning surface area with web consumption for different shapes of the propellants, like solid cylinder, tube, multi-tubular, multi-perforated, star-perforated etc. These shapes are used in many operational rockets of the world and have been point of discussion in many research articles. Complete mathematical expressions are derived and the same has been implemented in illustrated examples. The developed mathematical formulations can be used directly for the evolution of burning surface area and from that pressure time and thrust time profile for rocket can be derived. This section discusses predictive capabilities for different configurations of solid propellants for rockets. Design of complete rocket in different configurations for same given input parameters are discussed in Appendix – 2. This section may be called *configuration ballistic or geometrical ballistic* for solid propellants. This section is divided into five chapters, each describing typical rocket propellant configuration.

- Solid Cylindrical Configurations
- Tubular and Multi-Tubular Configurations
- Multi-perforated Configurations
- Internal Star and Fin Port Configurations
- Simple Configurations and Numerical

6

Solid Cylindrical Configurations

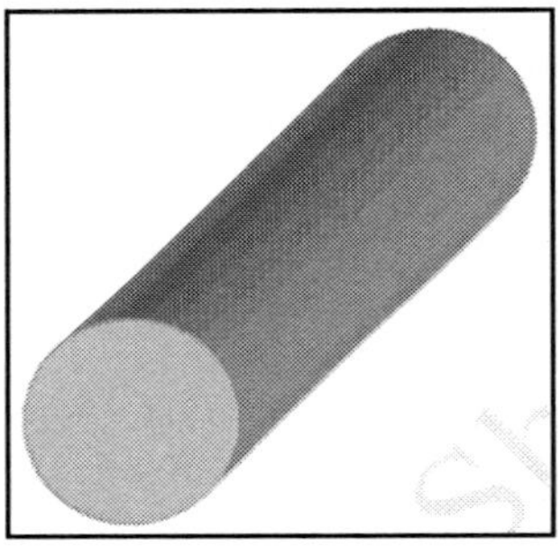

CHAPTER SUMMARY

Right Circular Cylinder is the simplest geometry used in rocket propulsion. This configuration is easy to manufacture and has potential for very high volumetric loading. The mathematical aspects of burning area variation with web consumption is explained in this chapter and different derived configurations by drilling hole or applying inhibition are also discussed, in this chapter. The chapter discusses the following:

- Features of solid cylindrical propellant configuration
- Configuration with cigarette mode of burning
- Uninhibited cylindrical configuration
- Cylindrical configuration with single blind hole
- Cylindrical configuration with annular blind hole

INTRODUCTION

Solid sustainer is one of the important rocket propellant configurations, as it gives the best neutrality and the highest solid loading. The solid sustainer, in cigarette-mode of burning, has circular burning surface which remains so, throughout the propellant combustion. Constant burning area for the entire burning duration is one of the important aspects from ballistic calculations and predictions point of view. This configuration resembles in burning with burning of a cigarette, if entire external surface is inhibited, except for one end (flat) surface. However, when practically evaluated, this results in some variations in burning area over time. This configuration is generally present in case bonded configuration or laterally inhibited cartridge-loaded configurations. The main features of this configuration are as follows.

- This configuration has the highest volumetric loading possible for the given chamber volume. No other configuration can match solid sustainer in filling the available chamber volume of rocket motor casing with energetic propellant formulations. This results in high mass ratio and probably higher ranges can be achieved.
- This configuration has inherent high neutrality and has been suitable for those applications where constant thrust is expected from rockets.
- Burning area of the propellant configuration is small, as compared to lateral burning propellant configurations (star, tube, fin, finocyl). So, for generating stable pressure, high burning rate is expected from employed propellant. This requirement is reiterated again in the chapter with practical examples.
- For stability of pressure, this type of propellant configuration desire smaller throat diameter. Smaller throat diameters are sensitive to variations in dimensional parameters and performance variations may give predictions, which are very much deviated from ideal behaviour.
- Practically, this type of propellant configuration have stringent inhibition requirement and secondary heating of propellants by conduction through rocket motor casing can be avoided only with adequate level of thermal isolation of propellants.
- This rocket propellant configuration is well suited for sustainer application, where low thrust is desired for prolonged duration. This configuration will not be suitable for booster, progressive, or variable thrust rocket motors.
- There is an ease of manufacturing in solid sustainer propellant configuration by either extrusion or casting routes. The process is simple and well established. It does not require any central core and can be automated easily. The process can be designed for a long solid sustainer grain from which lengths of required sizes

can be chopped off. The lateral inhibition requirements of propellants can be resolved by case-bonding or sleeve casting processes of manufacturing.

Cigarette Mode Burning

This propellant configuration is also called end burning configuration or solid sustainer configuration. Solid sustainer propellant is like stick, which can be initiated at one end. All lateral surfaces are inhibited and are not initiated during combustion of propellant. The burning surface, which is equal to cross-sectional area of the configuration, regresses along the length. The propellant web is equal to length of the propellant grain. The pressure-time as well as thrust time profile is neutral for this configuration, ideally. The same is reflected in burning area variation with web consumed curve shown as Fig. 31. This type of configuration has very large web for complete consumption of propellant and is generally used for providing small thrust for large duration. Due to small burning area, this type of configurations generates very small pressure in rocket motor chamber. Sometimes, the pressure is even lower than low pressure combustion limit of the propellant. Hence these propellants are generally designed with high burning rate propellant compositions.

For definition of the geometry of this configuration, only two parameters are needed. One is diameter (D) of propellant and another is length (L) of propellant. Burning area of the propellant for the entire burning duration is given by $\pi D^2/4$ and it is independent of web consumed and remains constant for the entire burning duration of propellant. From burning rate, throat diameter and other ballistic parameters of the propellant, operating pressure inside rocket motor chamber for steady state condition can be obtained. For the operating pressure, burning rate of propellant is calculated. Operating time or burning time of the propellant/rocket motor is calculated from division of length (L) by burning rate of propellant at the operating pressure. This calculation strategy is adopted for these propellant configurations.

For 200 mm diameter and 200 mm long propellant grain with standard condition depicted in this section, the burning area is calculated as 31415.93 mm^2. For the calculation of pressure, different ballistic parameters are enlisted below:

Characteristic velocity, C* = 1540 m/s

Density of propellant = 1750 kg/m^3

Throat diameter of nozzle = 50 mm

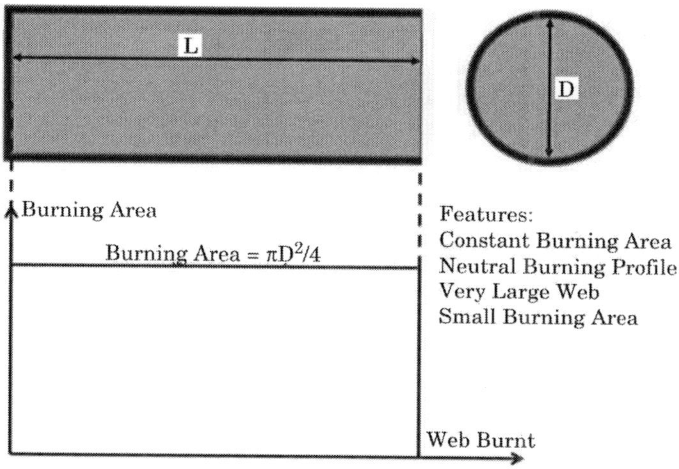

Fig. 31: Predicted variation of burning area with cigarette mode burning of propellant configuration

Burning rate of propellant at 7 MPa = 10 mm/s

Burning rate index of propellant = 0.3

These values are taken as standard values in further calculations, if values are not specified. The pressure generated is calculated as 0.9462 MPa. This pressure level is very small. At this small pressure, burning rate of the propellant is calculated as [10 x $(0.9462/7)^{0.3}$ =] 5.48 mm/s. The burning time of propellant is calculated as 36.5 seconds. Small pressure and large burning time makes such configuration suitable for long duration sustainers. However, if burning rate of propellant is raised to 40 mm/s at 7 MPa, then pressure generated is 6.856 MPa. This pressure level is generally utilized in rocket motors for various operations. For this burning rate, burning time is estimated as 5.03 seconds. Development of such high burning rate propellants is another domain of propellant development.

Another method to enhance operating pressure of rocket motor with solid sustainer configuration is reduction in throat diameter, but this leads to following problems:

- Low throat diameter has more probability of blocking by particulate matters.
- Low throat diameter may adversely affect thrust, depending on relative rise in pressure.
- Low throat diameter for same exit diameter may result in increased nozzle area ratio.

- Low throat diameter has higher throat erosion during operation of rocket.

These types of configurations are used as auxiliary propulsion rockets, as sustainer for retaining constant thrust after a given velocity is imparted to rocket. This is also used as small thrusters for attitude correction.

Another variation of this configuration comes, when cylindrical propellant configuration is inhibited at one end only. Compared to previous configuration, this configuration has higher burning surface area due to inclusion of lateral surface areas. For this configuration, maximum web for complete consumption of propellant is lower dimension of D/2 and L. In this case, during combustion propellant, length will reduce from one side, as previous configuration. Additionally, outer diameter will also reduce. The burning area will be high and nature of burning profile is regressive in nature. It is clear that propellant burning area will reduce with consumption of the propellant.

$$\text{Initial burning area} = \text{lateral surface area} + \text{surface area of one end face} = \pi \times 200 \times 200 + (\pi/4) \times 200^2 \text{ mm}^2 = 157079.6 \text{ mm}^2$$

As propellant web is consumed, burning area reduces. The variation of burning area with different web (w) consumed in course of burning is calculated from the equation given below. The burning area variation is plotted against web consumed as Fig. 32.

$$\text{Burning area, } A_b = \pi/4 \times [12 \times w^2 - 8 \times (D+L) \times w + D \times (D+4L)]$$

Using ballistic parameters given in the beginning of this section, pressure time profile of the configuration is also generated and the curve

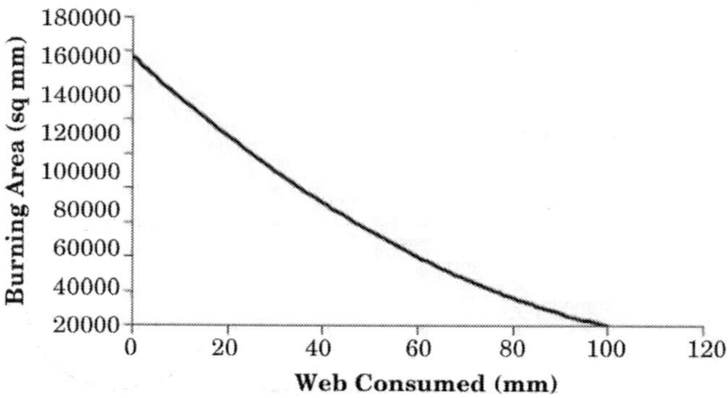

Fig. 32: Burning area variation for laterally burning solid cylinder configuration

is shown as Fig. 33. As low pressure continue to exist as per calculation for a very long time, as per customary, attainment of 10% of maximum pressure is taken as burning time of the configuration. As per this convention, burning time is calculated as 8.3 seconds. Although, propellant continues to burn for around 19 seconds, but after 8.3 seconds, burning may not be smooth and thrust generation will be inadequate. This time is much less than earlier predicted time (36.5 seconds) for end burning configuration with 10 mm/s burning rate at 7 MPa. However, it is observed that this configuration has lower burning time but higher pressure than end burning configuration. Overall, area under pressure time profile for this configuration and the end burning solid sustainer configuration remains same.

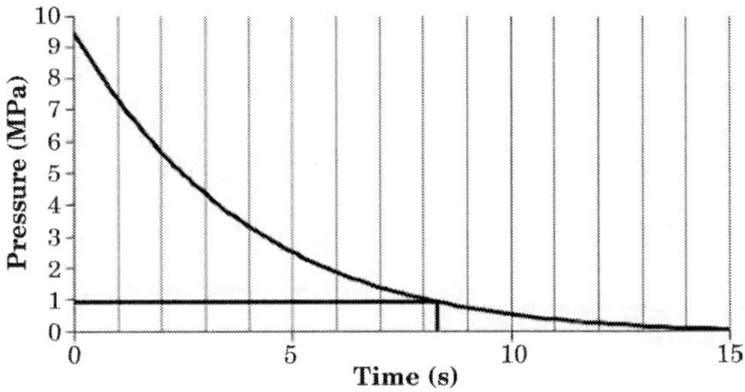

Fig. 33: Pressure time curve for laterally burning solid cylinder configuration

Uninhibited Configuration

Another variation of similar configuration is obtained in free standing mode of uninhibited cylindrical propellant grains. The geometry is fully defined by two parameters, namely Diameter (D) and Length (L). In such grains, burning area of propellant reduces with progress of combustion. This type of burning mode is called regressive burning. Initial burning area of propellant is the highest, which reduces with web consumption and time. The pressure and thrust from such propellant configuration also varies in regressive manner. This type of propellant grains can be bulk produced by extrusion. Since inhibition requirement is dispensed with, this propellant configuration has faster production rate. However, inherent regressivity limits use of such propellant configuration. In addition to regressive nature, burning at external lateral surface has following problems:

- Clearance provided at outer surface for proper initiation reduces volumetric loading of propellants in the rocket motor chamber.

- Combustion at outer surface leads to exposure of rocket motor casing to hot combustion gases for prolonged duration.
- Sometimes thermal insulation is needed at the inner surface of rocket motor casing, enhancing the inert weight of the rocket.
- Initiation all around simultaneously, needs specially designed igniters.
- Ensuring uniform combustion is also difficult to ensure in such configuration.
- Centering of such propellant inside rocket motor chamber is difficult.
- Chances of differential pressure may lead to crumbling of propellant at the end of burning, when very small web remains.
- Near end of propellant consumption, chances of ejection of unburnt propellant pieces from nozzle exist.

For a 200 mm diameter and 200 mm long propellant grain in uninhibited cylindrical configuration, the performance prediction needs development of equation for the burning area of propellant with web consumed.

$$\text{Initial burning area} = \text{lateral surface area} + \text{surface area of two end faces} = \pi \times 200 \times 200 + 2 \times (\pi/4) \times 200^2 \text{ mm}^2 = 188495.6 \text{ mm}^2$$

During combustion of propellant outer diameter and length of the propellant piece reduces. Maximum web of propellant in this case is given by D/2. Here maximum web to be consumed for complete consumption of propellant is 100 mm. For a given web (w) consumed, burning area of the propellant is calculated by the equation given below.

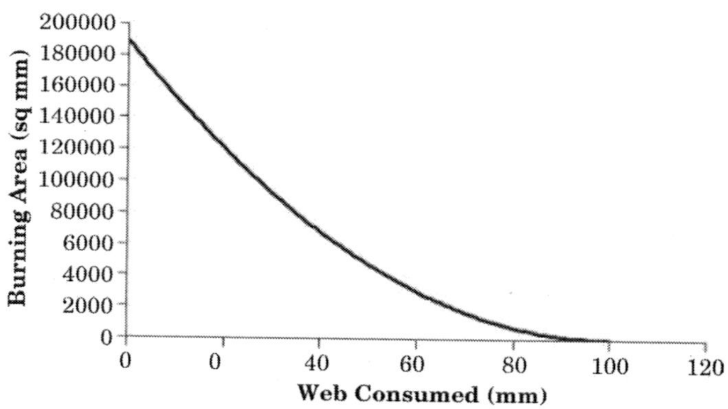

Fig. 34: Variation of burning area for uninhibited cylindrical propellant configuration

Burning area, $A_b = \pi/2 \times [12 \times w^2 - 4 \times (2D+L) \times w + D \times (D+2L)]$

Using equation above, burning area versus web burnt profile is generated for the configuration as Fig. 34. The initial burning area matches with the earlier calculated result. It is observed that burning area reduces in monotonic way from initial value to zero at web consumed of 100 mm. The variation is quadratic as depicted by the developed equation. This type of burning area reduction leads to poor utilization of propellant at the end of burning. Near end of propellant consumption, burning area becomes so small that pressure generated inside rocket motor chamber may not be sufficient to impart sufficient thrust. Additionally, the chamber condition may be non-steady under such circumstances. It is also clear that normal burning rate law may not be valid for these extreme situations and rather than following the burning rate law, propellant exhibit some lump-sum burning rate at low pressures.

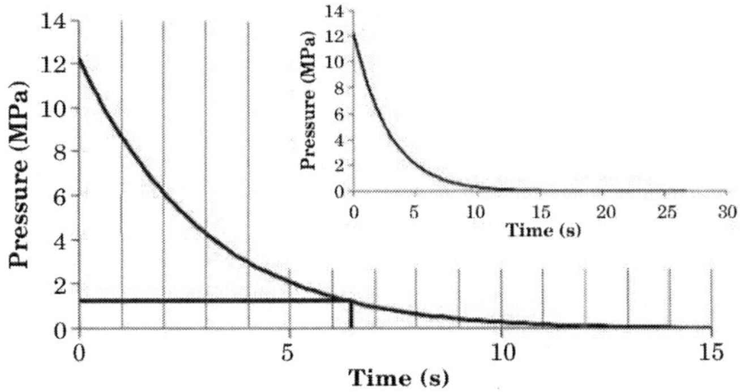

Fig. 35: Evolved pressure-time profile for the uninhibited cylindrical propellant configuration

From the ballistic parameters, given in the beginning of the section, pressure time profile is generated for the configuration and the same is depicted in Fig. 35. Maximum pressure is realized as 12.236 MPa in the beginning, which reduces continuously. If burning rate law is assumed valid till complete consumption of propellant, the burning time is around 31 seconds. However, propellant burning rate may not be very low as predicted by the used power law of burning rate ($r = ap^n$). In rocketry, it is customary to take 10% of maximum pressure as cut-off of burning time. Accordingly, in this case, 10% of maximum pressure is 1.2236 MPa. Horizontal line is drawn on the pressure-time plot and the burning time of propellant, considering attainment of 10% of maximum pressure as criteria, is 6.45 seconds.

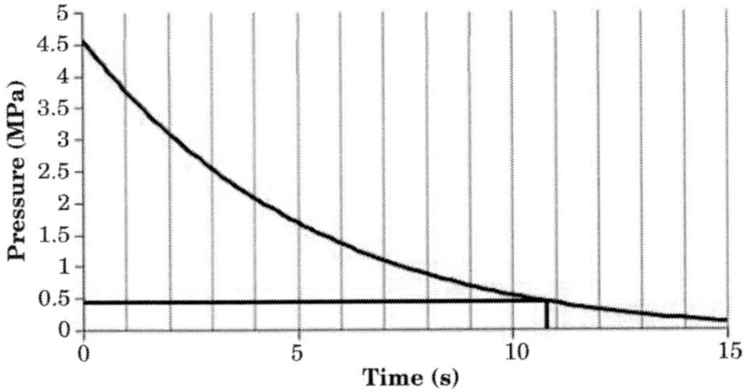

Fig. 36: Evolved pressure-time profile for alternate uninhibited cylindrical propellant configuration

Suppose, length of this propellant configuration is reduced to 50 mm, then web of the propellant for complete consumption of propellant is 50 mm (not 100 mm, as earlier). The initial and subsequent burning area of the propellant will reduce and so is the case with pressure and thrust values. However, quadratic nature of variation of burning area with web consumed is exhibited by this propellant configuration also. As pressure levels of this configuration will be lower than earlier configuration, burning rate of the propellant will be less. This may lead to higher burning time of the propellant. On calculation, initial burning area of this propellant configuration is obtained as 94247.8 mm^2. Initial pressure is 4.545 MPa and burning time on the basis of 10% of maximum pressure, is 10.8 seconds. The pressure time profile is shown in Fig. 36.

Blind Hole Configuration

With the help of inhibition and blind perforations in the solid tubular propellant configuration, it is possible to generate dual-thrust burning profile, where two distinct levels of thrust and pressure are achieved in course of burning by getting two different burning area domains. A typical configuration may have a blind central hole till around one-fourth length of the solid cylindrical propellant configuration. External surface at one side is inhibited from one end and remaining three-fourth distance on lateral side. The configuration is shown as Fig. 37.

This Propellant grain is characterized by 5 dimensional parameters only, contrary to most popular star configuration, where 7 geometrical parameters are needed for their complete definition. The propellant configuration is truly axis-symmetric and burning surface evolution is three dimensional in nature. Propellant grain radius (R_o) and total length (L_g) indicates overall dimension, while central hole radius (R_h),

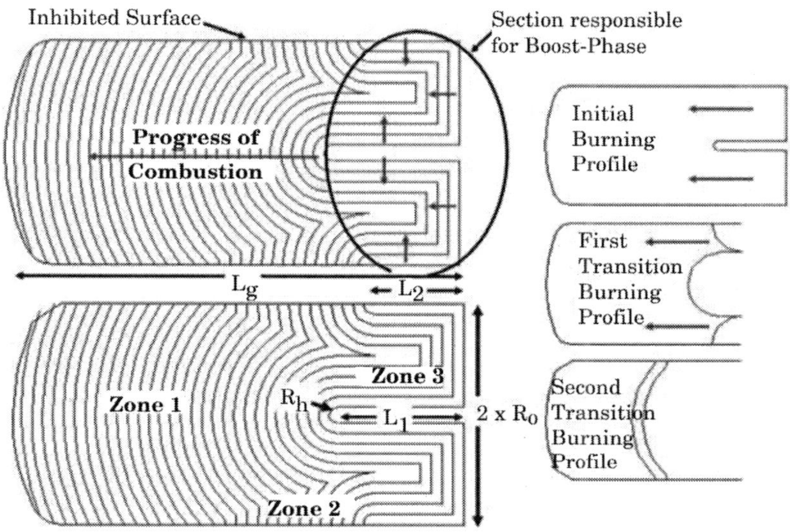

Fig. 37: A typical dual-thrust configuration

hole depth (L_1) and length of uninhibited outer surface (L_2) indicates essential features for evolution of dual thrust burning pattern. Hole depth (L_1) is always kept more than length of uninhibited lateral surface (L_2) to get smooth and gradual transition between two thrust phases. In this case variation of burning surface area is responsible for dual-thrust mode.

In the initial phases of combustion, higher surface area is available for burning from the uninhibited end. It gives higher pressure representing boost phase in thrust history. However, it gets consumed in the initial phase. This part is combination of three surfaces – two cylindrical and one end face of a cylinder. This is represented as 'Zone 3' in the Fig. 37. Junction of outer inhibited and uninhibited surface give rise to a concave burning cusp with center at the interface. Because of cyclic symmetry, surface area is obtained as a surface of revolution of evolved spherical section about axis of the solid propellant grain. This gives rise to surface specified by 'Zone 2' in Fig. 37. As $L_1>L_2$, a second transition in burning surface is observed.

The most significant part of burning pattern is imparted by the central hole which acts as a solid of revolution which is initially hemi-spherical, but later on it becomes spherical section with lower angles at the center of spherical cross-section. This is representing 'Zone 1' and is active throughout the combustion. Although in the initial phases, it contributes less in total burning profile but later low thrust profile is result of burning in this zone alone. This contributes in sustainer phase burning significantly.

The grain design is simple and processing of this geometry is also very easy. It is observed that propellant can be manufactured by both casting and extrusion methods. Longitudinal section of propellant configuration is considered (Fig. 38) such that solid of revolution about central axis can give the entire propellant configuration. It is clear that first parameter namely P_1 (= L_1-L_2) represents difference of hole depth (L_2) and outer uninhibited surface length (L_1). Second parameter P_2 (= "$R_o^2 + P_1^2$) represents central distance between two spherical surfaces. Burning surface evolution is characterized by several transitions. Complete propellant grain consumption is characterized by web burnt of ($L_g - L_1 - R_h$).

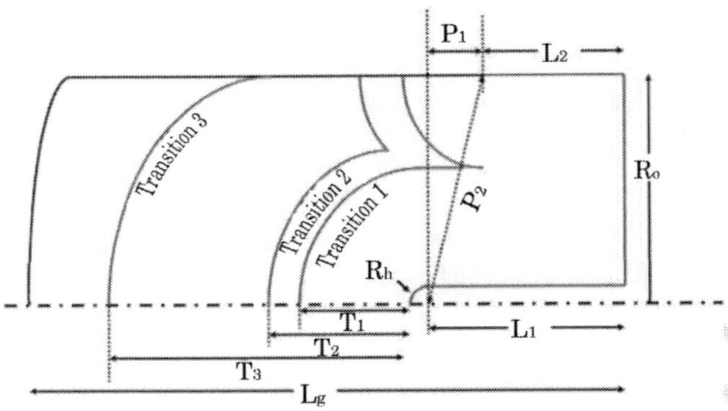

Fig. 38: Salient parameters for burning surface evolution

First transition of solid propellant grain is characterized by consumption of uninhibited section complete conversion of cup shape solid of revolution at the right into a single cylindrical surface. It is represented by web burn of T_1 and given by equation 1. Second transition occurs when cylindrical surface is completely consumed and is represented by T_2 in the Fig. 2 and mathematically expressed as equation 2. Third transition, characterized by T_3 is represented by complete consumption of outer spherical surface. After this transition, entire burning surface is made of central spherical section only and is represented by equation 3.

$T_1 = (R_o - R_h)/2$ Equation 1

$T_2 = [P_1^2 + (R_o\text{-}R_h)^2]/ (2R_o - 2R_h)$ Equation 2

$T_3 = (P_1^2 + R_o^2 - R_h^2)/ (2R_h + 2P_1)$ Equation 3

Most simple and well developed formulation is possible for zone 3 portion of the propellant configuration (Fig. 37), which is valid till transition T_1 is reached. For a given web burnt, w this surface is given by equation 4.

$$S_3 = 2p(R_h+w)(L_2-w) + 2p(R_o-w)(L_1-w) + p(R_h+ R_o)(R_o-R_h-2w)$$
Equation 4

Between transitions T_1 and T_2, this zone is reduced to a simple cylindrical section. Length of this cylindrical section (L_x) is given by equation 5. After transition T_2, this section does not contribute to the surface area. Surface area between transition T_1 and T_2 is given by equation 6.

$$L_x = P_1-\sqrt{[(R_o-R_h)(2w+R_h-R_o)]}$$ Equation 5
$$S_3 = 2p(R_h+w) L_x$$ Equation 6

Zone 1 is section of a spherical surface, with limiting angle given by. On integration (Fig. 39), the surface area is calculated in terms of this control parameter (a). However, value of a varies with transition status. Variation of a with web burnt is given by equation 7.

$$S_1 = 2\pi(R_h+w)^2(1\text{-Cos}a) \text{ where}$$
$$a = \pi/2 \text{ for } w < T_2$$
$$= \pi-\text{Tan}^{-1}(R_o/P_1)-\text{Cos}^{-1}[\{(R_h+w)^2+P_2^2-w^2\}/\{2(R_h+w)P_2\}] \text{ for } T_2 < w <$$
$$T_3 = \text{Sin}^{-1}\{R_o/(R_h+w)\} \text{for } T_3 < w < (L_g-L_1-R_h)$$
Equation 7

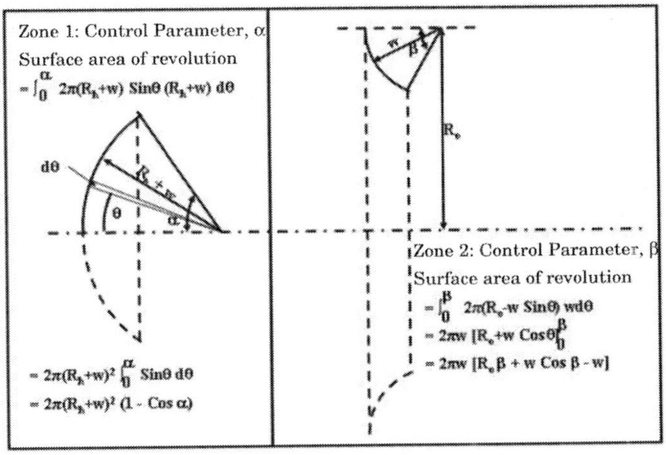

Fig. 39: Surface of revolution for different zones

Zone 2 is created at the external diameter of grain and is also in the form of a surface of revolution of a circular arc. But in this case axis of revolution is offset in the direction opposite to the center of curvature of arc. This surface is governed by a limiting angle named b (Fig. 39). This surface disappears, once transition T_3 is reached. The evolved surface is given by equation 8.

$$\begin{aligned} S_2 &= 2\pi w\{R_o\beta - w + w\cos\beta\}, \text{ where} \\ \beta &= \pi / 2 \text{ for } w < T_1 \\ &= \pi/2 - \sin^{-1}\{(P_1 - L_x)/w\} \text{ for } T_1 < w < T_2 \\ &= \sin^{-1}[\{R_o - (R_h + w)\sin\alpha\}/w] \text{ for } T_2 < w < T_3 \end{aligned} \qquad \text{Equation 8}$$

To validate the mathematical formulation, a propellant grain with salient dimensions and parameters as indicated in Table 12 is evaluated. Pressure time profile is predicted using mathematical formulation developed above.

Predictions are made without considering any ignition delay and tail-off portion obtained at the end of firing is also not included in the formulation. To understand effect of each of the salient parameters, parametric study is conducted for the given configuration. Since mathematical formulation developed above is restricted to burning surface area calculations, comparison and effect of various parameters are also depicted in burning area versus web burnt profiles only. It must be noted that pressure-time or thrust time profiles are in fact scaled versions of burning surface area and web burnt profiles. For different propellant properties, suitable pressure-time and thrust time profiles can be extracted.

For parametric study, outer radius (R_o) of the propellant is considered as base parameter. Total length (L_g) to outer radius ratio (L_g/R_o) of the propellant grain is taken as 5. Variation of volumetric loading fraction is plotted against ratio of L_1/R_o for different R_h/R_o in Fig. 40.

Table 12: Salient parameters for dual thrust configuration

Parameter	***Symbols***	***Value***	***Unit***
Outer radius of propellant	R_o	54	mm
Total length of propellant	L_g	213	mm
Hole radius	R_h	4	mm
Hole depth	L_1	59	mm
Uninhibited length at outer diameter	L_2	42	mm
Propellant density	r	1.58	g/cc
Characteristic velocity	C*	1380	m/s
Burn rate law	r	$6.6491 \times P^{0.1052}$	mm/s
Throat area of rocket motor	A_t	54.63	mm^2

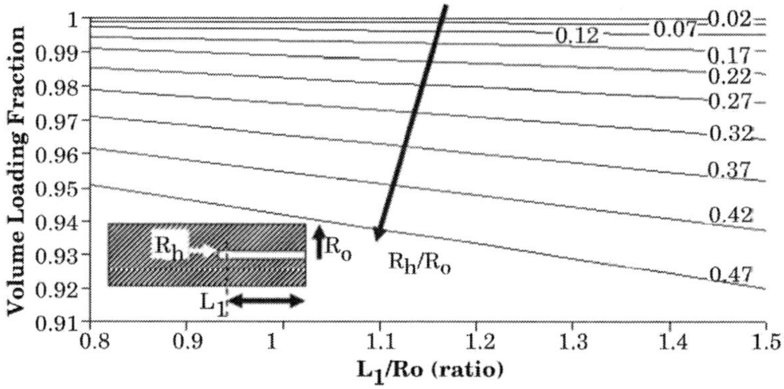

Fig. 40: Variation of volume loading fraction

It is clear that as L_1/R_0 increases volume loading fraction reduces. A similar trend is observed with increase in R_h/R_0. However even for a very unlikely hypothetical case of hole depth fraction equal to 1.5 and hole radius fraction of 0.47, volumetric loading fraction is more than 90%. It clearly indicates that prime requirement of high loading fraction is accomplished by this propellant grain configuration.

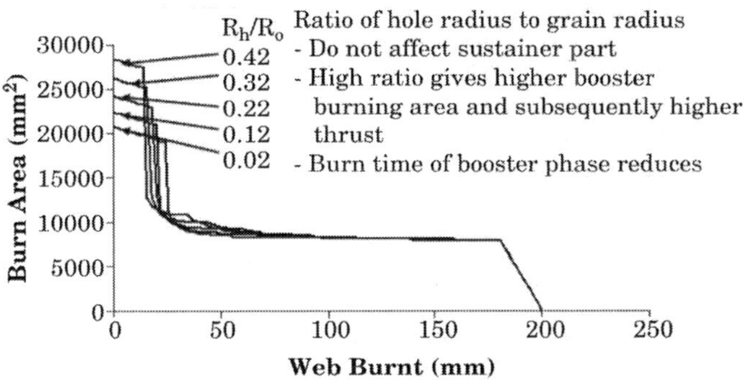

Fig. 41: Effect of varying hole-radius fraction

Effect of hole radius fraction (R_h/R_0) on the burning surface area versus web burnt profile is also studied. The variations are plotted as Fig. 41. It is clear that variation of radius of hole alone does not affect sustainer part or low burning area part of the curve. However, booster phase is significantly affected. Higher hole radius fraction indicates higher hole radius and ultimately higher initial burning surface area is available. However, burning time shows a reverse trend and higher hole radius invariably gives lower boost phase time. Transition is also smooth for lower value of hole radius.

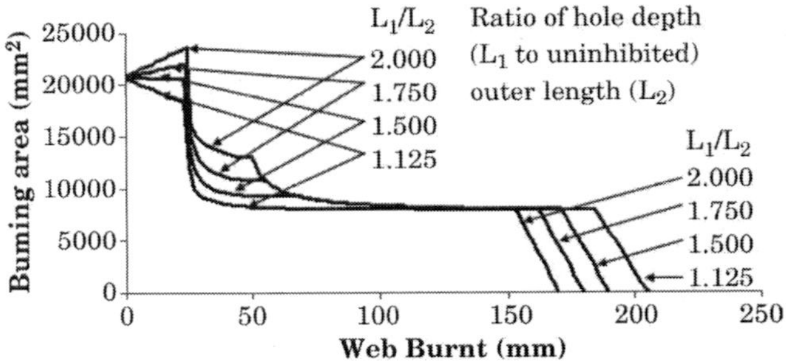

Fig. 42: Variation in performance profile with length ratio

In the propellant configuration, two important length parameters are defined. One is hole depth (L_1) and other is length of uninhibited portion at the outer periphery of the propellant grain (L_2). For the configuration under consideration, L_1 is always more than L_2. Effect of ratio of these lengths on performance prediction curves is also explored (Fig. 42).

As length ratio (L_1/L_2) increases boost phase profile changes from regressive to progressive profile. Sudden drop in burning surface area occurs almost at same duration. At high length ratio, total burning time reduces. Alternately, sustainer phase burning time reduces for higher length ratio. For smaller length ratio, transition is more-smooth. At higher length ratio, tendency to form a secondary higher burning area is observed. Single chamber, single grain, single nozzle, single propellant configuration is explored for realization of dual-thrust burning profile. Achievement of dual thrust by configuration alone is established. As

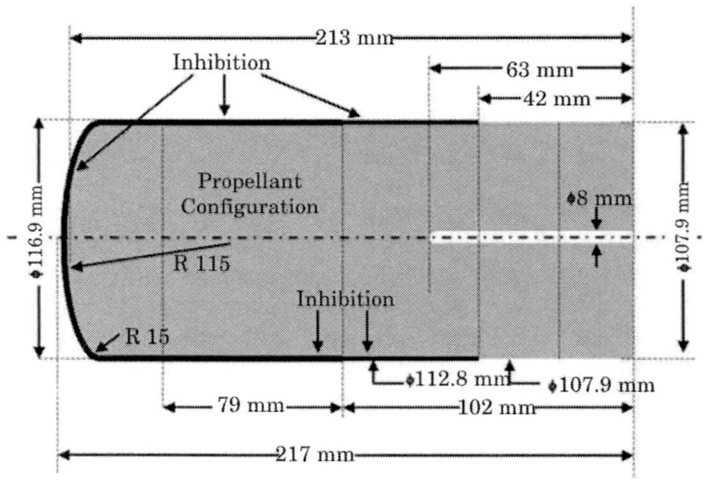

Fig. 43: Dimension of dual thrust configuration

number of parameters are less, better control and smooth repeatable performance is expected from the grain under consideration.

With mathematical formulation developed above, burn area versus web burnt for the propellant configuration can be obtained. For numerical ease, dimensions of a typical propellant grain are shown as Fig. 43 and generated burning area versus web burnt profile is shown as Fig. 44. The pressure time profile is shown as Fig. 45. Single blind hole configuration along with proper inhibition gives a dual thrust configuration, but if the hole is extended to other end the configuration becomes a tubular configuration, which is explained in the next chapter.

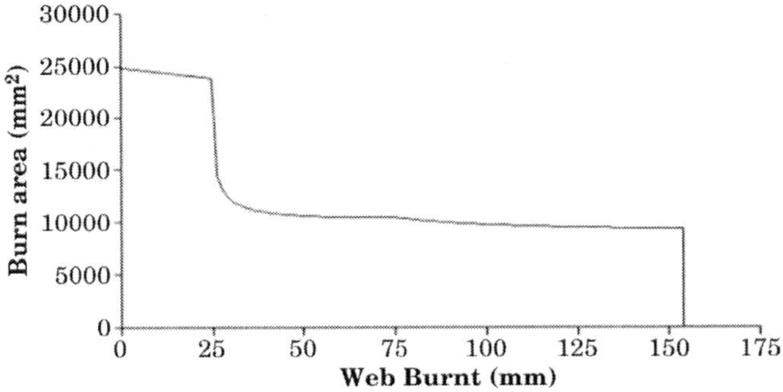

Fig. 44: Burning area versus web burn profile

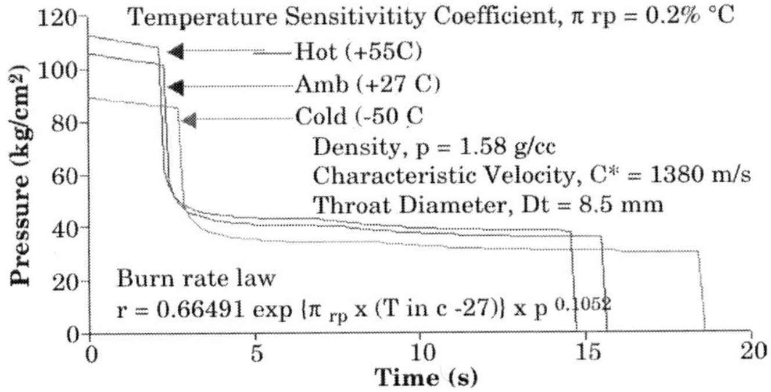

Fig. 45: Pressure-time profile for dual thrust configuration

Blind Annular Hole Configuration

Another dual thrust propellant configuration is possible with annular blind hole in inhibited solid cylindrical propellant grain. This type of propellant configuration has 4 independent dimensional parameters. The booster phase is created by the annular section, where end face and lateral faces of annular ring gives a higher burning surface area for so called booster phase, while later on propellant burns in sustainer mode. The length (L) of propellant grain and that (l) of the annular hole are two independent variable dimensions. The hole width (2t) is another independent dimension. The outer diameter (D) of propellant is also independent dimension. This propellant configuration (Fig. 46) is inhibited from all side except one and resembles solid sustainer configuration. The uninhibited side contains annular ring in such a way that outer and inner webs are matching and vanishes at the same time. The inner diameter (d) of the propellant configuration to give a perfect dual thrust propellant is given by d = D/2.

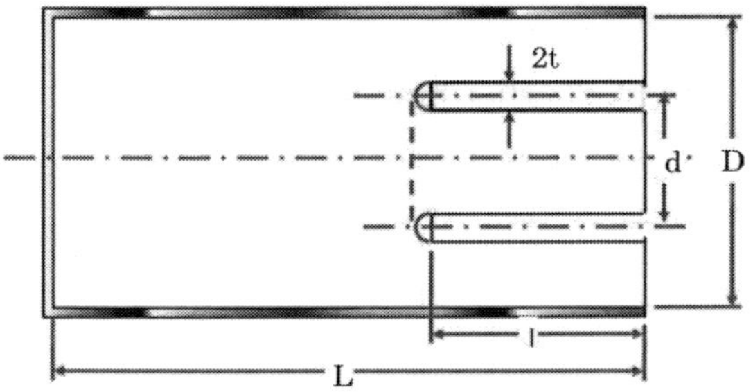

Fig. 46: Solid sustainer with blind annular hole

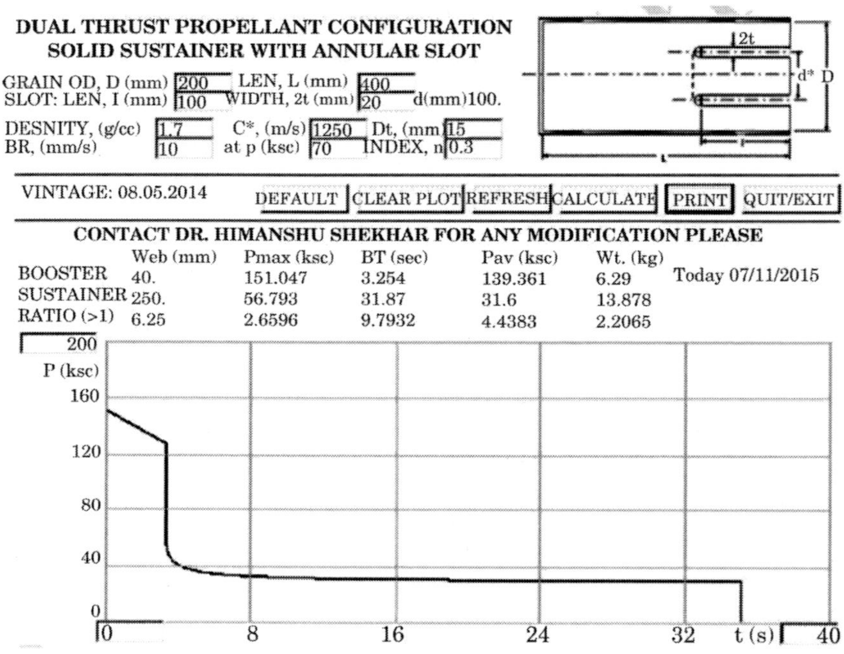

Fig. 47: Dual thrust solid sustainer with annular hole

The propellant configuration has only one web transition at the conversion of booster to sustainer. This web transition is indicated as w_1 and is represented by w_1 = (d/2-t). After this sustainer phase continues. The complete consumption of propellant can be assumed to take place, when blind hole touches inhibited end of the propellant length. This is represented by w_2 and is mathematically expressed as w_2 = L-l-t. The equation of burning surface area can be defined for both the phases of burning by the following expression.

$$A_b = \pi D^2/4 + 2\pi d(l\text{-}t\text{-}2w) + \pi^2 d(t+w) \text{ for } w < w_1$$
$$= 2\pi d(t+w) \, Sin^{-1}[(D\text{-}d)/\{2(t+w)\}] \text{ for } w_1 < w < w_2$$

Using above simplified equation burning area versus web burnt profile can be generated for the configuration and then with known propellant properties, pressure time and thrust time profile using global analysis can be ascertained. As the calculation is simple, the calculation strategy is coded and front page of the trial result for a given configuration is shown as Fig. 47.

REVIEW QUESTIONS

1. Why is simple cylindrical configuration known as solid sustainer?
2. What are advantages of simple cylindrical solid propellant configuration?
3. What are disadvantages of using simple cylindrical solid propellant configuration?
4. How cigarette mode of burning is achieved in simple cylindrical configuration?
5. Compare cigarette mode of burning with uninhibited solid cylindrical configuration.
6. Why is cigarette mode of burning used in rockets?
7. What are repercussions of removing lateral inhibition from cigarette mode of burning propellant?
8. What are advantages and disadvantages of using inhibition on propellant?
9. What are limitations of uninhibited solid cylinder propellant configuration in rockets?
10. What is meaning of dual thrust configuration?
11. What are features of blind hole solid cylinder configuration with inhibition?

12. What controls the higher thrust level in the dual thrust configuration obtained by (i) blind central hole (ii) blind annular hole?
13. How annular blind hole helps in achieving dual thrust configuration?
14. Whether area during sustainer phase of dual thrust configuration, derived from solid cylindrical configuration matches with burning area of cigarette mode of burning propellants?
15. Whether regressive burning is inherent feature of solid cylinder configurations?
16. Derive expression for the burning area of uninhibited solid cylinder configuration.
17. Derive expression for the initial burning area of annular blind hole configuration.
18. What is total burning area of uninhibited solid cylinder with 50 mm diameter and 1000 mm length? [Ans: 161006.6 mm^2]
19. What is initial burning area of cigarette mode of burning propellant with diameter of 40 mm and length of 700 mm? How does burning area changes, if lateral inhibition is removed? [Ans: 1256.6 mm^2, 89221.2 mm^2]
20. Rocket propellant, designed to burn in cigarette mode has diameter of 30 mm and length of 400 mm. Find burning area of the propellant. If burning time of the propellant is 10 seconds, what is burning rate of the propellant at the operating pressure of the rocket?[Ans: 706.86 mm^2, 40 mm/s]

7

Tubular and Multi-tubular Configurations

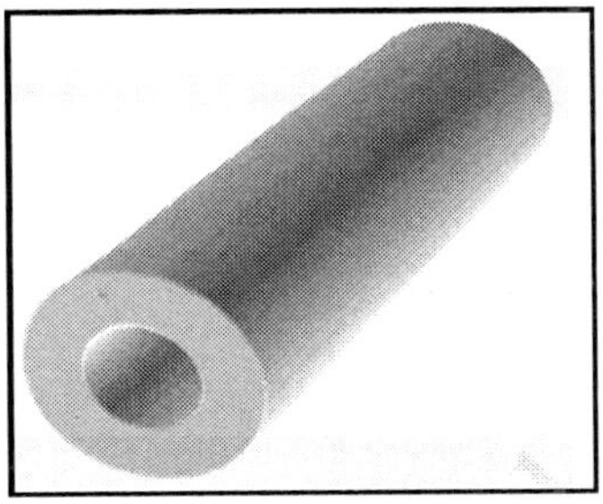

CHAPTER SUMMARY

This chapter discusses another important solid propellant configuration, where central through perforation in the solid cylinder is made. This propellant configuration has ballistic importance and is used for the assessment of burning rate law of solid propellants for rockets. The mathematical derivation of burning area and generation of pressure time profile is explained for various configurations, derived from simple tubular configuration. The chapter discusses the following:

- Features of tubular propellant configuration
- Bare tubular propellant configuration
- Inhibited tubular propellant configuration (progressive and regressive)
- Taper tubular propellant configuration
- Offset port tubular configuration
- Multi-tubular propellant configuration

INTRODUCTION

Tubular configuration of solid propellant grain is basically solid cylinder with central concentric perforation. Compared to solid sustainer propellant grains, this configuration is generally employed universally in rocket and missile applications due to various features, offered by this geometrical identity:

- From manufacturing point of view, this type of propellant configuration can be made in bulk quantity by extrusion technique. The consolidation (high density) in this configuration is better than solid cylinders. However, outer diameter of tubular propellant grain is limited by capacity of extrusion press and associated enhanced hazards with higher dimensions.
- For larger diameter tubular propellant grains, casting technique is generally applied, where propellant slurry is cast in a cylindrical container around a concentric mandrel or core. Both double base propellants and composite propellant grains are processed in tubular configuration by casting routes. In this route, pressing hazards are relatively less, but friction hazards during decoring of central mandrel become a major challenge from safety point of view.
- This type of propellant grains are utilized in main motor in both mono-tube and multi-tubular configurations and it was extrusion of double base matrix and bulk production possibility, which makes these grains useful for almost all rockets of earlier vintage (before 1970s).
- This propellant grain has unique feature of providing all type of burning profiles namely progressive, neutral and regressive, by suitably inhibiting the different surfaces.
    - o If tubular grain is allowed to burn on all surfaces (both ends, inner and outer lateral surfaces), the burning profile will be regressive, *i.e.*, burning area (and pressure) will reduce with web consumed (or time).
    - o If ends of the tubular grain are inhibited (prohibited from burning), and propellant is allowed to burn in radial direction from both inside and outside, this results in neutral burning profile.
    - o If tubular propellant grain is allowed to burn from inner lateral surface in radial outward direction, by inhibiting both the ends and outer lateral surface, the burning profile is progressive. Burning area (and pressure) increases with web consumed (or time).
- This propellant grain is utilized for determination of burning rate of solid rocket propellants in ballistic evaluation motors (BEM). All three modes of firings are adopted and generation of burning rate data is an informative mathematical ease from such firings.
- Burning area of tubular propellant grains can be tailored by having more number of control parameters (outer diameter, inner

diameter, length, inhibition) than solid sustainers (diameter and length). In tubular configurations, burning area can be controlled by changing outer diameter, inner diameter and length of the propellant grains. Since burning area utilizes length of propellant grain, burning area of tubular propellant grains is higher than solid sustainer propellant grain in same volume.

- o This leads to reduced burning rate requirements for tubular propellant configuration than that for the solid sustainer propellants configuration.
- o Tubular grains provide higher pressure levels than identical solid sustainer propellant (same burning rate, characteristic velocity, density) in identical rocket motor (same throat diameter).
- o Tubular grain has less burning time as compared to solid sustainer propellant grain because maximum web to be consumed is much lower for tubular grains.

- For suitable tailoring of variation in burning area, inhibition conditions and tube dimensions are optimized.
- Further variations in performance can be obtained by providing taper in the central port section of the propellant configuration.
- Volumetric loading of propellant in this case, is of the order of 70 to 85%.

Bare Tubular Propellant Grain

Tubular propellant grain is obtained by extrusion or casting and it resembles a thick cylinder (Fig. 48). After ignition, propellant burns at inner diameter, outer diameter and also along the length. Outer diameter and length of the propellant reduces during combustion of propellant, while inner diameter enlarges. The variation of propellant configuration during combustion is depicted in Fig. 48. It is apt to express and analyze the propellant geometry in terms of non-dimensional parameters, represented in terms of outer diameter of the propellant grain (D). The inner diameter in the beginning is given by sD, while length is given by hD. Max web of the propellant that is available for the propellant configuration depends on relative value of non-dimensional parameter 'h' and 's'.

When propellant grain is smaller in length or $2h+s < 1$, maximum web is given by half of the length or hD/2. When propellant is sufficiently long or $2h+s > 1$, maximum web is given by (1-s)D/4. The reduction in outer diameter and length, during combustion makes this grain regressive. However, enhancement in inner diameter counters this reduction in burning area and reduces regressivity of the propellant

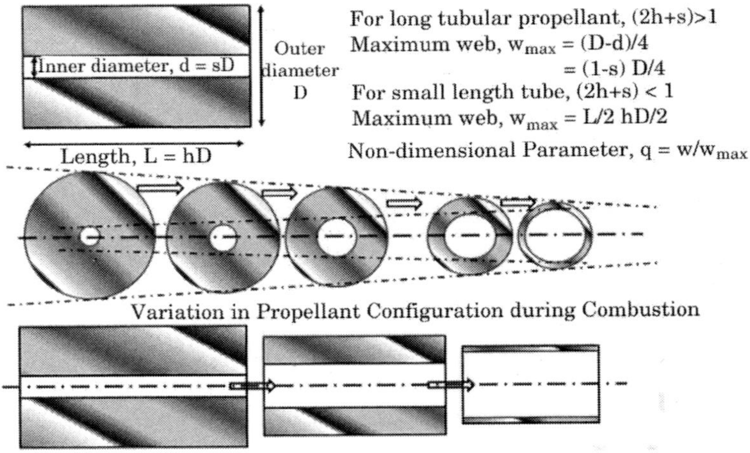

Fig. 48: Salient features of bare tubular propellant grain

grain during combustion. But overall, the burning area reduces with consumption of propellant. The expression for the burning area is given below:

$$\text{Burning area, } A_b = \pi D^2/2 \text{ x } (1+s) \text{ x } \{(1-s) \text{ x } (1-2q) + 2h\}$$

Variation of burning area with web consumed is main input for ballistic prediction and for analysis a tubular propellant grain with outer diameter of 120 mm is considered. Parameters, 'h' is taken as 1.5 and 's' is taken as 0.5. For this configuration, burning area is normalized and ratio of current burning area to initial burning area is calculated for plotting. Initial burning area using derived equation is obtained as 118752.2 mm^2. Similarly, this propellant is long propellant grain because 2 x h + s = 3.5, which is more than 1. So, maximum web of the propellant

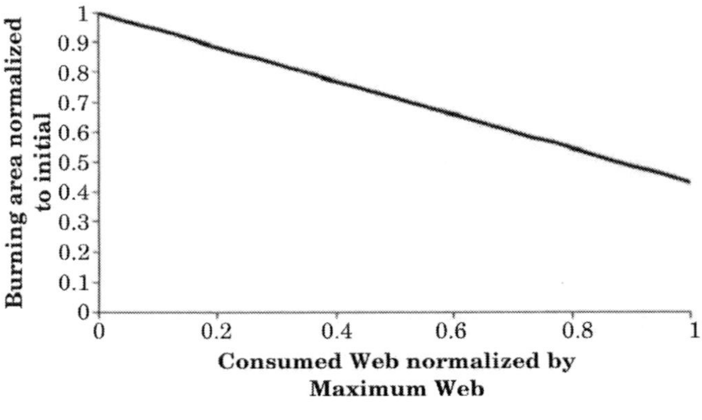

Fig. 49: Normalized variation of burning area with consumed web

grain is given by (1-s) x D/4 = 120 x 0.5/4 mm = 15 mm. Instantaneous value of web consumed is divided by maximum possible web to normalize the web parameter. Normalized plot of burning area versus web is depicted in Fig. 49. The regressive nature of curve is obvious and the variation seems to be a straight line. The equation for such normalized curve is given below.

$$A_b / A_{bi} = 1 - [4 \times (1\text{-}s)/\{(1\text{-}s)+2h\}]\ q$$

The slope of the curve is dependent on parameters 'h' and 's', only. Form the equation, it is clear that slope is negative (second term) and slope is independent of propellant outer diameter. At complete combustion of propellant, percentage of burning area remaining is also a function of these two parameters only. This is given by putting q=1 in the above derived equation.

Fraction of burning area at the end of propellant consumption

$$= 1 - [4 \times (1\text{-}s)/\{(1\text{-}s)+2h\}] = 1-[4 \times (1\text{-}0.5)/\{(1\text{-}0.5)+2 \times 1.5\}]$$

$$= 1 - (2/3.5) = 0.4285$$

This indicates that burning area at the end of combustion of 42.85% of the initial burning area. The values may vary, depending on selection of parameters. Outer diameter does not play any role in the calculation of (i) normalized burning area (ii) slope of variation of normalized burning area with normalized web burnt (iii) fraction of burning area available at the end of propellant consumption.

If practical situation is considered, then bare tubular propellant grain burns at outer surface also and for initiation and flow of hot gases, some clearance between outer diameter of propellant and inner diameter of rocket motor must exist. Because of this, propellant loading in the rocket motor reduces. The empty space available at a cross-section for the flow of gases is called port area. For tubular propellant configuration, port area is space between rocket motor and propellant as well as at the inner diameter. Both these area increases in course of combustion of propellant. Ratio of burning area to port area is one of the importance parameter (K_I) for establishing pressure values. Abnormalities may be observed, if this area ratio is improper. If internal diameter of rocket motor (D_m) is expressed as p x D, then value of burning area to port area is given by following expression.

$$K_I = 2\ (1+s)\{(1\text{-}2q)(1\text{-}s)+2h\}/[p^2\text{-}\{(1\text{-}s)^2(1\text{-}q)\}]$$

For a typical situation of h = 1.5, parameter 's' of a tubular propellant configuration is varied from 0.0 to 1.0. Zero represents solid sustainer

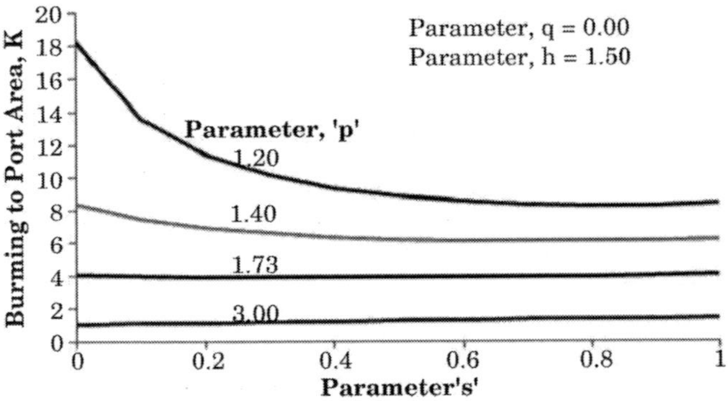

Fig. 50: Parametric variation for burning to port area

type of propellant grain, which is permitted to burn in radial outward direction from center also. Unity value of parameter represents equality of inner and outer diameter, resulting in virtually zero web and zero mass propellant configuration. Simultaneously parameter 'p' is varied from 1 to 3. Unity value represents a case where outer diameter of propellant is equal to inner diameter of rocket motor. Although, mathematically this case can exist, practically absence of gap for flow of gases at the outer surface of propellant makes this situation non-realizable. As value of 'p' increases, gap appears between propellant and motor. Ratio of burning area to port area (K_I) is plotted for initial condition (q = 0) against parameter 's' for constant values of parameter 'p' as Fig. 50.

As internal diameter of the configuration (s) increases, K_I reduces initially at a higher pace for relatively lower value of parameter p (gap between rocket motor and propellant). This slows down as internal diameter of tubular grain is increased for a given outer diameter. As value of parameter 'p' is increased, the flat portion of variation captures entire domain. For a peculiar value of parameter 'p' (=1.732), the value of K_I for s=0 and s=1 is same. Although in between slightly lower value of K_I is realized. For higher value of parameter 'p' (= 3.00), the value of K_I has very little variation with 's' but it increases with increase in 's'. This reversal of trend (from reduction to increment in K_I for small s) occurs at $p = \sqrt{3} = 1.732$ for given situation (q = 0, h = 1.5).

Although, Fig. 50 represents variation of area ratio (burning to port) for initial firing condition at zero web consumed, when combustion starts the variation also changes and at intermediate web burnt of 30%, the same variation is shown as Fig. 51. It is clear that, now all the curves are progressive in nature for higher value of parameters 'p'. This means inner diameter of rocket motor is many times bigger than outer diameter

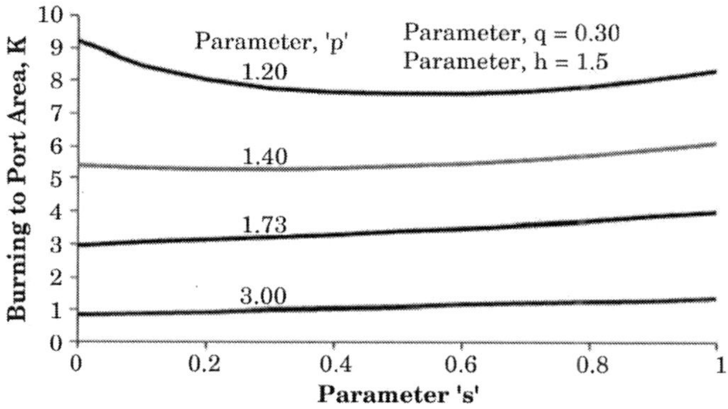

Fig. 51: Parametric variation for burning to port area

of propellant, increase in inner diameter of propellant grain leads to increased burning area compared to port area. However, both the figures represent the variation for completely different propellant dimensionally. High value of parameter 'p' represent non-practical case, as volumetric loading of propellant in the rocket motor casing reduces. To get better insight, same propellant configuration (same 'h' and 's') is considered and variation of K_I against web consumed parameter 'q' is plotted as Fig. 52. Here parameter 'p' is taken as 1.2, which is a more practical situation encountered in the rocket firings.

From the figure, it is clear that the burning profile is regressive in nature. Constant value of 'h' is representative of constant length of the propellant. Parameter 's' represents inner diameter and a lower value represents lower value of inner diameter. For smaller value of 's' (near zero), the burning area to port area is more regressive. For larger values of 's', the burning area to port area ratio is almost invariant with web consumed. This is because, at larger value of 's', the inner and outer diameter are very near to each other and variation in burning area is small with web consumed. However, for constant lower value of parameter 'h' (length of propellant), the value of burning area to port area is equal in the zone, where q = 0.35 to q = 0.50. However, for larger values of propellant length (large 'h'), the curve attains a more or less neutral curve.

Inhibited Tubular Propellant Grain

Inhibition is selectively restricting the burning surface of propellant to tailor the burning pattern. As uninhibited tubular propellant grain always leads to slightly regressive profile with regressive nature dominating for lower length of the propellant, neutrality of propellant

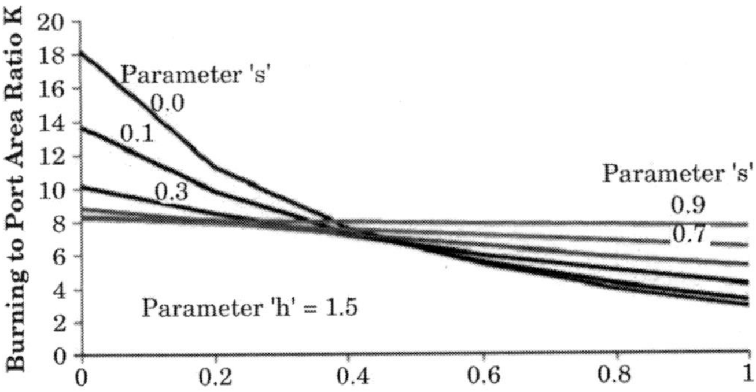

Fig. 52: Effect of inner diameter of propellant on burning to port area ratio

is achieved with end inhibition of the propellant surface. This allows propellant to burn at inner and outer diameters, only. Thus gain in burning area by increment in inner diameter is compensated by reduction in burning area from outer diameter reduction. The propellant configuration is shown in Fig. 53A.

However, the requirement of a gap between rocket motor and propellant outer surface is retained by this configuration also. This type of configuration has a practical difficulty of pressure equalization in the empty spaces. One port is between, rocket motor case and outer diameter of the propellant, while another port lies at the inner diameter of the propellant. The balance of gas generation and port volume increment is responsible for equality of pressure. In case of imbalance of pressure, propellant is subjected to differential pressure loading in

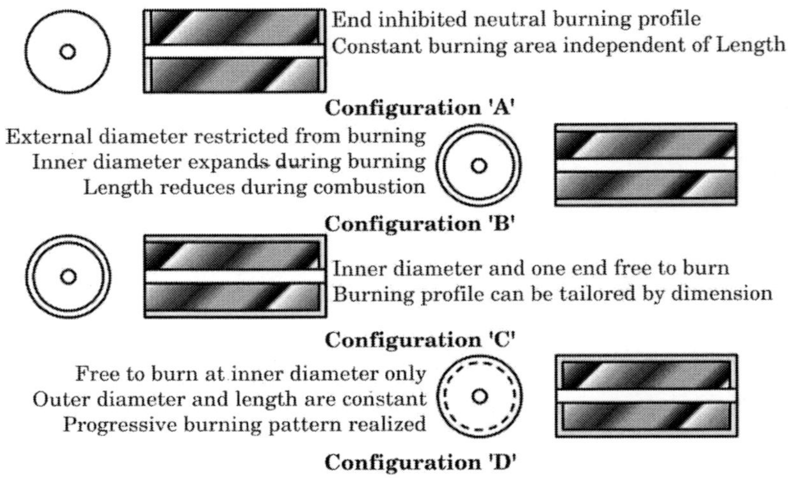

Fig. 53: Different inhibited tubular configurations

radial direction. The burning pattern is same as depicted in Fig. 48 except for the reduction of length. Burning area for such configuration is constant and is independent of web consumed. It is given by the following expression.

$$\text{Burning area} = \pi D^2 h\,(1+s)$$

It is clear that the expression is similar to the bare tubular configuration, if first term under curly bracket in earlier configuration is eliminated. Absence of any dependence of burning area on 'q' indicates that burning area is constant during burning of propellant. This gives a perfectly neutral burning profile. Maximum web in this case is given by $w_{max} = D(1\text{-}s)/4$. Effect of length is completely eliminated in such configurations.

Another configuration has inhibition at the external surface, only. This type of propellant is free to burn from inner diameter and along length. The configuration is shown in Fig. 53B. The inner diameter enlarges in course of burning and length reduces from both the ends for such propellant configuration. Increase of inner diameter is responsible for progressivity, while reduction in length is responsible for regressive nature. For long propellant grains, the end effects (of length) are negligible and the propellant burns in progressive manner. Such propellant configuration has relatively faster production rate, as the propellant can be directly cast in long inhibition sleeves and can be cut into required dimensions for evaluation and use. However, dimension of such propellants can be selected to tailor the progressivity. In these propellant configurations also, maximum web is given by same expression ($w_{max} = D\{1\text{-}s\}/2$) as explained earlier. With non-dimensional parameters described earlier (s = d/D, h = L/D, q = w/w_{max}), burning area of the propellant in terms of non-dimensional parameters is given by expression below.

$$\text{Burning area} = [\pi D^2/2] \times [2hs + (1\text{-}s) \times \{1+s+q(2h\text{-}4s\text{-}3(1\text{-}s)q)\}]$$

In the above expression, q = 0, represents initial burning area and q = 1 represents final burning area. Non-dimensional form of the burning area can be obtained by taking initial burning area as normalizing factor. For numerical appraisal, tubular propellant grain in specified configuration (Fig. 53B) is considered, with outer diameter of 120 mm. value of 'h' and 's' are taken as 1.5 and 0.5 respectively. Then w_{max} = (1-s) x D/2 = 30 mm. When parameter 'h' is varied and parameter 's' is kept constant, higher burning area is obtained for higher values of parameter 'h'. However, for understanding relative worth of the configurations, non-dimensional form of burning area and web is considered. As the non-dimensional web consumed increases, burning area (normalized to initial burning area) shows different form of burning patterns (Fig. 54). For higher values of 'h' (= 5), the variation rises monotonically. At the end of burning, burning area is around 56% higher than the initial burning area. For smaller values of 'h' (= 0.5), the non-dimensional burning area shows regressive nature and at the end of burning, burning area reduces to zero. It is clear that in between, for some assorted value of 'h', the burning area at end (q = 1) and at the beginning (q = 0) of combustion is almost equal. Mathematically, non-dimensional parameter can be derived for such situations.

Initial Burning area = $[\pi D^2/2] \times [2hs + (1\text{-}s) \times (1+s)]$
Final Burning area = $[\pi D^2/2] \times [2hs + (1\text{-}s) \times \{1+s+2h-4s-3(1\text{-}s)\}]$
So, $1 + s = 1 + s + 2h - 4s - 3(1\text{-}s)$
And $h = (s + 3) / 2$

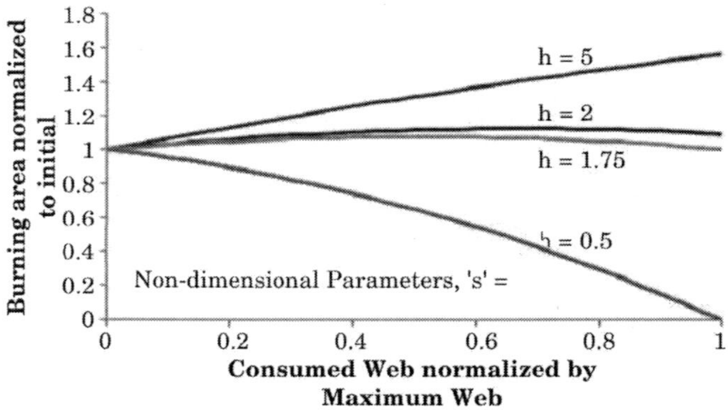

Fig. 54: Variation of non-dimensional burning area with non-dimensional web

For the given example, where 's' is assigned a value of 0.5, the value of 'h' can be obtained as 1.75 (= 0.5 + 3/2). Burning area variation for this typical situation is also shown in Fig. 7. The variation is more or less neutral and initially, burning area increases to around 7.50% of initial burning area and then regresses to original burning area at the end of burning. The value of non-dimensional web at which burning area attains a maximum can be obtained by differentiation of expression for the burning area.

$$\text{'q' for maximum burning area} = (1/3) \times (h\text{-}2s)/(1\text{-}s)$$

For the configuration, where initial burning area is equal to final burning area, the value of 'q' for the maximum burning area is obtained as ½. For other configurations, other values are obtained. For the given example, when 'h' is 5, 'q' for maximum burning area is 8/3. So burning area is maximum after complete web is consumed (q=1), indicating a progressive burning profile. For 'h' equal to 2, q for maximum burning area is 2/3 and the burning profile is progressive in the beginning following by a regressive curve. The maximum burning area is 12% higher than initial burning area and final burning area is around 9% higher than initial burning area.

As a critical case, value of parameter 'h' is calculated, when maximum burning area is achieved at q=1. This is possible, if h = 3-s. For the considered example, this is possible at h = 2.5. For any value of 'h' above 2.5, the burning profile will be progressive for the considered example. Similarly, for 'h' equal to 1, q for maximum burning area is zero and burning profile is maximum in the beginning. This leads to regressive burning profile. For any value of 'h', less than 1, burning profile is regressive. For any value of 'h', between 1 and 2.5, the burning profile is a combination of progressive and regressive burning. Typical examples are shown in Fig. 55. It is clear that for h = 1.2, the curve is containing less of progressive and more of regressive part, while for h = 2.2, the curve is predominantly, progressive. For h = 1.2, maximum value of burning area is obtained at q = 2/15 (= 0.1333) as 0.68% higher than initial burning area. However, at the end of web consumption, the burning area for this configuration reduces to around 71.79% of initial burning area. For h = 2.2, the maximum area is obtained at q = 4/5 (= 0.8) and it is 16.27% higher than initial burning area. Since the configuration has less regressive tendency, burning area at the end of combustion comes to 15.12% higher than initial burning area. A compromise and balance of burning area variation is obtained for h = 1.75 and both progressive and regressive tendency of burning area variation is equalized. So it is possible tailor the dimensions of configuration (Fig. 53B) to give progressive, regressive or any other pattern of their combinations.

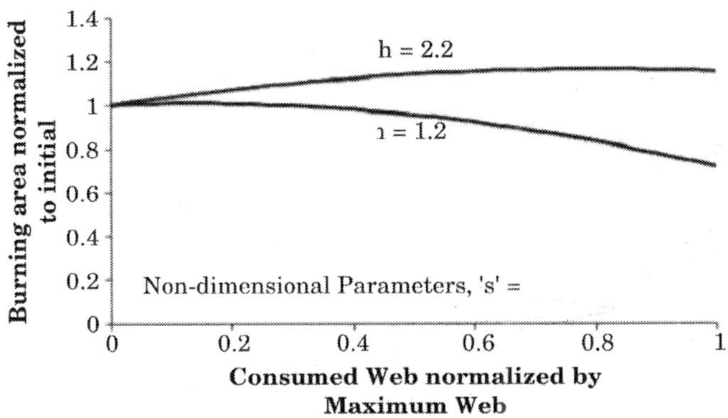

Fig. 55: Typical burning profiles for externally inhibited tubular propellants

A variant of this configuration has additional inhibition at one of the end faces (Fig. 53C). This configuration resembles previous configuration (Fig. 53B) in almost all respects. However, requirement of end inhibition reduces the production rate of such propellant configurations. This configuration can also be tailored to give all forms of burning profiles. In this case, propellant burns at inner diameter and at one of the end faces. Compared to configuration in Fig. 53A, Fig. 53B gives lower regressivity. Fig. 53C reduces regressivity, due to end burning, further. The burning area variation with web consumed for such propellant configuration is given below.

Burning area = $[\pi D^2/4] \times [4hs + (1\text{-}s) \times \{(1+s) + q(4h\text{-}4s\text{-}3(1\text{-}s)q)\}]$

For 120 mm outer diameter and 's' equal to 0.5, burning profile is generated using equation given above. Different values of 'h' are considered and it is observed that the curves resembles more or less with Fig. 54, but with different values of 'h'. For higher values of 'h', the curves are progressive, while for lower values the curves are regressive. For some of the values, the curves may be near neutral. For getting maximum value of burning area, the expression of burning area is differentiated with respect to 'q', and expression for 'q' for maximum burning area is obtained.

Expression for q for maximum burning area = $(2/3) \times (h\text{-}s)/(1\text{-}s)$

For the example, under consideration, where s is equal to 0.5, the expression of 'q' for maximum burning area is $(2/3) \times (2h\text{-}1)$. For 'h' equal to 2, maximum burning area is obtained for $q = 2$ (*i.e.*, after complete consumption of propellant, $q = 1$). So the nature of the curve is progressive. The burning area increases continuously to 2.5 times initial burning area for the given example. For 'h' equal to 0.4, maximum burning area occurs after complete consumption of the propellant. The rise in burning area, at the end of propellant web consumption, compared to initial burning area is 11.36%. Reduction in value of 'h', results in reduction in increase in burning area. For critical case, when maximum burning area lies at complete consumption of the web, the value of h is equal to 0.875. This is obtained by putting $q = 1$ and $s = 0.5$ in the expression for q for maximum burning area. This indirectly says that for any value of h, higher than 0.875, the nature of the curve is progressive. The maximum burning area in this case is 7.5% higher than initial burning area.

At the other side, value of 'h' can be obtained for which maximum burning area lies before initial burning area. This is possible, when value of q for maximum burning area is less than zero or negative. For

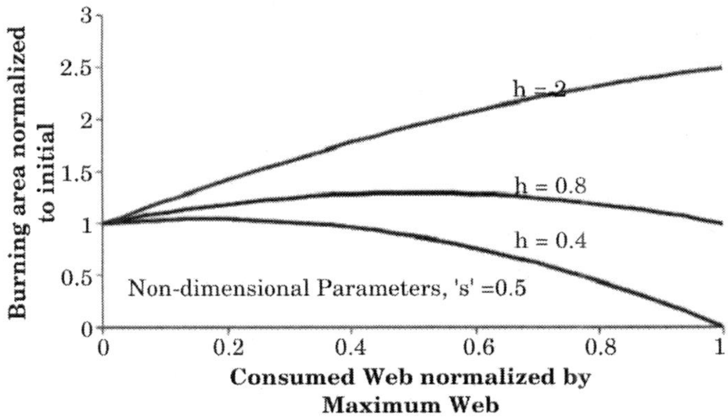

Fig. 56: Burning profiles for tubular configuration with one end and outer diameter inhibited

such situations, burning profile is entirely regressive. This is possible for all those values of 'h', which are less than 0.5. In between h equal to 0.5 and h equal to 0.875, the burning profile contains both progressive and regressive curves. One of the important considerations is that value of 'h', for which, the initial and final burning areas are equal. This situation occurs, when value of h = (s + 3)/4. For the given example, this occurs at h = 0.6875. The same curve is depicted in Fig. 56. However, this must be kept in mind that increasing value of h increases propellant weight and burning area of the propellants. The curves are normalized representation of the actual situation.

The variation of inner diameter can also be plotted for the given configuration. As an alternative, value of h for the given configuration is taken as 1.0. This represents a situation of outer diameter being equal to the length of the propellant. Curve is reproduced from Fig. 9 for h = 1.0 and s = 0.5 in Fig. 57. Since value of parameter 's' will always be less than 1, two more values, one higher (s = 0.8) and another lower (s = 0.2) than this reported value is considered. Higher value of inner diameter (or s) gives a lower propellant weight and the value of web reduces. However, in the plot the variation on web is normalized to maximum web. As value of 's' is lowered, the burning profile becomes more progressive. This is because, for lower inner diameter, more rise in diameter or burning area is possible during combustion. Contrary to this, if value of 's' is near 1, there is very little margin for increase in burning area and progressivity of burning profile is not observed. In fact for hypothetical case of s = 1, the curve is neutral but no web of propellant will be available for burning.

Another variant has only inner diameter, free to burn, rest all surfaces are restricted from burning by inhibition. This type of propellant

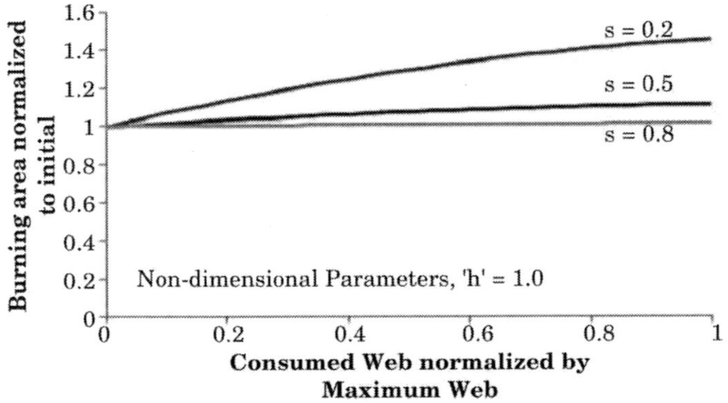

Fig. 57: Burning profiles with variation of inner diameter

configuration can be case-bonded easily. Better utilization of chamber volume is also observed, as space for flow of combustion gases is not needed between propellant and rocket motor casing. Another advantage of this configuration is that propellant length remains constant and propellant can be held firmly with the help of grids at ends. Such provisions cannot be adopted, if propellant length is reducing with time or propellant consumption. Since propellant is free to burn only on the inner diameter (Fig. 53D), the burning profile is progressive. Maximum web in this case is given by $w_{max} = (1 - s) \times D/2$, which is two time higher than earlier cases. Length has very little effect on the burning profile and same percentage rise in burning area is observed for different length of propellant grains. If parameter 's' is smaller, more percentage rise in propellant burning area is observed. The equation for burning area is very simple in this case.

$$\text{Burning area} = [\pi D^2] \times [s + (1\text{-}s) \times q] \times h$$

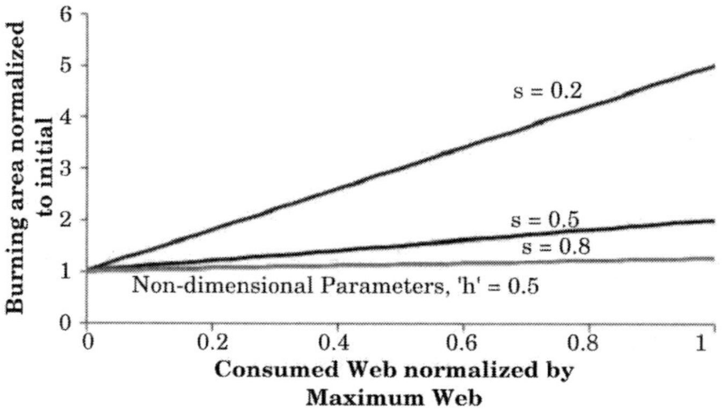

Fig. 58: Burning profiles of progressive tubular propellant configuration

For different inner diameter, percentage change in burning area is plotted during consumption of web in the Fig. 58. It is clear that curve is more progressive for lower values of 's' (inner diameter). In fact, from the equation of burning area, it is clear that percentage change in burning area for progressive tubular propellant configuration is given by 1/s. lower the value of s, higher will be value of percentage change in area. For s = 0.2, burning area becomes 5 times (1/0.2 = 5) original burning area. For s = 0.8, burning area becomes only 1.25 times (1/0.8 = 1.25) original burning area. Since the variation of burning area can be tailored and wide ranges of pressure values are observed in single firing, this type of propellant grain is useful for evaluation of burning rate law of the propellant. The calculation step for burning area evolution during combustion is simple and propellant is safer to evaluate.

Taper Port Tubular Propellant Grain

One of significant variation in burning profile is achieved in tubular propellant grains by making inner diameter in tapered form for certain length of the propellant. This type of configuration may be inhibited on external surfaces and one end. The propellant is free to burn at inner diameter and taper length. The dimensions can be tailored to give neutral burning profile. However, other forms of burning profiles are also realized by such propellant configurations. The initiation of such propellant is better. Inhibition requirement is also reduced to one side only. The tapering can be achieved by several means like taper mandrel and taper turning of tubular port. The manufacturing of such grains is also possible by mandrel insertion or carthias casting technique. The fact extruded tubular propellant grains can be taper turned easily. This configuration gives a very high port to throat area. The chances of erosive burning are also lower. The propellant grain can be case-bonded for better utilization of rocket motor space. However, enlargement of central port may lead to reduced volumetric loading. The configuration is shown as Fig. 59. The benefits of such propellant grains are enlisted below.

- The configuration is simple robust and the fabrication of mandrel will be easier.
- The propellant configuration can be realized by mandrel insertion technique.
- The propellant can be realized by taper turning of tubular sleeve cast propellants.
- The requirement of inhibition remains for lateral surface and only one end surface.
- Generation of configuration by pressure casting will be simpler.

- The configuration has very large port area at nozzle end resulting in less accumulation of combustion gases in the port.
- The port to throat area ratio will be very high.
- The chances of erosive burning will be less.
- The grains will have very small stresses due to less discontinuity in port area.
- Propellant grain will result in higher web, which is advantageous for higher burning rate propellants.
- The configuration will have lower tail-off.

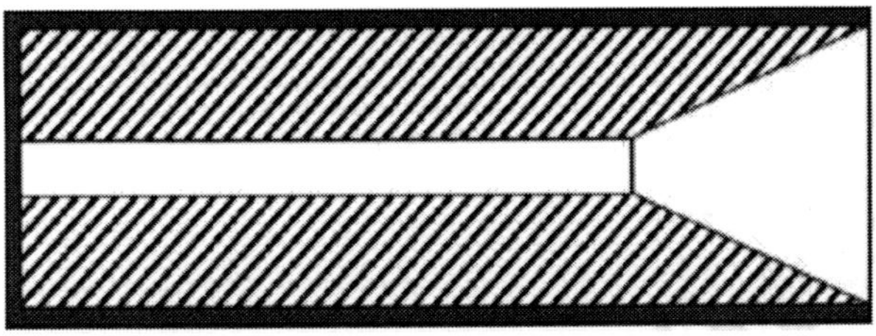

Fig. 59: Propellant configuration with funnel shaped port

In addition to earlier defined parameters 's' (d/D), 'h' (L/D), 'q' (w/w_{max}), this configuration needs one more non-dimensional parameter, which is ratio of slant length to total length. This is nothing but fraction of total length, which is tapered. The ratio can be stated to be 'p'. Using these parameters, the angle between slant section and the axis of the propellant grain can be given by 'q', whose expression is given below.

$$\text{Tan } q = (1\text{-}s)/(2xpxh)$$

The burning area can be calculated in terms of these parameters. The expression for burning area is given below. With q = 0, initial burning area can be obtained. Expression for the same is also given. The burning area normalized to initial burning area gives a better representation of parametric variations in geometrical parameters.

$$\text{Burning area} = \pi D^2 x\{[s+q(1\text{-}s)][h(1\text{-}p+qp)\text{-}\{qph/\text{Cos}q\}]+ph(1\text{-}q)/\text{Cos}q\}$$

$$\text{Initial Burning area} = \pi D^2 \text{ x } \{s[h(1\text{-}p)]+ph/\text{Cos}q\}$$

For a trial case, outer diameter of the propellant grain (D =) is taken as 200 mm. The value of s, h and p were assumed to be 0.4, 5 and 0.2, respectively. This gave inner diameter of the propellant grain as 80 mm, length of the propellant grain as 1000 mm and length of slant section as 200 mm. The value of slant angle (q =) is calculated to be 16.7^0. The maximum web available for complete consumption of propellant is obtained as (200-80/2 mm =) 60 mm. The initial burning surface area is calculated as 332258.7 mm^2. For checking correctness of the formulation, parameter 'p' is made zero, resulting in a pure tubular propellant grain burning in progressive manner (lateral and longitudinal surfaces inhibited). For this case, value of 'q' is 90° and initial burning area is found as (π dL = π x 80 x 1000 mm^2 =) 251327.41 mm^2. From the expression of the burning area, if p and q are made zero, the burning area = $\pi D^2 hs$ = π x sD x hD = π dL. Burning area versus web consumed profile for this configuration is shown as Fig. 60. The burning profile is progressive in nature.

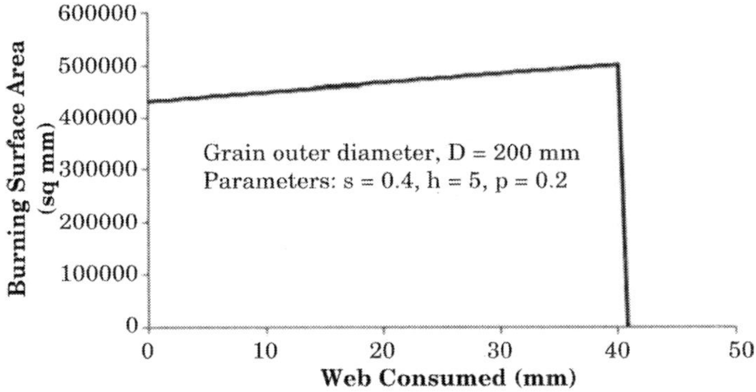

Fig. 60: Burning area variation for taper tubular propellant grain

For parametric study, one of the parameters is varied at a time and burning profile is obtained. If value of 's' is increased, inner diameter increases. This reduces maximum web available for complete consumption of propellants. This result in lower variation of burning surface area and corresponding lower range of pressure covered in single firing for same propellant composition. Normalized web burnt to normalized burning area (to initial burning area) is shown in Fig. 61.

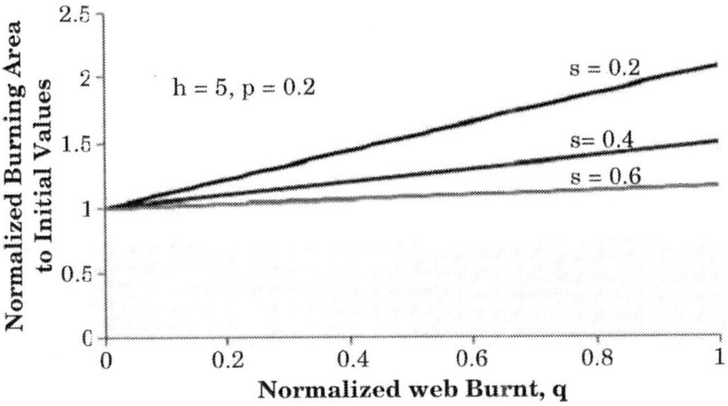

Fig. 61: Variation of 's' for taper tubular propellant grain

It is clear from the figure that initial burning area is normalized to 1. However, lower values of 's' represents a lower values of initial burning area. In other words, for smaller values of 's', the combustion starts with a lower pressure values and higher available web gives wider coverage of pressure range. It is found that variation in final burning area is relatively smaller. The same is depicted in normalized curve of Fig. 61. Variation of parameter 'h' may not lead to any significant change in the burning profile. Only value of burning area will be higher for higher 'h'-values.

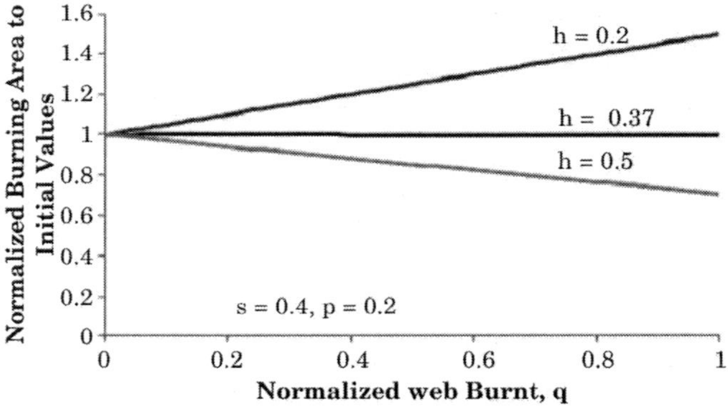

Fig. 62: Variation of 'h' for taper tubular propellant grain

The most significant parameter for taper tubular configuration is 'p', which governs slant length. This has power to convert a progressive profile into regressive profile and for an optimum condition, near neutral profile is also possible. For lower values of 'p', the burning area variation is progressive in nature and effect of slant section is not felt by the configuration. However, as value of 'p' is increased, the regressivity picks up and burning profile becomes regressive. The same is depicted in Fig. 62.

It is clear that for a particular value of 'h' (in this case 0.37), the variation of burning area is minimal and almost neutral burning profile is possible. The expression of optimal values of 'p', for which neutral burning profile is possible, has following expression.

'p' for neutral burning = $(1\text{-}s) / [(2/\cos\theta) - s]$

Using this expression, the value of 'p' for the trial case is 0.368975 (~0.37).

Offset Port Tubular Configuration

Sometimes abrupt reduction in thrust at the end of propellant consumption is not preferred and gradual reduction is recommended.

For concentric central port in tubular configuration, thrust termination is theoretically sudden but in practical situation, some variations are observed. To introduce tail-off in the burning profile deliberately by design, offset port tubular configuration is conceived, which burns in progressive manner. This shape of propellant is referred as moon-burning propellant grain, also. Continuing with the earlier notation, the outer diameter of the propellant grain is 'D'. The hole diameter is taken as sD and length of the propellant grain is hD. The central port is offset by pD. This means that center and outer diameter and that of hole diameter is pD. Propellant combustion takes place in radial outward direction from the hole surfaces. This type of propellant grain is depicted in Fig. 63.

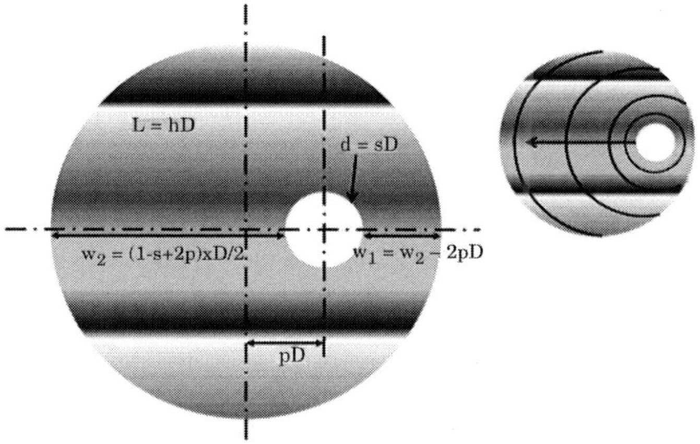

Fig. 63: Moon burner configuration

This propellant configuration has two transition web burnt. First web transition (w_1) occurs when circular burning profile is converted to sectorial burning profile and second web transition (w_2) indicated complete web consumption. The expressions of these web transitions are given below.

$$w_1 = (1\text{-}s\text{-}2p)xD/2 \text{ and } w_2 = (1\text{-}s\text{+}2p) \text{ x } D/2$$

Since maximum web is w_2, the parameter 'q' is expressed as ratio of web consumed at any instant to w_2. The expression for 'q' is given below. The burning surface area for the configuration is given by expression below.

Web fraction, $q = w/w_2 = 2w/[(1\text{-}s\text{+}2p)D]$

Burning surface area = $hD^2\theta\,[s+(1\text{-}s\text{+}2p)q]$,

where $\text{Cos}\theta = \pi$ for $w \leq w_1$,

$= 0$ for $w > w_2$

$= [1\text{-}4p^2\text{-}\{s+(1\text{-}s+2p)q\}^2]/[4p\{s+(2\text{-}s+2p)q\}]$ for $w_1 < w < w_2$

For the configuration under consideration, if the offset is zero (p=0), the propellant configuration resembles progressive burning tubular propellant grain. For such configuration, value of θ is equal to π and burning area is equal to $hD^2\pi[s+(1\text{-}s)q] = \pi \times hD \times [sD+(1\text{-}s)qD] = ð \times L \times [d+2w]$.

For analysis, outer diameter for the tubular propellant grain with offset port is taken as 200 mm. The value of s and h are taken as 0.4 and 5 respectively. The offset conditions, depicted by 'p' are changed. The value of 'p' cannot be independently, but is a function of other parameters. Maximum value of 'p' is given by (1-s)/2. If 'p' higher than this value is taken then hole circumference will touch outer diameter in the beginning itself and burning profile is moon-shaped right from the beginning. Ideally, the highest value of 'p' can be 0.5. In such case, the hole center will be at the periphery of the propellant grain and hole will not be a complete circle. The value of 'p' can be higher than 'p' and can reach upto (1+s)D/2, but under such situations (p > 0.5), the hole is like a small carving from the solid sustainer propellant grain. The Fig. 64 depicts burning area variation with web burnt for different values of 'p' for offset port tubular propellant grains.

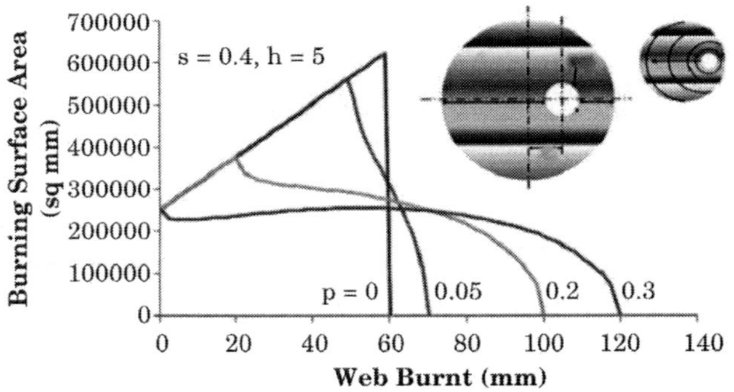

Fig. 64: Burning area variation for offset port tubular configuration

When, offset parameter 'p' is zero, the configuration resembles tubular configuration burning in progressive mode. Both the web transitions become equal and in the case being discussed, the value is 60 mm. when 'p' is changed to 0.05, first web transition occurs at 50 mm and complete web consumption is depicted at 70 mm. The sharp fall at the end of burning is replaced with a tail-off from 50 mm to 70 mm web burnt. However, the burning profile before first web transition remains unaltered. When 'p' is equal to 0.2, the web transitions are calculated to be 20 mm and 100 mm. this clearly indicates that the web transition is symmetrically varying about the web depicted for corresponding tubular propellant configuration. Higher value of 'p' results in higher tail-off. However, for the highest value of p (= 0.3), first web transition is immediately realized at the beginning and the hole periphery touches outer diameter of the propellant in the beginning of burning. Now entire web of 120 mm is available for combustion in moon-burning fashion. In fact for the higher values of 'p', the tail-off portion becomes main burning profile.

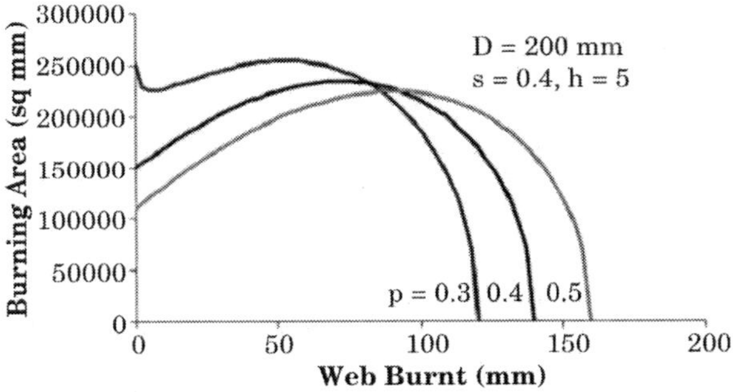

Fig. 65: Burning area variation for moon burner configuration

For values of 'p' higher than 0.3, propellant configuration in moon-shaped, right from the beginning. For 'p' equal to 0.3, 0.4 and 0.5, the variation of burning area is plotted against web burnt as Fig. 65. The trend of higher web availability for higher value of 'p' continues. The contribution of progressive tubular propellant grain is completely eliminated and the burning profile is entirely due to moon-burner configuration. The kink in burning profile at the transition from progressive tubular to moon-burner is not present. The burning profile is progressive in the beginning and is regressive at the end. Obviously, there exists a maximum burning surface area in all such configurations. For p = 0.5, the maximum burning area occurs at web burnt of 89.6 mm. For p = 0.4, the maximum burning area lies at web burnt of 72.8 mm. For p = 0.3, it occurs at 52.8 mm of web burnt.

One of the peculiarities of these profiles is that as value of 'p' increases, the initial burning surface area reduces. Higher 'p' gives higher web available, lower initial burning area, higher web for maximum burning area and maximum burning area for the configuration. The value of 'p' can acquire any value between zero and (1+s)/2. Between zero and (1-s)/2, the propellant burning has traces of progressive burning tubular propellant grains. This disappears for higher values of 'p' [between (1-s)/2 and (1+s)/2] and the burning profile is progressive initially, which becomes regressive at later stages. Obviously, there exists a maximum value of burning area for such configurations.

Although, this type of propellant configuration is not suitable in cartridge loaded configuration and unsymmetrical burning leads to thermal protection layer unevenly distributed at the inner surface of the rocket motor, the configuration may have some academic interest. The lateral shift of center of gravity is another concern for such configurations. In fact, a near neutral burning profile is possible, when the propellant is designed with higher offset of central port.

Multi-tubular Propellant Configurations

At several occasions, many tubular grains are bundled inside a single rocket. Such configuration enhances propellant burning profile. If progressive burning propellant grains are used, progressivity is enhanced by using multiple grains. However, multi-tubular propellant configurations are generally utilized in regressive burning mode without any inhibition. Sometimes, end inhibited propellant grains are also utilized. The main features of multi-tubular configurations are enlisted below.

- Multi-tubular propellant configuration is exact replica of their tubular configuration counterpart.
- Multi-tubular configuration gives very high burning area for available volume within rocket.
- If same pressure is maintained inside rocket, multi-tubular propellant grain require higher throat diameter.
- Higher throat diameter leads to less nozzle erosion.
- Multi-tubular propellant configuration gives higher thrust without enhancement of burning time or operating pressure.
- The web of propellant is lower, so this configuration is suitable for booster configurations requiring higher thrust and lower burn time.
- Practically, this configuration is susceptible to combustion instability and propellant grain crumbling near end of burning.
- Symmetric, simultaneous and uniform burning of all the tubular propellant grains is difficult to achieve.
- The pressure time or thrust time profiles of this configuration is not generally smooth, because of interacting flame fronts of different propellant grains.
- Since higher throat diameter is permitted, suitable grid to hold propellant in rocket, at the end of burning, is needed.

- Volumetric loading for multi-tubular configuration is lower than tubular configuration.

For analysis and understanding some of the salient features of multi-tubular propellant configuration, following assumptions are made. These assumptions are made to understand the utility and behavior of the multi-tubular configurations. The actual burning profile with multi-tubular configuration is analyzed a little later.

- All propellant grains are identical in geometry, shape, size, weight and density.
- All propellant grains have same web and burning surface.
- All propellant grains are inhibited at end (lateral surface).
- Propellant burning takes places on both internal and external longitudinal surfaces.
- The burning surface area is constant for the propellant grains during rocket evaluation.
- The burning profile for the propellant grains is neutral.
- The web available for complete consumption of propellant is taken as representative of burning time of the rocket.
- The burning periphery represents burning surface area and in turn a measure of operating pressure of the rocket.

An assessment of volumetric loading can be made on the assumption of different number of propellant grains inside a rocket motor. Since this assumed configuration is two dimensional in nature, the cross-sectional loading is considered. The ratio of inner diameter (d) to outer diameter (D) of a single propellant grain is taken as s (= d/D). For single tubular grain inside a rocket, the cross-sectional loading is $(1\text{-}s^2)$. If s = 0.4, the cross-sectional loading can be maximum 84%. If two propellant grains are positioned inside rocket, the cross-sectional loading is $(1\text{-}s^2)/2$. For s = 0.4, maximum cross-sectional loading with 2 propellant grains is 42%, only. If 3 propellant grains are positioned inside rocket, then maximum cross-sectional loading is given by $9(1\text{-}s^2)/(2+\sqrt{3})^2$. For s = 0.4, the value of cross-sectional loading is 54.28%. If s = 0.2, the cross-sectional loading for 3 tubular propellant grains has a maximum value of 62%. For hepta-tubular configuration, 7 propellant grains are positioned inside rocket chamber. The arrangement of propellant grains is one central propellant and 6 at periphery of the central grain. For this configuration, cross-sectional loading is $7(1\text{-}s^2)/9$ and the value for s = 0.4 is 65.33%. Similarly for cluster of 19 propellant grains in multi-tubular configuration, the cross-sectional loading can reach a maximum value of $19(1\text{-}s^2)/25$. For s = 0.4, the cross-sectional loading can attain a maximum value of 63.84%. None of the values acquire the cross-sectional loading of single grain (84%). So the sequence of cross-sectional (volumetric) loading is single grain > 7-tube > 19-tubes > 3-tubes > 2 tubes.

Burning area of the propellant set-up in multi-tubular configuration can be compared, assuming an end-inhibited tubular propellant grain, burning from inner and outer longitudinal sides. The rocket inner diameter for accommodating propellant is fixed as D. As propellant length is invariant during burning, burning periphery is compared. For single grain, burning periphery is given as $\pi D(1+s)$. If two propellant grains are accommodated in same rocket, then diameter of each grain is D/2. Total burning periphery is $\pi D(1+s)$. There is no change in the burning periphery. If three propellant grains are accommodated, then diameter of each propellant grain is $D\sqrt{3}/(2+\sqrt{3})$ and the burning periphery is given by $[3\sqrt{3}/(2+\sqrt{3})] \times \pi D(1+s) = 1.3923\ \pi D(1+s)$. So, burning periphery increases by around 39.23%. If 7 propellant grains are accommodated, then diameter of each propellant grain is given by D/3. The burning periphery is given by $(7/3) \times \pi D(1+s) = 2.333\ \pi D(1+s)$. This is a rise of around 133%. If 19 propellant grains are accommodated, then diameter of each propellant grain is given by D/5. The burning periphery is given by $(19/5) \times \pi D(1+s) = 3.8\ \pi D(1+s)$. This is a rise of around 280%. So, the sequence of change of burning periphery (or burning surface area) for constant diameter ratio is single grain = 2-tubes < 3-tubes < 7 tubes < 19 tubes.

For each of the given configuration, web available can be compared. For single grain web is D(1-s)/4. For two grain configuration, this value is also halved (multiplier: 0.5). For three grain configuration, the value is web of single grain multiplied by $[\sqrt{3}/(2+3)] = 0.464$. The multiplier for cluster of 7 grain in multi-tubular configuration is 0.3333. The multiplier for cluster of 19 grains is 0.25. It is clear that, as number of propellant grains inside a rocket chamber is increased, the burning periphery (or burning surface area or pressure) increases, but web available for complete consumption of propellant (or burning time) reduces. The above mentioned comparison is for constant value of 's', which is ratio of inner diameter to outer diameter of the propellant.

Similar simulation can be carried out for those conditions, where total weight of the propellant is kept constant. The outer diameters of the propellant are maximum possible for rocket of given available diameter 'D'. Inner diameter is adjusted to give same total weight. The value of 's' will change for this governing condition. Effect of length is neglected by assuming inhibition at the lateral surfaces. For matching weight, propellant cross-sectional area is to be matched. Suppose single grain is replaced with two identical tubular propellant grains. Now if value of 's' for single grain is less than 0.707 (√0.5), even if the two propellant grains are made in solid cylinder (without central hole or s = 0) form, the weight cannot be matched. If value of 's' for single propellant grain is above this number, then two tubular grains could be realized. If the value of diameter ratio (inner to outer diameter) for cluster of 2-grains is denoted by 't', then it is related to 's' for single grain by the relation $(1\text{-}t^2) = 2(1\text{-}s^2)$. If 's' is 0.8, then t = 0.529. Web available for single grain is 0.05D and the same for two grains is [(1-t)D/8 =] 0.05887D. The burning periphery (equivalent to burning area) is 1.8ðD for single grain and is [(1+t)ðD =] 1.529 ðD for cluster of 2 grains. So, to maintain same propellant weight, cluster of 2 propellant grains has lower diameter ratio, higher web availability and lower burning area as compared to single propellant grain.

If value of 's' for single grain is 0.9 and cluster of 2 grain in equivalent weight is perceived, then values of parameters will change. If 's' is 0.9, then t = 0.7874. Web available for single grain is 0.025D and the same for two grains is [(1-t)D/8 =] 0.026575D. The burning periphery (equivalent to burning area) is 1.9ðD for single grain and is [(1+t)ðD =] 1.7874 ðD for cluster of 2 grains. So, to maintain same propellant weight in cluster of 2 propellant grains, if diameter ratio of single grain is increased, diameter ratio of cluster of 2 grains also increases, web availability reduces and burning area increases.

Similar calculations are repeated for cluster of 7 propellant grains. Now the condition for value of 's' becomes less stringent. For solid sustainer cluster of 7 grains, weight matching to single propellant grains is not possible, if 's' for single propellant grain is less than 0.4714 (√2/9). If the value of diameter ratio (inner to outer diameter) for cluster of 7-grains is denoted by 't', then it is related to 's' for single grain by the relation $(1\text{-}t^2) = 9(1\text{-}s^2)/7$. If 's' is 0.9, then t = 0.8693. Web available for single grain is 0.025D and the same for cluster of 7-grains is [(1-t)D/12] 0.01089D. The burning periphery (equivalent to burning area) is 1.8πD for single grain and is [7(1+t) πD/3 =] 4.3617 πD for cluster of 7-grains. So, to maintain same propellant weight, cluster of 7 propellant grains has lower diameter ratio, lower web availability and lower burning area, as compared to single propellant grain.

The value of 't' for cluster of 2 is 0.7874 (for cluster of 7, t = 0.8693). Web available for cluster of two grains is 0.026575D (0.01089D for cluster of 7). The burning periphery (equivalent to burning area) is 1.7874 ðD for cluster of 2 grains (4.3617 ðD for cluster of 7). So, to maintain same propellant weight, cluster of 7 propellant grains has higher diameter ratio, lower web availability and higher burning area as compared to cluster of 2 propellant grains.

Similar calculations are repeated for cluster of 3 propellant grains. Now the condition for value of 's' becomes less stringent than cluster of 2 or 7 grains. For solid sustainer cluster of 3 grains, weight matching to single propellant grains is not possible, if 's' for single propellant grain is less than 0.5948. If the value of diameter ratio (inner to outer diameter) for cluster of 3-grains is denoted by 't', then it is related to 's' for single grain by the relation $(1-t^2) = \{2+\sqrt{3}\}^2(1-s^2)/9$. If 's' is 0.9, then t = 0.84. Web available for cluster of 3-grains is 0.0185D. The burning periphery (equivalent to burning area) is 1.9ðD for single grain and is 2.562 ðD for cluster of 3-grains. So, to maintain same propellant weight, cluster of 3 propellant grains has lower diameter ratio, lower web availability and higher burning area, as compared to single propellant grain. Compared to other cluster configuration, this has highest diameter ratio, highest burning periphery (or burning surface area) and lowest web available (or burning time). Some of the cluster configurations are shown as Fig. 66 for ready reference.

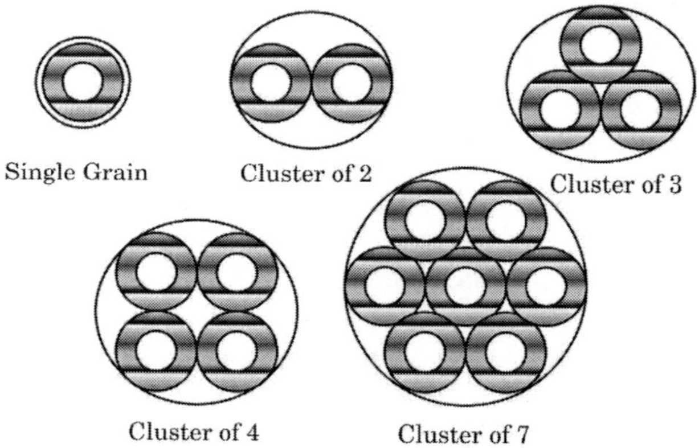

Fig. 66: Cross-section of multi-tubular configurations

Various expressions can be written for the cluster configurations, under assumption of lateral inhibition and outer diameter equal to 'D'. The diameter ratio of single grain is 's', while for the cluster configuration, it is 't'. For cluster configuration, 's' is replaced by 't'.

$(1-t^2) = K(1-s^2)$ or $t = \sqrt{[1- K(1-s^2)]}$

Maximum web available = D(1-dia.ratios)/P

Burning periphery = Q π D (1+dia.ratio)

Lower Limit of 's' for feasible design, $M = \sqrt{[1-(1/K)]}$

The constants in expressions for different cluster configurations are tabulated below.

Cluster of	*Grain dia*	*K*	*M*	*P*	*Q*
1	D	1	0	4	1
2	D/2	2	0.7071	8	1
3	$\sqrt{3}D/(2+\sqrt{3})$	$(2+\sqrt{3})^2/9$	0.5948	$(8+4\sqrt{3})/\sqrt{3}$	$3\sqrt{3}/(2+\sqrt{3})$
4	$D/(1+\sqrt{2})$	$(1+\sqrt{2})^2/4$	0.5601	$4(1+\sqrt{2})$	$4/(1+\sqrt{2})$
7	D/3	9/7	0.4714	12	7/3
19	D/5	25/19	0.4899	20	19/5

As number of propellant grains in cluster configuration increases, the value of 'K' reduces till 7-tube multi-tubular configuration. In fact, the feasible range of diameter ratio for single grain weight equivalent is also the lowest for cluster of 7 grains. This gives a better flexibility with cluster of 7-tube multi-tubular configuration.

The parameter 'P' represent maximum web available. Higher values of 'P' indicate lower value of maximum available web for complete consumption of propellant. It is natural that as number of propellant grains in cluster configuration increases, the maximum web available reduces. Although from cluster of 2 to cluster of 3, the enhancement 'P' (or relative value of maximum web available) is marginal (from 8 to 8.62). However, for any cluster configuration, the value of diameter ratio also plays a major role, which changes with number of grains in cluster configurations.

The parameter 'Q' represents burning periphery (or burning area and pressure realized) and higher value of 'Q' means higher burning area and pressure realized. As number of propellant grains in cluster configuration increases, the value of available burning periphery also

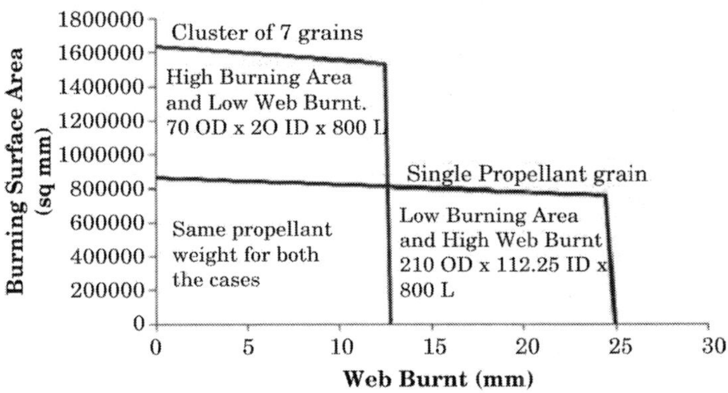

Fig. 67: Comparison of burning surface area variation

increases. It is interesting to note that multiplier for burning periphery (Q) does not change if single grain is replaced with two identical tubular propellant grains, having same total weight, accommodated in same rocket (fixed rocket diameter). However, since diameter ratio for 2 propellant grains is lower than single grain, overall, burning area reduces.

From the analysis depicted above, it seems that for cluster configuration, 7-tube multi-tubular configuration is probably an optimum design, considering web available and burning surface area availability. Further analysis of uninhibited multi-tubular propellant grain is carried out for cluster of 7 tubular propellant grains. The burning profile is regressive and this configuration is utilized for practical application.

The propellant dimensions are taken as 70 mm outer diameter, 20 mm internal diameter and 800 mm length. The web burnt for complete consumption of the propellants is 12.5 mm. Initial burning surface area is 1632843 mm^2, while final burning surface area is 1533883 mm^2. This leads to reduction of around 6.06% in burning surface area. The burning surface area versus web burnt profile is shown in Fig. 67.

The minimum internal diameter of the rocket casing to accommodate 7-tubular propellant grains is 210 mm. The same is considered as outer diameter of an equivalent single propellant grain. Single grain is assumed to have same length (800 mm). To match the propellant weight, internal diameter of equivalent single propellant grain is 112.25 mm. matching total weight of propellant automatically assumes equal cross-sectional loading of the propellant. The web burnt for complete consumption of the propellants is 24.44 mm. Initial burning surface area is 859382 mm^2, while final burning surface area is 760421.8 mm^2. This leads to reduction of around 11.51% in burning surface area. This numerical calculation clearly indicates that multi-tubular configuration makes burning profile less regressive. However, this reduction in regressivity may be due to lower web for multi-tubular propellant grain. If same web burnt is considered for single propellant grain, the burning surface area at web burnt of 12.5 mm is 808763.1 mm^2. This is a reduction by 5.89%, which is less than single propellant grain configuration. One of the significant deduction is that cluster of 7 propellant grain has a very high burning surface area as compared to single propellant grain of equal weight.

To simulate the actual situation, rather than equating weight of propellants in single and cluster of 7 tubes, attempt is made to equate initial burning area and total available web for the propellants. Cluster of 7 tubular propellant grains with dimensions as 70 mm outer diameter, 20 mm inner diameter and 800 mm length is considered. This has cross-sectional loading of 71.42% for the inner diameter of rocket chamber of 210 mm. However, if equivalent propellant grain with same maximum web availability (12.5 mm) and initial burning surface area (1632843 mm^2) is conceived, the outer diameter of single propellant grain is calculated as (4D+3d = 4x70+3x20 =) 340 mm. The inner diameter of the propellant grain is (340–50 =) 290 mm to maintain same web availability of 50 mm. Length of the propellant grain remains invariant. For such configuration, the variation of burning surface area with web burnt is identical to that for the cluster configuration and no change is observed in numerical values at each of the web burnt. Identical variation of burning surface area versus web burnt profile ensures equality of weight of propellant for both the cases. However, for single propellant grain the cross-sectional loading is reduced drastically to 27.25%. This is because of enhancement in outer and inner diameters of single propellant grains, exorbitantly (even beyond 210 mm).

A generalized relation for diameters of single propellant grain to replace same burning area, maximum web available and propellant weight of a multi-tubular combination can be obtained. Length can be assumed invariant but combustion is taking place at the lateral surfaces and length of propellant reduces in course of burning thus giving a regressive burning profile. 'w' may be considered as web burnt at any instant. If n-tube multi-tubular propellant grain of outer diameter 'D' and inner diameter 'd' is considered, the outer diameter (D_1) and inner diameter (d_1) can be obtained by the relation given below.

Maximum web available, $w_m = (D\text{-}d)/4 = (D_1\text{-}d_1)/4$, or $4w_m = (D\text{-}d) = (D_1\text{-}d_1)$

Outer diameter of single grain, $D_1 = n(D+d)/2+2w_m = (n+1)D/2 + (n\text{-}1)d/2$

Inner diameter of single grain, $d_1 = D_1 - 4w_m = (n\text{-}1)D/2 + (n+1)d/2$

Expression for burning area = $n\eth(D+d)(L+2w_m\text{-}4w) = \eth(D_1+d_1)(L+2w_m\text{-}4w)$

Second invariant of the configuration, $n(D+d) = (D_1+d_1)$

Expression for Volume of propellant = $n\eth(D+d)(w_m\text{-}w)(L\text{-}2w)$

Expression for volumetric loading = $n(D^2\text{-}d^2)D_r^{\ 2}$, where D_r = Rocket diameter

It is clear from the above expressions that difference of outer and inner diameters for the configurations are same thus giving constant value of maximum web available. Another invariant is sum of diameters times number of propellant grains. Expressions for burning surface area and instantaneous volume of propellants give same numerical values at each of the web consumed. So burning area versus web burnt profile is identical. The pressure time profiles for both the configurations are similar. However, volumetric loading for single grain configuration will be very low.

REVIEW QUESTIONS

1. What are main features of tubular propellant configuration?
2. What area advantages of tubular configuration over solid cylinder configuration?
3. How are all three modes of burning obtained in tubular configurations?
4. What are different inhibition conditions for tubular propellant configuration?
5. Derive equation of burning surface area for uninhibited tubular propellant configuration for different web consumed?
6. How burning area to port area varied for tubular propellant configuration?
7. Compare end inhibited tubular configuration with solid sustainer configuration.
8. Derived expression for maximum web available for different inhibition conditions for tubular configuration.
9. What are features for taper tubular port configuration?
10. Can neutral burning be obtained from taper tubular port configuration?
11. What are features of offset-port tubular configuration?
12. Why is multi-tubular configuration used in rocketry?
13. What are features with multi-tubular configuration?
14. What are assumptions for analysis of multi-tubular configuration?
15. Derive expression for ratio of maximum individual grain diameter to inner diameter of rocket for cluster of (i) 2 (ii) 3 (iii) 4 (iv) 7 tubular propellant grains.

8

Multi-perforated Configurations

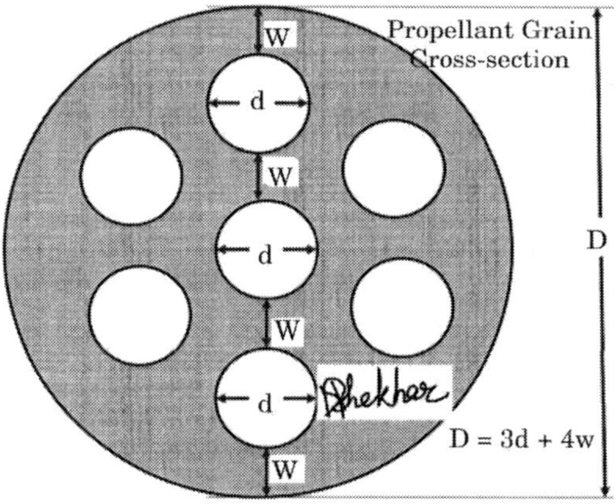

CHAPTER SUMMARY

This chapter concentrates on highly progressive multi-perforated propellant configuration, which is mostly utilized in gas generating situations and is rarely used in main rockets for propulsion. The multi-perforated configuration has more importance for gun propulsion field and the geometrical aspects, relevant for gun propulsion are included in this chapter. In rocketry, this configuration is used in all side inhibited condition and the same is explained in this chapter. The chapter discusses the following.

INTRODUCTION

Multi-perforated solid propellant configuration is derived from gun propellants area. Compared to other configurations like cord (here solid cylinder), tubular, slotted tube, flakes, ribbon etc used in gun propulsion, multi-perforated configurations evolved due to progressive nature of burning pattern, even in uninhibited condition. In gun propellants, inhibition is never applied and propellant burns at all the exposed surfaces. Getting a progressive burning profile during combustion is fulfilled by this multi-perforated configuration. In rocketry also, high progressivity requirement compels use of amulti-perforated configurations. The application of inhibition at the external surfaces and restricting burning at the perforations, only may enhance progressivity to a very higher value and in course of burning pressure rises by more than 4-5 times in certain specific cases of multi-perforated configurations. The salient features of this configuration are enlisted below:

- This configuration (with inhibition on all external surfaces) has very high progressivity ratio or ratio of final to initial burning surface area in the initial phases of burning.
- This progressivity is reflected in pressure-time and thrust-time profiles also.
- This propellant configuration is utilized for cartridge actuation devices (CAD), impulse cartridges and ejection devices.
- This type of propellant has smaller web availability and burning time is generally very short.
- This propellant configuration is suitable for small devices and accordingly dimensions are limited. Use in bigger devices may lead to very high rise in pressure.
- This propellant configuration is used with inhibition at all the longitudinal and lateral burning surfaces. So, this can be manufactured in case-bonded configuration.
- Since burning profile is progressive in nature, the propellant configuration can be used for practical evaluation of burning rate from pressure time profiles, by static evaluation.
- This propellant configuration has very high volumetric loading (~ 90-95%) and chamber volume is sufficiently filled with propellant.
- This propellant configuration has longer tail-off, due to sliver burning at the end of propellant consumption.
- Sliver formation gives sharp pressure fall and during sliver burning, unburnt pieces can come out of nozzle giving pressure kinks. This configuration needs suitable grid to hold propellant grains till sliver burning.
- This propellant configuration cannot be used as main propulsion device due to high pressure at the end of burning and long tail-off.
- Manufacturing this configuration require perforations, which are created by either mandrel or drilling.

- The production rate of this configuration is slow, due to more manufacturing processes and inhibition needs.
- Ignition system design for this configuration becomes critical, as propellant gases are to be distributed even through many perforations. Since perforations are invariably not inter-connection, the flame spread in each of the perforations takes places independently.

In this chapter, expressions for burning surface area of various multi-perforated configurations are evolved. The first configuration is uninhibited hepta-perforated configuration, which is utilized, mostly in gun propulsion area. Similar configuration can be used in rocketry to tame the level of progressivity. All the subsequent configurations are inhibited on all the lateral and longitudinal surfaces. Combustion takes place through cylindrical perforations. Number of holes is increased gradually to attain various type of burning pattern. But whatever may be number of holes, the configurations are characterized by very fast increase in burning surface area and longer tail off for the sliver burning phase.

Uninhibited Hepta-perforated Configuration

Uninhibited hepta-perforated configurations have more significance for gun propulsion, as they give progressive burning profile comfortably. In guns, entire energy is imparted to projectile in the beginning and propellant combustion after projectile is ejected out of gun barrel has no meaning. The gases created at the base of projectile due to combustion of propellants has to execute two activities simultaneously – (i) increasing pressure behind the projectile to impart high acceleration and (ii) fill the empty space created by movement of projectile with gases to maintain and sustain the high pressure. This definitely needs a propellant configuration, where amount of gas generation continues to rise. High gas evolution is possible only when burning surface area continues to rise. This needs a progressive burning profile. Since gun propellants are manufactured by extrusion technique and their outer diameters are also very small (less than 10 mm), the progressive nature of burning can be attempted by configuration alone. The flexibility of adding inhibition does not exist for gun propellants. So, the uninhibited hepta-perforated configuration evolved. This configuration is characterized by two diameters – one is grain diameter (D) and other is hole diameter (d). The unique features of this configuration are as follows:

- This gives progressivity in uninhibited conditions.
- The progressivity can be enhanced by applying inhibition.

- The tail-off can be reduced by reducing hole diameter as compared to outer diameter of the propellant grain.
- Higher tail-off can be obtained when hole diameters are increased giving a low rise in pressure.
- The web between any two nearest holes and between any outer hole and periphery is same. This gives a sudden end of progressivity.
- Since no inhibition is used, some gap between propellant and rocket casing is needed to continue combustion at the external surface comfortably.

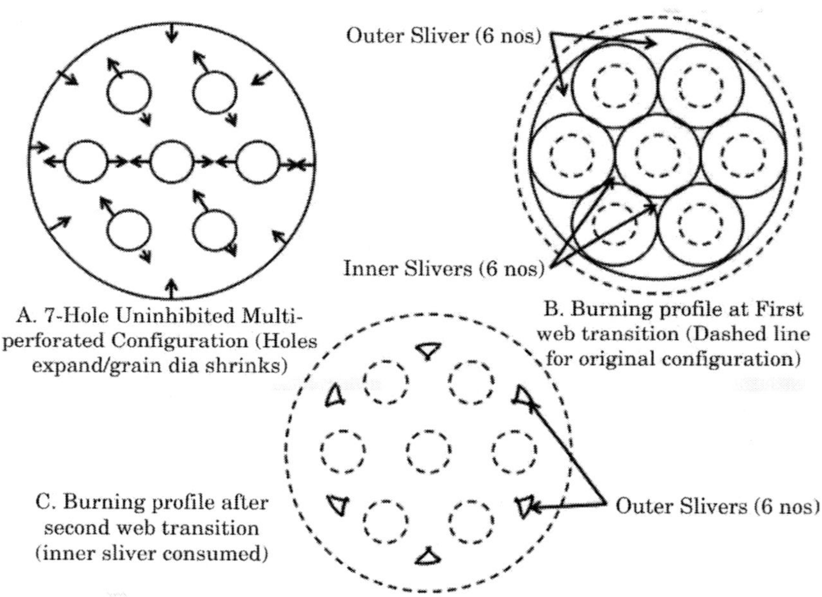

Fig. 68: Burning of hepta-perforated uninhibited propellant

A typical hepta-tubular propellant configuration during combustion inside rocket motor is depicted in Figs. 68 and 68A, shows an initial configuration, which contains one central perforation and six perforations on a pitch circle. During combustion, the outer diameter of propellant shrinks and hole diameters expands. The directions of combustion of different burning surfaces are shown in Fig. 68A. All six holes are equidistant from each other and also with the central hole. The outer diameter shrinks and meets external 6 holes at the same time when central hole touches the external 6 holes (depicted in Fig. 68B). This is first web transition. The simultaneity of convergence of all the burning surfaces restricts various dimensions. Only 2 cross-sectional parameters namely (i) outer grain diameter (radius, R) and (ii) hole diameter (radius, r) are needed and pitch circle diameter of external 6 holes can be calculated from these two variables. The propellant is consumed in length direction also. Till first web transition, propellant is a single piece, but after this propellant is divided in unconnected 12 pieces – 6 inner slivers and 6 outer slivers. Inner slivers are consumed first and this is denoted as second web transition. After second web transition, only outer slivers remain (Fig. 68C). At third web transition, entire propellant is consumed.

So, for hepta-perforated propellant grain, 3 web transitions are defined – first is beginning of sliver formation, second is consumption of inner sliver (convex triangles), third is consumption of outer sliver (convex-concave triangles). Rather than using diameter, the calculations can be simplified, if radii are considered. Outer radius (R = D/2) and length (L) of the propellant grain reduces during burning, while hole radii (r = d/2) increases in course of burning. If propellant density (ñ) is considered variation of weight of propellant in course of burning can also be obtained.

Outer radius of the propellant grains = R

Hole radius of the propellant grains = r

First web transition, $w_1 = (R-3r)/4$

Second web transition, $w_2 = (R-r)/(2\sqrt{3}) + r$

Third web transition, $$w_3 = \frac{(R^2-r^2)+(r+w_1)[4r+4w_1-2\sqrt{3}R]}{2(R+r) - 2\sqrt{3}(r+w_1)}$$

The burning area and volume of propellant grain can be obtained between each web transitions with the help of certain auxiliary variables.

For $w < w_1$

Weight of Propellant, $W = \rho[\pi(R-w)^2-7\pi(r+w)^2][L-2w]$

Burning area, $A_b = 2\pi[R+7r+6w][L-2w]$

The calculation after first web transition is separated into two streams – one for internal sliver and other for the outer sliver. Since, this sliver formation starts after web burnt is exceeding w_1, another variable 'H' is defined as $(r+w_1)$. One more variable (r+w) is defined as 'S'. It is worth consideration that 'H' is a constant for the configuration, while 'S' changes with the web consumed. One more variable (L-2w) can be taken as 'P', which also changes with consumption of propellant. The instantaneous outer radius is defined by 'Q' and is given by expression (R-w). Internal slivers exist till second web transition. When web exceeds w_2, internal sliver is completely consumed giving zero value of weight as well as burning area. For internal sliver, web varying from w_1 to w_2 is relevant.

For $w_1 < w < w_2$

Intermediate variable, $A1 = Cos^{-1}[H/S]$

Weight of internal sliver, $W_1 =$

$$\frac{\rho PH^2[6\sqrt{3}Cos(A1)-18Sin(A1)-3\pi+18(A1)]}{Cos(A1)}$$

Burning area of internal sliver, $A_{b1} = S\ [6\pi-36(A1)]$

Outer slivers exist till third web transition. When web exceeds w_3, outer sliver is completely consumed giving zero value of weight as well as burning area. For outer sliver, web varying from w_1 to w_3 is relevant.

For $w_1 < w < w_3$

Intermediate variable, $A2 = Sin^{-1}[H/S]$

Intermediate variable, $A3 = Cos^{-1}\ [Q^2-S^2-4H^2]/4HS$

Intermediate variable, $A4 = (\pi/6)-Sin^{-1}\ [S\ x.sin(A3)/Q]$

Wt: outer sliver, $W_2=\pi P\{12HS[Sin(A3)-sin(\pi/6+A2)]+6(A4)Q^2-S^2[6(A2+A3)-5\pi]\}$

Burning area of internal sliver, $A_{b2} = p\{12Q(A4)+S[12(A2+A3)-10\pi]\}$

Total weight of propellant remaining is obtained by addition of W_1 and W_2. Same is case for burning area. To understand the variation of propellant weight and burning area, numerical analysis is carried out with outer diameter of 200 mm, hole diameter of 20 mm and length (L) of 1000 mm. The propellant density is taken as (π=) 1760 kg/m^3. The grain radius and hole radius are (R=)100 mm and (r=)10 mm, respectively.

Initial propellant weight = $\rho\ [\pi R^2L - 7\ \pi r^2L]$ = 51.42 kg.

Initial burning surface area $=2\pi L\ (R+7r)$ = 1.0681415 m^2

First web transition, $w_1 = (R-3r)/4$ = 17.5 mm

Second web transition, $w_2 = (R-r)/(2\sqrt{3}) + r$ = 21.75 mm

Third web transition, $w_3 =$

$$\frac{(R^2-r^2)+(r+w_1)[4r+4w_1-2\sqrt{3}R]}{2(R+r)-2\sqrt{3}(r+w_1)} = 27.25\ \text{mm}$$

Before 17.5 mm web is consumed, all the holes are completely circular. Propellant weight and burning surface area can be calculated in same way as initial values are calculated above, but instantaneous values of

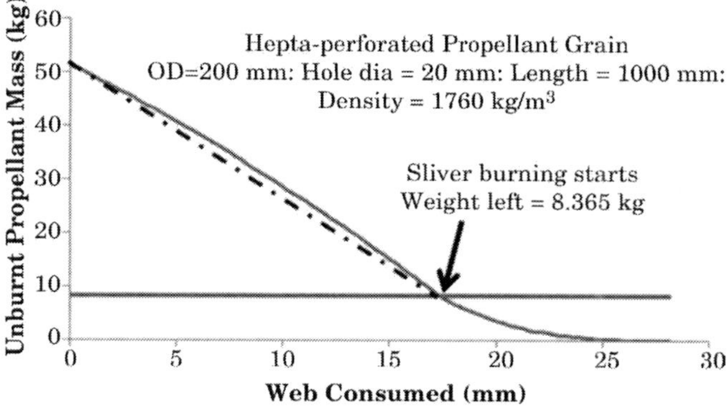

Fig. 69: Variation of mass of propellant with web consumed

radii and length are to be considered. Calculation is repeated till complete consumption of web and Fig. 69 represents variation of weight of propellant with consumption of propellant web.

Propellant combustion till beginning of slivers is quadratic in nature and is given by expression $\rho\pi P(Q^2\text{-}7S^2)$. However, the variation can be simulated as a straight line variation also as depicted in the Fig. 69 by dashed line. The linear variation of weight with web consumed is lost after sliver formation. Burning of propellant as sliver starts at 17.5 mm, where mass of propellant left is around 8.365 kg. This is 16.27% of the initial propellant weight taken. Burning during sliver burning phase may be irregular as they are unconnected pieces and their movement through nozzle may result in pressure kinks, in practical evaluations. From variation of mass, it is difficult to predict or understand progressive nature of the propellant grain. So variation of burning surface area is plotted against web consumed as Fig. 70.

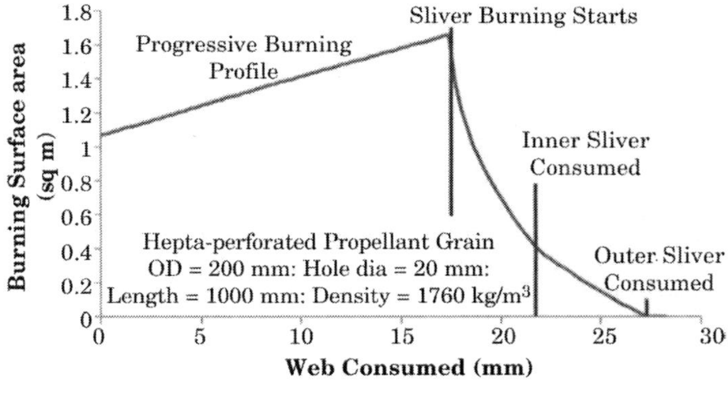

Fig. 70: Variation of burning surface area of propellant with web consumed

The burning surface area increases continuously till combustion of propellant in sliver burning mode starts. This makes the burning pattern of propellant grain progressive in the initial phases. Although web for progressive burning is smaller, but burning surface area enhances by around 1.8 times. This leads to higher change in pressure levels. Once sliver burning starts, burning surface area reduce at a faster pace. When inner sliver is consumed, the rate of variation of burning surface area reduces and burning surface area attains zero value when outer sliver is completely consumed at third web transition of 27.25 mm. The web transition points are also depicted in the Fig. 70.

When propellant length is changed, neither web transition points are changed nor is point of maximum burning surface area altered. For initial length of 1000 mm, the enhancement in burning surface area over initial burning surface area is 155.79%. If initial length is changed to 500 mm, the enhancement in burning surface area over initial burning surface area is 150.18%. So, percentage change in burning surface area increases with rise in propellant length. This is because of relatively smaller value of first web transition (w_1) with respect to length of propellant grains for higher propellant lengths. In burning surface area, term 'P' (=L-2w) is responsible for contribution of length. When 'L' is larger, the effect of 'w' will not be felt that significantly and reduction effects are reduced. So for larger lengths, higher change in propellant burning surface area is observed. The upper limit of burning surface area enhancement over initial burning surface area is given by [5(R+r)/ 2(R+7r)]. For the current case the upper limit of burning area enhancement is 161.76% over initial burning surface area. This occurs at a hypothetical propellant grain of infinite length.

Two Hole Multi-perforated

Two-hole multi-perforated configuration is analyzed for inhibited combustion on all external surfaces. This means propellant outer diameter and length does not change during combustion due to application of inhibition and combustion takes place at the holes surfaces only. The location of holes can be fixed arbitrarily, but for analysis purpose, two holes located at known pitch circle is considered. Again pitch circle is to be selected in such a way that both circles touch each other at the same time, when they touch the outer diameter. This leads to formation of two convex-concave slivers. As depicted in previous section, grain outer radius, length and hole radius are taken as 'R', 'L' and 'r', respectively. Out of the three parameters, only 'r' changes with combustion of propellants. Other two parameters are fixed throughout the combustion of propellants. A typical configuration and burning profile is shown in Fig. 71.

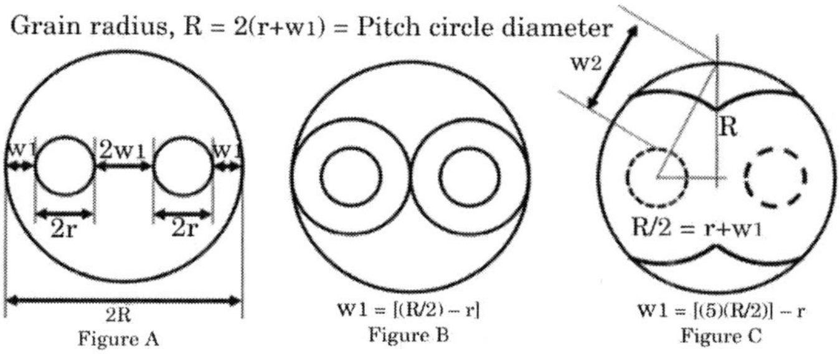

Fig. 71: Two hole inhibited propellant configuration

The configuration is depicted by two web transitions. At first web transition (w_1), both the holes touch each other and also the outer grain diameter (Fig. 71B). The expression for the first web transition is also shown in Fig. 71B. After first web transition, propellant burning continues as two separate propellant grains, which is terminated at complete web consumption at second web transition (w_2) as depicted in Fig. 71C. Expression for the calculation of burning area till first web transition is given by expression below.

$$\text{For } w<w_1, A_b = 4\pi(r+w)L$$

Since, value of 'w' increases continuously, the burning area is increases continuously. However, after web transition, circular burning areas of both the holes are disturbed. The burning area is calculated for different web consumed using two auxiliary intermediate parameters A1 and A2, which are defined for the webs after first web transition.

For $w_1<w<w_2$:

$A1 = \text{Cos}^{-1}\{(r+w)^2 - (3/4)R^2/[R(r+w)]\}$
$A2 = \text{Cos}^{-1} R/[2(r+w)]$
$A_b = 4(A1\text{-}A2)(r+w)L$

There are certain other geometrical factors, which are given by expressions below. They are utilized to analyze various altered configuration for parametric studies.

Slope of burning area profile in progressive zone:
$dA_b/dw = 4\pi L$ for $w<w_1$ and $dA_b/dw = 4(A1\text{-}A2)L$ for $w_1<w<w_2$
Progressive web, $w_1 = (R/2)\text{-}r$
Regressive web, $w_2 - w_1 = (\sqrt{5}\text{-}1)R/2$
Fraction of progressive web to total web $=(1\text{-}2r/R)/(\sqrt{5}\text{-}2r/R)$
Fraction of regressive web to total web $= (\sqrt{5}\text{-}1)/(\sqrt{5}\text{-}2r/R)$
Progressivity ratio, $A_{b\,max}/A_{b\,initial} = R/2r$

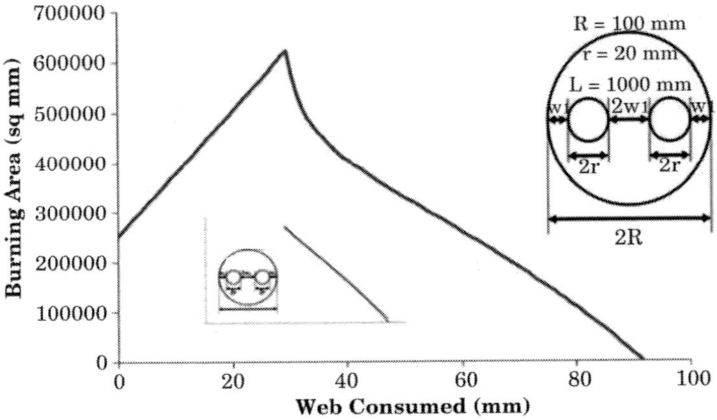

Fig. 72: Burning area versus web consumed profile for 2-hole configuration

After second web transition, propellant is completely consumed. For analysis using above developed formulation a two-hole inhibited propellant configuration with outer radius of 100 mm, hole radius of 20 mm and length as 1000 mm is selected. First web transition is calculated as 30 mm. The second web transition, which indicates complete consumption of propellants, is calculated to be 91.8 mm. Burning area versus web burnt profile is generated as Fig. 72. Initial burning area is calculated as 251327.4 mm^2, which can be calculated independently, for verification as 4ðrL. Progressive burning profile is obtained in the beginning till first web transition (30 mm). At first web transition the burning area is 628318.5 mm^2. This gives a progressivity in burning area of around (R/2r =) 2.5. This progressivity is amplified further, when pressure time profile is generated. For burning rate index of 0.3, the ratio of maximum to initial pressure is calculated as 3.7. This means that if initial parameters for rocket firing are set to give a pressure of 4 MPa in the beginning, then maximum pressure can reach14.8 MPa. Although high level of progressivity is obtained but lower web at first transition, leads to lower burning time till achievement of peak pressure.

After first web transition, sliver burning starts and burning profile becomes regressive. This regression continues till complete consumption of propellant (web = 91.8 mm). However, the rate of regression is steep in the beginning, which later becomes more or less linear. This is depicted as insert in Fig. 72. In fact the tail-off portion is very long and is around two-thirds of the maximum available web for the propellant grain. In fact, such tail-off sections are not appreciated in main motors, but if action is completed at the achievement of maximum pressure, this configuration is very suitable.

This configuration has only 3 geometrical parameters, which can be tailored to get different durations for progressive and tail-off periods. Although length does not play a major role in changing nature of the burning area versus web consumed (or pressure-time) profiles, it does affect the numerical values. The curve is only scaled on y-axis for different lengths of propellants. Propellant outer radius and hole radius affects both the web transitions (w_1, w_2) and difference of web transitions (w_2-w_1). Both the web transitions depend on both the radii, while difference between web transitions [($\sqrt{5}$-1)R/2] is a function of propellant outer radius only. When hole radius (r) is increased, the progressive portion of the curve reduces, but difference of web consumed between first and second transitions remains constant. Reverse situation occurs for reduction in hole radius. When outer radius is increased, values of both the web transitions increase. The difference of web transitions also increases, giving a longer tail-off. This indicates that although both the web transitions are shifted to later instances, but shift of second web transition is by larger amount for increased propellant outer radius.

To numerically display the effects of various parameters, reference propellant configuration with 100 mm propellant outer radius, 20 mm hole radius and 1000 mm propellant length is considered. Length of propellant is changed from 1000 mm to 800 mm, keeping outer radius and hole radius as 100 mm and 20 mm, respectively. The burning area variation shifts downwards (Fig. 73A), but web transition points are unaffected.

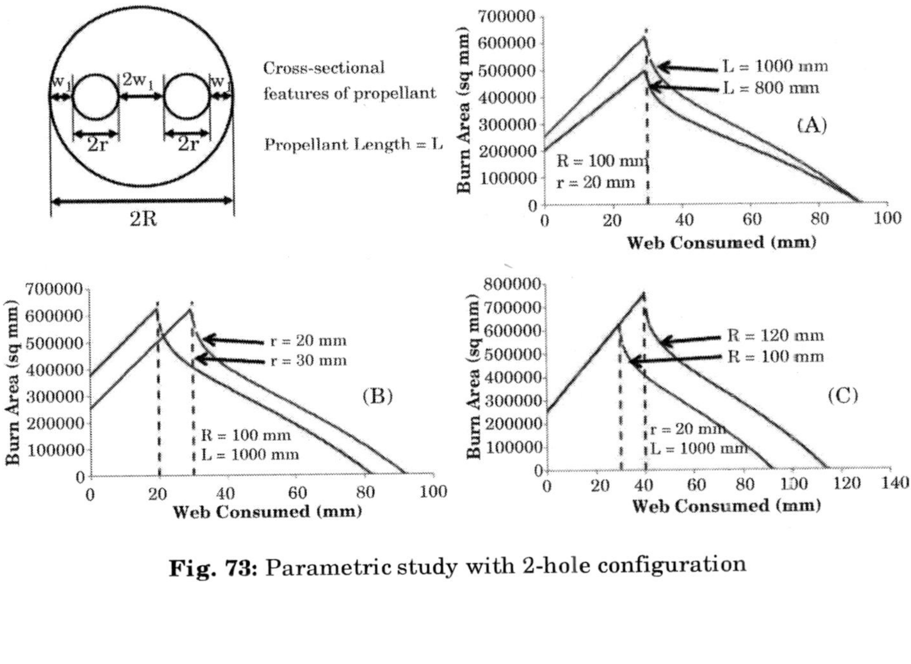

Fig. 73: Parametric study with 2-hole configuration

Both the web transitions occur at 30 mm and 91.8 mm, respectively. Initial burning area reduces for reduced length of propellant but the ratio of maximum burning area to initial burning area remains constant (2.5). Since slope of burning area versus web consumed profile for the progressive phase is ($dA_b/dw = 4\eth L$) a function of length of propellant, and so is the situation for the regressive phase of burning [$dA_b/dw = 4$ (A1-A2)L], the slope is found different for the two curves. For the curve of reference propellant geometry, the slope of progressive section is 12566.37 mm, while same for the propellant with reduced length is 10053.1 mm. For the reference propellant configuration, initial burning area is 251327.4 mm^2, which increases in 30 mm web consumed by (30 x 12566.37 mm =) 376991.12 mm, giving a maximum burning area of 628318.5 mm^2. For the propellant configuration with reduced length, initial burning area is 201061.9 mm^2, which increases in 30 mm web consumed by (30 x 10053.1 mm =) 301593 mm, giving a maximum burning area of 496392.6 mm^2.

When hole radius of the propellant is increased from 20 mm to 30 mm, both the web transitions are shifted (increment in hole radius) 10 mm earlier. It is as good as offsetting the burning area axis by 10 mm rightward. The first and second web transitions occur at 20 mm and 81.8 mm, respectively. As a result, initial burning area is increased, but maximum burning area remains same. The nature of the curves seems to be identical in all respect (Fig. 73B), as both progressive and regressive phases show parallel variations. However, the progressivity ratio (R/2r) reduces to 1.667.

Similarly, in the reference configuration, outer radius of the propellant is changed to 120 mm (Fig. 73C). The first web transition occurs at 40 mm, which is (10 mm: from 30 mm to 40 mm) half the increase in outer grain radius (20 mm: from 100 mm to 120 mm). It is worth noting that increase in hole radius by certain amount increases first web transition by same amount, but the same of outer radius increases first web transition by half the increment amount. This is clear from the equation of first web transition (=R/2 - r) also. Since initial burning area is a function of hole radius, changing outer radius does not alter this. The slope of burning area versus web consumed profile is a function of length and it remains constant, even if propellant outer radius is changed. However, larger first web transition extends burning area increase up to 40 mm of web consumed giving a maximum burning area of 753982.3 mm^2. This gives a progressivity ratio of burning area (maximum to initial) as 3. Curves for reference configuration and altered configuration are same till 30 mm web consumption. As far as second web transition is concerned, it increases by [($\sqrt{5}/2$) x 20 mm =] 22.36 mm and becomes 114.16 mm (from 91.8 mm). This is proportional rise, but with certain factor. The curves of regressive portion for both the configurations are parallel.

This configuration is suitable for limited progressivity. The tail-off portion is generally very long as compared to progressive portion. Out of the total web available, fraction of regressive web is given by [($\sqrt{5}$-1)/(5-2r/R)]. For the reference configuration, this is calculated to be 67.32 %. For limiting condition of zero hole radius, minimum value of this fraction is 55.27%. So, regressive web cannot be reduced beyond 55% of total available web. For getting higher progressivity and lower tail-off, smaller and smaller hole radius are needed for this configuration. This profile should not be selected, when neutral burning profile is needed. At most a regressive burning profile can be obtained when holes touch outer radius of grain in the beginning itself.

There are several variations possible for this configuration and some neutrality in the pressure time profile can be attained, if all webs are not consumed at the same instant. Following options may be adopted for getting differential webs:

1. One hole is at the center and other at certain distance as sun-planet arrangement.
2. Holes touch each other at different instant than they touch outer diameter.
3. One hole is nearer to outer diameter than the other.
4. Both the holes are of different diameters.
5. Holes are not located diametrically opposite to each other.

Each of these configurations needs specialized treatment. Some of them are illustrated with numerical examples. If one of the holes is shifted to center of the propellant grain, another hole will be at any distance from the central hole. The configuration may be treated as sun-planet system. Central hole acts as sun and the peripheral hole act as planet. Both the holes may be of same diameter and location of planet-hole is maintained in such a way that it touches sun-hole and outer periphery at the same time. In the mentioned configuration, no extra information, other than what is given earlier, is needed. However, web transition values are changed (lowered). First web transition (w_1) occurs at [(R/3) – r] and not at [(R/2) – r]. Till first web transition, burning area variation with web consumed is same as earlier case of symmetrical configuration. Complete consumption of propellant grain occurs when web (w_2) is equal to difference of propellant and hole radius (R-r). The burning area for the configuration can be given by the following relation.

Intermediate variables:

A1 = 0 if $w < w_1$ else A1 = $\text{Cos}^{-1}[(r+w_1)/(r+w)]$

A2=π if $w<w_1$ else A2 = $\text{Cos}^{-1}[\{R^2-(r+w)^2-(r+w_1)^2\}/\{4(r+w_1)(r+w)\}]$

Burning area, $A_b = 2(\pi+A2\text{-}2xA1)(r+w)L$ for $w<w_2$ else 0

To get burning nature of propellant configuration, numerical example is considered. Propellant radius, hole radius and propellant length are taken as 100 mm, 20 mm and 1000 mm, respectively. First web transition occurs at 13.33 mm and complete consumption of propellant occurs at 80 mm. Both these values are lower than the same for the earlier configuration. Using expression, derived earlier, burning area versus web consumed profile is generated and shown as Fig. 74.

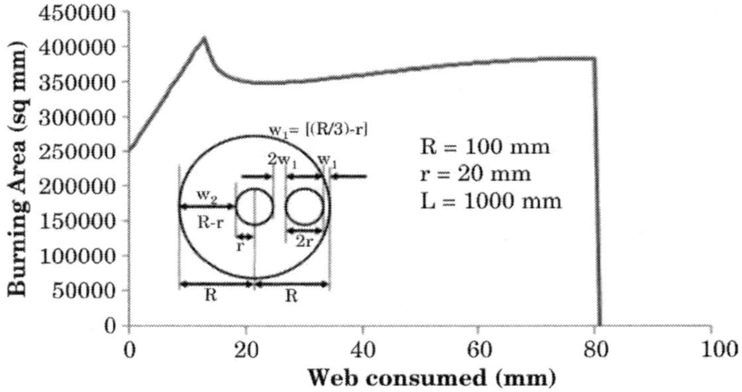

Fig. 74: Burning profile for sun-planet type 2-hole configuration

It is clear from Fig. 74, that initial progressive burning has same numerical value as for previous configuration, depicted in Fig. 5. However, the progressivity stops at much lower web consumption of 13.33 (as compared to 30 mm in Fig. 70) mm. This restricts abnormal rise in the burning area to high values. After first web transition, the hole diameter continues to rise, but the part of hole relevant for the calculation of burning area (angle) reduces. This restrict rise in burning area and very good neutrality is observed. If somehow first web transition is increased (either by increasing outer radius of propellant or reducing hole radius), the initial progressivity will rise and neutrality at relatively higher burning area will be attained. In addition to neutrality, this configuration does not generate any sliver or unconnected propellant pieces in course of burning. So in actual firing, pressure kinks, due to passage of these small unconnected pieces through the nozzle is avoided. However, one major drawback of this configuration is propellant burning surface touching outer diameter at the early stages. For a case-bonded configuration, this leads to higher thermal insulation layer requirements at the exposed sites (towards planet-hole). The rocket motor for the numerical example will be exposed from 13.33 mm web consumed level to 80 mm web consumed. As far as restriction on hole radius is concerned, for retaining initial progressivity, the hole radius should be smaller than one-third outer radius of the propellant grain. For this limiting situation, the planet hole periphery touches outer periphery of the propellant grain.

For the symmetrical hole configuration, separation between the holes may be increased or decreased to have a different configuration and burning profile. For this type of configuration, one more parameter, *i.e.*, separation of both the hole centers is needed as input. Let use define it as 2p. If hole separation is increased, during combustion of propellant, first holes will touch outer diameter of the propellant, then holes will touch each other. Sliver burning will start after this web transition. The complete consumption of propellant will occur at third web transition. The web transitions for this configuration are calculated by the expression given below.

First web transition, w_1 = R-r-p
Second web transition, w_2 = p-r
Third web transition, $w_3 = \sqrt{R^2 + p^2} - r$

For the calculation of burning area, following expressions are derived with the help of certain intermediate parameters.

$A1 = \pi$ if $w<w_1$ else $A1 = \text{Cos}^{-1}[\{R^2-(r+w)^2-(r+w_1)^2\}/\{4(r+w_1)(r+w)\}]$
$A2 = 0$ if $w<w_2$ else $A2 = \text{Cos}^{-1}[(p)/(r+w)]$
Burning area, $A_b = 4(A1\text{-}A3)(r+w)L$ for $w<w_3$ else 0

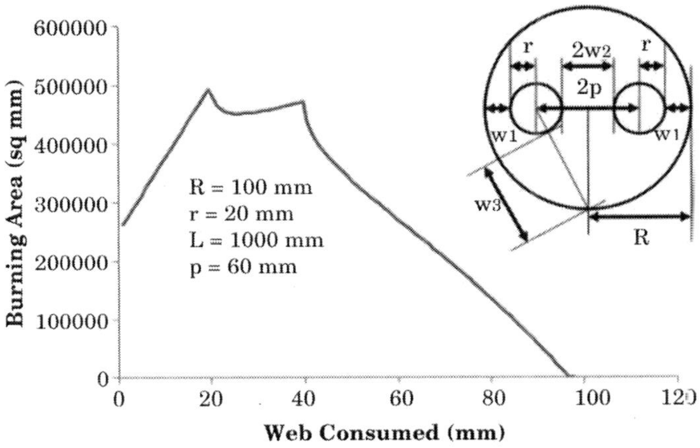

Fig. 75: Burning profile for 2-hole configuration in three webs

For numerical derivation, propellant configuration with 100 mm outer radius, 20 mm hole radius and 1000 mm length is considered. If separation between the hole centers is taken as 100 mm, the configuration resembles the symmetrical configuration, which is shown in Fig. 71. Separation of hole center is changed to (2p=) 120 mm.

First web transition occurs at (R-r-p =) 20 mm and at this point, holes will touch outer periphery of the propellant. Second web transition occurs at (p-r =) 40 mm, when holes touch each other. Third web transition occurs at (w_3 =) 96.62 mm. it is worth noting that the maximum web available is increased as compared to symmetrical configuration defined in Fig. 71. The burning area versus web consumed profile for the propellant configuration is shown in Fig. 75.

As discussed earlier, the burning profile has a longer tail-off. Progressivity of burning profile (increase in burning area) continues till first web transition. The first web transition may be when holes touches outer periphery or when holes touch each other. In the numerical example considered, holes touches propellant outer periphery at first web transition. The progressivity stops there and more or neutral burning profile is observed till second web transition. The triangular tip of Fig. 72 is flattened in Fig. 75. Rest of the curve behaves in similar fashion as Fig. 72.

As compared to sun and planet arrangement, rather than putting one hole at the center, it can be placed at different distance from center on the same diameter on which first hole lies. This is an extension of the previous two configurations. Compared to sun and planet, one hole is not at the center and compared to differential web configuration, both the hole are not at the same distance from the center of the propellant grain. This is a versatile configuration, from which all the earlier configurations can be derived. However, number of input parameters will increase to account for many variations. In addition to R, r and L, two more parameters are required to define the configuration completely. Both indicates distance of hole center from propellant center and are denoted by p_1 and p_2, respectively. For this configuration, web transitions are defined below.

First web transition, $w_1 = R\text{-}r\text{-}p_1$
Second web transition, $w_2 = R\text{-}r\text{-}p_2$
Third web transition, $w_3 = (p_1+p_2-2r)/2$
Fourth web transition, $w_4 = \sqrt{R^2 + p_1 p_2} - r$

Number of intermediate parameters also increases and burning area calculation has the following expression.

$A1 = \pi$ if $w<w_1$ else $A1= Cos^{-1}[\{R^2-(r+w)^2-(r+w_1)^2\}/\{4(r+w_1)(r+w)\}]$

$A2 = \pi$ if $w<w_2$ else $A2= Cos^{-1}[\{R^2-(r+w)^2-(r+w_2)^2\}/\{4(r+w_2)(r+w)\}]$

$A3 = 0$ if $w<w_3$ else $A3= Cos^{-1}[(p)/(r+w)]$

Burning area, $A_b = 2(A1+A2-2xA3)(r+w)L$ for $w<w_4$ else 0

To get a feel of the burning profile of such configuration, grain radius, propellant length and hole diameter is taken as 100 mm, 1000 mm and 20 mm, respectively. First hole center is taken at (p_1 =) 30 mm from propellant center, while second hole center is taken at a distance of (p_2 =) 70 mm. First web transition (w_1) occurs at web consumption of 50 mm, when first hole touches outer periphery. As second hole is very near to periphery, the web transition indicating this event occurs at (w_2 =) 10 mm. both the holes meet at (w_3 =) 30 mm of web consumption. Complete consumption of propellant occurs at web consumed of (w_4 =) 90 mm. It is worth noting that the number of web transitions does not occur in same sequence, as their cardinal numbers are given. The sequential web transitions have no role in the burning area calculations and webs can be tackled as and when burning surface crosses a particular web transition. In current situation, first second hole meets outer periphery, then both the holes meet each other and after that first hole meets the outer periphery. The burning area versus web consumed profile is shown in Fig. 76.

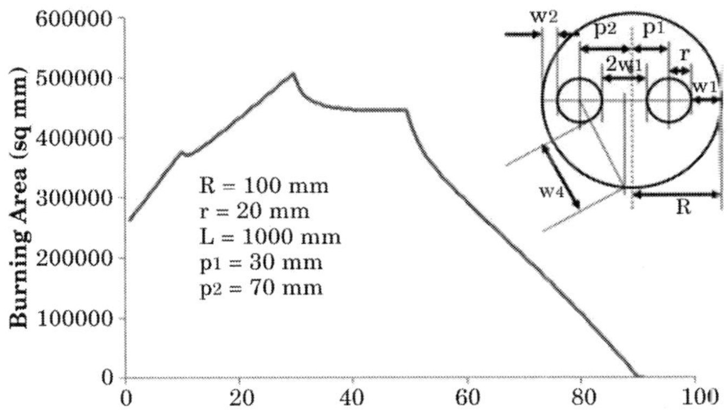

Fig. 76: Burning profile for 2-hole configuration in four webs

From the figure, it is clear that propellant burning profile has web transitions clearly marked on the burning surface area. Initial progressivity till web equal to 10 mm is faster, because both the holes are contributing in progressivity. After this web transition, one hole progresses at a slower rate, resulting in change of slope of the burning profile towards flatter side. At second web transition of 30 mm, the progressivity is lost and grain burning moves in the near neutral burning phase. After 50 mm, the burning profile is regressive and tail-off portion is seen in the burning profile. With this type of configuration, if difference between both the hole distances from propellant center (| p_1-p_2 |) is increased, the neutral portion of the burning profile can be enhanced. This results in reduction of tail-off portion also. Burning area profile for a 2-hole configuration with reduced tail-off is shown in Fig. 77.

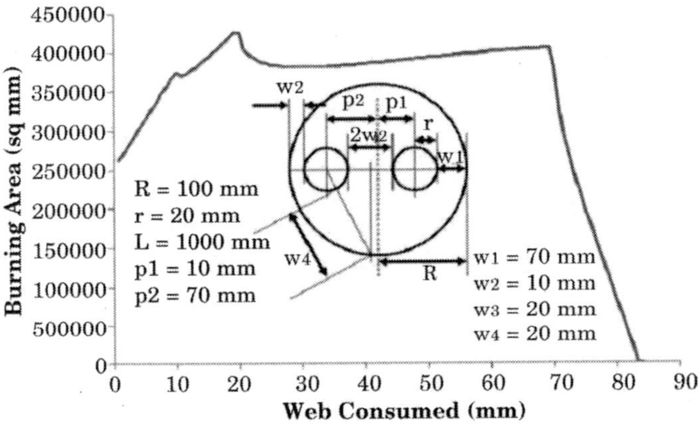

Fig. 77: Burning profile for four web configuration with reduced tail-off

Holes of different diameters can be introduced in the configuration to give different shapes of burning profile. Similarly propellant configuration can be made such that holes are not at same diameter. These configurations are specific and they need special treatment to arrive at the burning area profile by mathematical treatment. This becomes very generic configuration, which can be utilized for generating all the previous configurations with suitable selection of parameters. The mathematical expression for burning surface area profile is derived below for the generic configuration. The burning area profile for a typical numerical input is given as Fig. 78.

Input variables

Propellant Radius = R

Propellant length = L

Hole radius 1 = r_1

Hole radius 2 = r_2

Hole center 1 from propellant center = p_1

Hole center 2 from propellant center = p_2

Web transitions

Hole 1 meets propellant periphery, $w_1 = R\text{-}p_1\text{-}r_1$
Hole 2 meets propellant periphery, $w_2 = R\text{-}p_2\text{-}r_2$
Both holes meet each other, $w_3 = (p_1+p_2-r_1-r_2)/2$
Parameter 1, $H = 2(p_1r_2+p_2r_1)/(p_1+p_2)$
Parameter 2, $S = [(p_1r_2^2+p_2r_1^2)/(p_1+p_2)] - [R^2+p_1p_2]$
Complete consumption of propellant, $w_4 = \sqrt{H^2 - 4S} - H/2$

BURNING AREA CALCULATION

Intermediate parameters:

$A1 = \pi$ if $w<w_1$ else $A1= \text{Cos}^{-1}[\{R^2-(r+w)^2-(r+w_1)^2\}/\{4(r+w_1)(r+w)\}]$
$A2 = \pi$ if $w<w_2$ else $A2= \text{Cos}^{-1}[\{R^2-(r+w)^2-(r+w_2)^2\}/\{4(r+w_2)(r+w)\}]$
$A3 = 0$ if $w<w_3$ else $A3= \text{Cos}^{-1}[(r_1+w_3)/(r+w)]$
$A4 = 0$ if $w<w_3$ else $A4= \text{Cos}^{-1}[(r_2+w_3)/(r+w)]$
Burning area, $A_b = 2[(A1\text{-}A3)(r_1+w)+(A2\text{-}A4)(r_2+w)]L$ for $w<w_4$ else 0

It is observed that the 2-hole configuration can give progressive-regressive burning or near-neutral burning profile or 'M'-shaped burning profile. The number of input geometrical parameters can be increased to tailor the configuration. Control of tail-off and progressivity through geometry is possible.

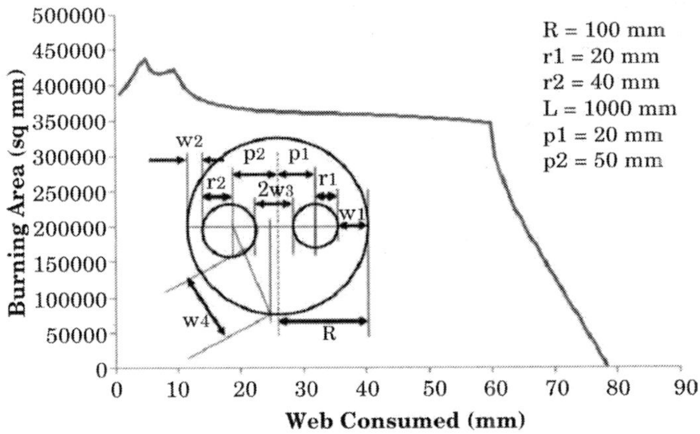

Fig. 78: Burning profile for generic 2-hole configuration

When number of holes is increased, two forms of configurations are possible. In first type of configuration center of one of the holes is concentric to propellant center. In those configurations, central unconnected piece of propellant at the end of burning is not formed. So, sliver burning starts at a much later stage. In second type of configuration, all the holes are non-concentric to propellant center. In this case central sliver piece is created. Rather than discussing number of holes as basis for section heading, the multi-perforated configurations are generalized in two types and analyzed. For each type, different number of holes is analyzed. Next two sections concentrates on various non-central hole and concentric hole multi-perforated configurations.

Multi-perforated Configuration without Central Hole

If number of holes is less, multi-perforated configurations can be conceived with non-central hole configurations. Many configurations are possible with different number of holes. For all these configurations, all the holes are of same radius. They lie at a pitch circle and are placed at equal distances on pitch circle. Although non-equidistance holes may exist, but that is specific configuration, which cannot be analyzed as a general case. They can be analyzed on case-to-case basis. The input parameters for the generation of burning profile for these configurations are listed below.

Input parameters

Propellant outer radius = R

Propellant length = L

Number of holes = n

Hole radius = r

Pitch circle radius = p

The web transitions for the configuration have four important stages. First web transition (w_1) occurs when hole touches outer periphery of the propellant configuration. Second web transition (w_2) occurs when holes touch each other. Third web transition (w_3) is when holes reach center of the propellant configuration. Fourth web transition (w_4) occurs when propellant is consumed completely. These web transitions are not sequential and any one of them may be bigger than the other. Pitch circle for these configurations is selected in such a way that the highest web is realized, when holes are consumed at the outer periphery and not when they are consumed as the internal unconnected pieces. This clearly indicates that highest web transition value is w_4. The expressions for these web transitions in terms of input parameters are given below.

Web transitions

Holes meet propellant periphery, w_1 = R-p-r

Holes meet one another, $w_2 = p\sin(\pi/n) - r$

Holes reach center of propellant, w_3 = p-r

Complete consumption of propellant, $w_4 = \sqrt{R^2 + p^2 - -2RpCos(\pi/n)} - r$

Calculation of burning surface area requires certain intermediate parameters and the expressions can be derived as given below.

Intermediate parameters:

$A1 = \pi$ if $w < w_1$ else $A1 = \text{Cos}^{-1}\,[\{p^2+(r+w)^2 - R^2\}/\{2p(r+w)\}]$

$A2 = 0$ if $w < w_2$

$= \text{Cos}^{-1}\,[(r+w_2)/(r+w)]$ if $w_2 < w < w_3$

$= (\pi/2) - (\pi/n)$ if $w > w_3$

$A3 = 0$ if $w < w_2$

$= \text{Cos}^{-1}\,[(r+w_2)/(r+w)]$ if $w > w_2$

Burning area, $A_b = 2nL(A1\text{-}A2\text{-}A3)(r+w)$ for $w<w_4$ else 0

The configuration can be tailored to give all types of burning profile including 'M'-shaped and near-neutral profile. For same input parameters, only number of holes is changed and changes in web transitions are tabulated in Table 13. Propellant geometry is defined in terms of propellant radius, propellant length, hole radius and pitch circle radius. Numerical values are shown in the first line of the table and they are consistent with the earlier values of these parameters. The values are same as discussed for two hole configuration in Fig. 72.

Table 13: Web transitions for non-central hole multi-perforated propellant configuration

Control Variables: R = 100 mm, L = 1000 mm, r = 20 mm, p = 50 mm					
n	2	3	4	5	6
W1	30	30	30	30	30
W2	30	23.30127	15.35534	9.389263	5
W3	30	30	30	30	30
W4	91.8034	66.60254	53.68129	46.40655	41.96568

It is clear from the table that if propellant radius, hole radius and pitch circle radius is not changed, the value of first web transition (w_1) does not change. For, all the configurations, the value of this web transition is 30 mm. Similarly, third web transition (w_3) is also same for all the configurations, as it is function of pitch circle radius and hole radius only.

As number of holes increases, the web for complete consumption of propellant reduces. Basically, increase in number of holes leads to reduced tail-off. For most of the situation, difference of w_4 with next highest web transition gives tail-off. Since pitch circle periphery is same, more number of holes, mean less distance between the hole centers. This gives reduced w_2 with increased number of holes. After getting an idea about trend in web transitions, burning profile variation is plotted as Fig. 79.

For higher number of holes, initial burning surface area is higher. In the Fig. 79, the highest initial burning surface area is obtained for 6-hole configuration. Higher number of holes has higher progressivity and initial rate of rise of burning surface area depicts this fact. However, second web transition (w_2) occurs early for more number of holes and despite higher initial burning surface area and higher rate of rise of burning surface area, more number of hole may not give corresponding rise in the burning surface area. It must be noted that second web transition (w_2) occurs before first web transition (w_1). Maximum burning surface area is restricted due to lowering of web transition values. First and third web transitions are deliberately made equal by proper selection of geometry.

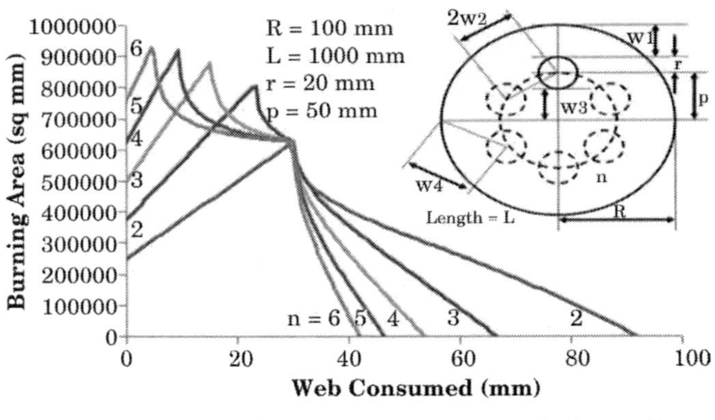

Fig. 79: Burning profile for non-concentric multi-hole configuration

One of the surprising facts is that burning surface area is same for all the configurations at first web transition (w_1). First web transition (w_1) occurs when holes meet outer periphery of the propellant and this occurs later that second web transition (holes meeting one another, w_2) for the given configurations. In the expression for the burning surface area at w_1, value of $2L(r+w_1)$ remains constant. The numerical value of the burning surface area becomes a factor of n(A1-A2-A3). For all these configurations, value of A1 remains π at w_1. Values of A2 and A3 are equal and each is given by $\cos^{-1}[(r+w_2)/(r+w_1)] = \text{Cos}^{-1}[p\sin(\pi/n)/(R-p)]$. For the given geometry, R = 2p and value of $\cos^{-1}[p\sin(\pi/n)/(R-p)]$ is equal to $[(\pi/2)-(\pi/n)]$. So, the burning surface control parameters n(A1-A2-A3) is equal to $n(\pi-\pi+2\pi/n) = 2\pi$, which is independent of 'n' or number of holes. So, burning surface area for this configuration at first web transition (w_1) is equal to $4\pi L(R-p) = 4\pi pL = 2\pi RL = 628318.53$ mm^2. This is depicted for all the configurations in Fig. 79. It is worth noting that same burning area, independent of number of holes, at web transition w_1 is observed only when R=2p. This must be treated as a special case and not a general rule.

For another numerical example, value of 'p' is changed to 60 mm and burning surface area plot is obtained as Fig. 80. Other parameters are kept same. For comparison, only 2-hole and 5-hole configurations are depicted in the Fig. 80. Again values of w_1 (= 20 mm) and w_3 (= 40 mm) are same for both the configurations, due to reason cited earlier. As number of holes increases, the value of w_2 and w_4 reduces. For 2-hole configuration, w_2 (= 40 mm) coincides with w_3 and w_4 (= 96.62 mm) has much higher values. The relative higher values of w_4 leads to longer tail-off portion. For 5 hole configuration, value of w_2 (= 15.27 mm) is very small and this is depicted as first kink in the burning surface profile. After this, regression of web starts. Another kink is observed when web attains value equal to that of w_1. The value of w_4 (= 42.38 mm) is point of complete consumption of web, which gives a lower tail-off. Higher initial burning surface area, higher rate of rise of burning surface area, lower tail-off, lower web for complete consumption of propellant, higher burning surface area at web transitions are general characteristics of more number of holes in multi-perforated configuration and the same is depicted adequately in the Fig. 80.

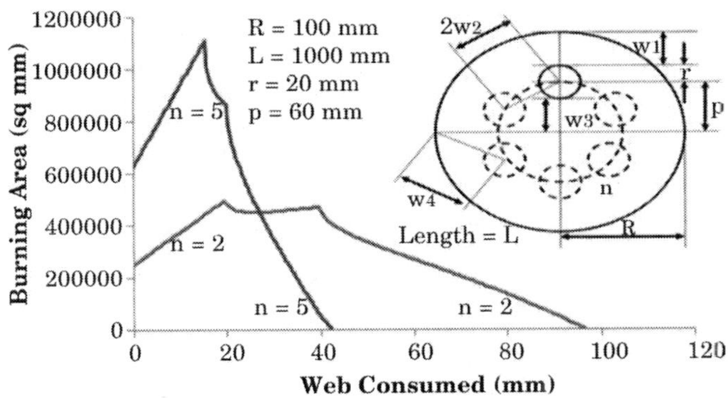

Fig. 80: Burning profile-2 for non-concentric multi-hole configuration

A general mathematical formulation is developed for multi-perforated propellant configuration, where none of the holes are concentric to propellant outer radius. The effect of number of holes has been studied in the text and the mathematical formulation can be extended to any number of holes, but at a single pitch circle.

Multi-perforated Configuration with one Concentric Hole

A variation in multi-perforated configuration can be analyzed, if one of the holes is concentric to the propellant outer periphery. This situation gets rid of central sliver portion and has a better control on performance of propellants. This is discussed for a two hole configuration as Fig. 74, where progressivity is dispensed with and a near-neutral burning profile is realized. Like in previous section, a generic mathematical formulation can be developed for the assessment of burning surface area variation with web consumption of the propellant. The input parameters of the propellants are same as depicted in previous section.

Input parameters

Propellant outer radius = R

Propellant length = L

Number of holes on a pitch circle = n

Total number of holes = n+1

Hole radius = r

Pitch circle radius of holes = p

Web transitions are also similar to that discussed in previous section. However, the value of w_3 has to change and it represents the web, when concentric hole touches outer holes. The central hole touches outer diameter for a hypothetical web consumed of w_5. Ideally, this should not be encountered in any propellant configuration. When configuration with one central and 'n' other holes on a pitch circle, is considered, one more web becomes important. That web is represented by circum-center of three holes – one central and any two adjacent holes at the pitch circle. This is represented by w_6 and has relevance only when total number of holes is four (one central and 3 on a pitch circle). For higher number of holes, this has no relevance. The formulation is developed for those situations, where total number of holes is at most 7 (one central and 6 at a pitch circle). When number of holes exceeds this, the distance between any two adjacent holes at a pitch circle is less than distance between central hole and any hole on the pitch circle. As a result, in course of burning, the central hole is incompletely consumed and unconnected pieces called slivers are generated. So, mathematical formulation is developed, for maximum 7 holes configurations.

Web transitions

Holes meet propellant periphery, w_1 = R-p-r

Pitch circle holes meet one another, $w_2 = p\sin(\pi/n) - r$

Concentric hole meets pitch circle holes, $w_3 = (p-2r)/2$

Complete consumption of propellant, $w_4 = \sqrt{R^2 + p^2 - -2RpCos(\pi/n)} - r$

Central hole touching outer periphery, w_5 = R-r

Circum-center: central hole with 2 holes of p.circle, $w_6 = p/[2Cos(\pi/n)] - r$

Calculation of burning surface area requires certain intermediate parameters and the expressions can be derived as given below.

Intermediate parameters

$A1 = \pi$ if $w < w_1$ else $A1 = Cos^{-1} [\{p^2 + (r+w)^2 - R^2\}/\{2p(r+w)\}]$

$A2 = 0$ if $w < w_3$

For n = 3

$A2 = Cos^{-1} [(p/2)/(r+w)]$ if $w_3 < w < w_6$

$A2 = (\pi/2) - (\pi/n) + Cos^{-1} [psin(\pi/n)/(r+w)]$ if $w > w_6$

For n = 4, 5 and 6

$A2 = Cos^{-1} [(p/2)/(r+w)]$ if $w_3 < w < w_2$

$A2 = (\pi/2) - (\pi/n) + Cos^{-1} [pSin(\pi/n)/(r+w)]$ if $w > w_2$

$A3 = 0$ if $w < w_3$

For n = 3

$A3 = A2$ if $w_3 < w < w_6$

$A3 = \pi/n$ if $w > w_6$

For n = 4, 5 and 6

$A3 = A2$ if $w_3 < w < w_2$

$A3 = \pi/n$ if $w > w_2$

Burning area, $A_b = 2L[\pi-n(A1-A2-A3)](r+w)$ for $w<w_4$ else 0

For a typical 4-hole configuration (n = 3), burning area profile is shown in Fig. 81. The dimensions are consistent with the earlier configurations. The nature of curve is 'M'-shaped and it has a long tail-off. When pitch circle is increased, first peak of 'M' is achieved at lower web consumed (earlier) and second peak occurs at relatively higher web consumed (later). This gives a broader peak. The maximum available web (w_4) is also reduced, thus reducing the tail-off. So, overall, the control of web under different phases of 'M' can be tailored.

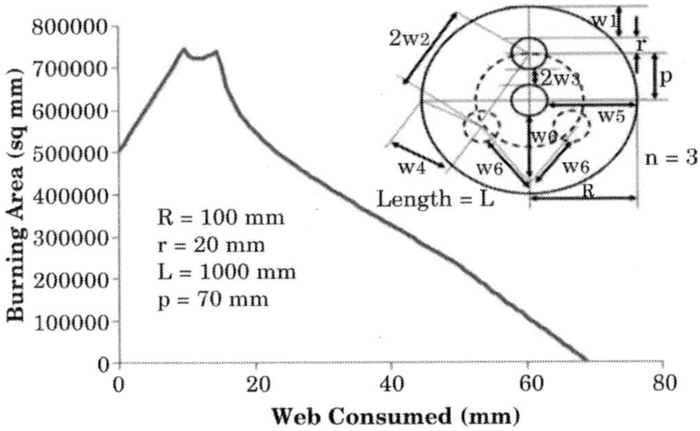

Fig. 81: Burning profile for 4 hole concentric multi-perforated configuration

Table 14: Calculated web transitions for concentric multi-perforated configurations

n →		*3*	*4*	*5*	*6*
Webs	***Unit***				
w_1	mm	10	10	10	10
w_2	mm	40.62	29.50	21.14	15
w_3	mm	15	15	15	15
w_4	mm	68.88	50.71	39.78	32.68
w_5	mm	80	80	80	80
w_6	mm	50	29.50	23.26	20.41

With same dimensional parameters as depicted in Fig. 71, number of holes is raised and various web transitions are shown in Table 14. The values of w_1, w_3 and w_5 are independent of number of holes and are not changing. Since w_2 is dependent on number of holes on a pitch circle, higher number of holes results in lower distance between the holes. Thus w_2 reduces with increase in number of holes. Same is situation with w_4, which also reduces with increase in number of holes. The value of w_6 is relevant for 4 hole configuration (n = 3), only.

Burning area variation with web consumed is shown in Fig. 82 for all the configurations. As number of holes increases, the rate of rise of initial phases increases and slope of burning area profile in the initial stages is higher for higher number of holes. Higher burning area is realized for higher number of holes and at first web transition (w_1) of 'M', the value of burning area is higher for higher number of holes. The second peak of 'M' is seen in all the curves at same web consumed (w_3). For higher number of holes, a sudden decrease in burning area is encountered, whereas for lower number of holes the reduction in subsequent burning area is relatively slower.

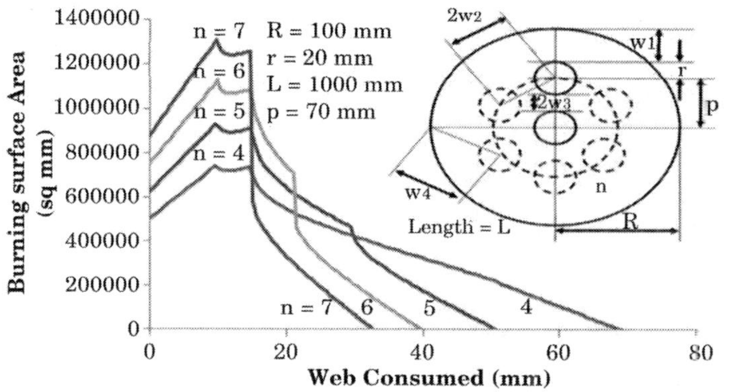

Fig. 82: Burning profile for concentric multi-perforated configuration

Multi-pitch-circle Multi-perforated Configuration

Multi-pitch circle multi-perforated configuration becomes a specific case for the calculation of burning area profile and it should be handled on case to case basis. The propellant grain is assumed to be inhibited on all external surfaces and burning is assumed to take place on the hole surfaces, only. A typical case is described with 132.75 mm propellant radius and 149 mm length. It contains 10 holes of 9.25 mm each. The propellant grain configuration is such that it has a central hole. Three holes are placed at pitch circle radius of 48.5 mm. Another six holes are placed at pitch circle with higher radius.

Propellant outer radius, R = 132.75 mm

Propellant length, L = 149 mm

Hole radius, r = 9.25 mm

Number of holes, n = 10

Radius of inner pitch circle, p_1 = 48.5 mm

Initial burning area = 10 x 2 x ð x 9.25 x 149 mm^2 = 86598 mm^2

Web transition, when central hole touches holes at inner pitch circle, $w_1 = (p_1 - 2r)/2 = (48.5 - 2 \times 9.25)/2$ *mm = 15 mm.*

If this web is maintained for the holes at outer pitch circle to touch outer periphery, then pitch circle radius of outer circle, $p_2 = (R\text{-}w_1\text{-}r) =$ *132.75 – 15 – 9.25 mm = 108.5 mm.*

Holes at the outer pitch circle are positioned in three pairs and pairs are such that line joining grain center with mid-point of holes of any pair is also bisecting line joining any two holes on pitch circle radius of p_1. To maintain web consumption for meeting a pair of holes as (w_1 =)

15 mm, the angle subtended at the center by a pair of holes at the outer pitch circle is given by 2 x $\sin^{-1}$ $[(r+w_1)/p_2]$ = 25.8296°. To ensure that pair of holes at the outer pitch circle meets each other after w_1, the angle is taken on higher side. Angle between pair of holes on outer pitch circle is arbitrarily chosen as *(q=) 47.5°*. This ensures that all the holes are intact till web consumption of 15 mm and variation of burning area is linear till this web. The propellant configuration is shown in Fig. 83.

The propellant configuration has many web transitions. First web transition occurs when central hole meets any hole on the pitch circle of radius p_1. This is denoted by w_1 and has a value of 15 mm. Since radius of outer pitch circle is selected in such a way that holes at this pitch circle also meets outer periphery at same web consumption, the value of second web transition w_2 is also 15 mm. All holes are intact till this web consumption and the value of burning area will rise in linear fashion till this point. *At this web consumed, hole will enlarge to a radius of (9.25 + 15 mm) = 24.25 mm. Burning area at web consumption of 15 mm = 10 x 2 x π x 24.25 x 149 mm² = 227027.19 mm².*

The third web transition (w_3) occurs when hole at inner pitch circle meets nearest hole at the outer pitch circle.

$$2(r+w_3) = \sqrt{p_1{}^2 + p_2{}^2 - 2p_1p_2Cos[(\pi/3) - (\theta/2)]} = 75.0805 \text{ mm}$$

$$\text{So, } w_3 = 28.29 \text{ mm.}$$

One of the salient angle for calculation is the angle between line shown with $2w_3$ and $2w_1$ in figure 16. If this angle is denoted by a, then it is calculated as follows.

$$\cos a = [4(r+w_3)^2+4(r+w_1)^2-p_2{}^2] /[8(r+w_3)(r+w_1)];\ a = 121.294^\circ.$$

Fourth web transition is w_4, depicting overlapping of nearest holes at the outer pitch circle. The value is calculated as follows.

$$\text{Web transition, } w_4 = p_2 \sin(q/2) - r = 34.448 \text{ mm.}$$

Fifth web transition (w_5) is a peculiar one, which may not be visible directly in the figure. This indicates web when contribution of the central hole is completely eliminated. This occurs when hole increases to such an extent that their radius surpasses circum-radius of centers of central hole and any two adjacent holes on inner pitch circle. As holes at inner pitch circle are separated by 120°, an equilateral triangle is created as shown in Fig. 83 at the center of circum-circle with any two other holes. So, $w_5 = p_1 - r = 39.25$ *mm.*

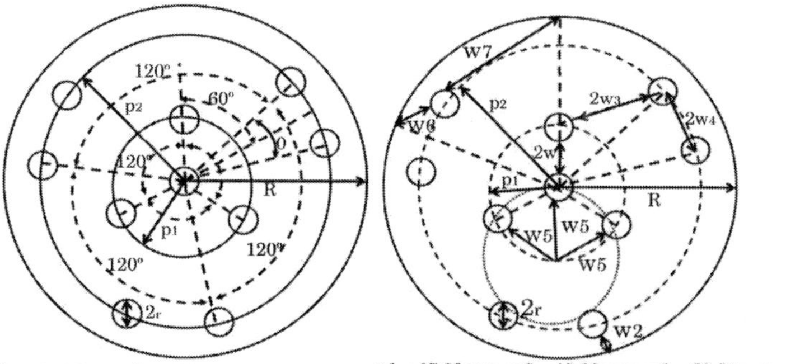

R = 132.75 mm, L = 149 mm, r = 18.5 mm, p1 = 48.5 mm, p2 = 108.5 mm θ = 47.5°

w1 = 15.00 mm, w2 = 15.00 mm, w3 = 28.34 mm, w4 = 34.19 mm, w5 = 39.25 mm, w6 = 45.49 mm, w7 = 69.25 mm

Fig. 83: 10-Hole multi-perforated configuration

Sixth and seventh web transitions are observed when holes at outer pitch circle vanishes one either side as depicted in Fig. 16.

$$\text{So, } r+w_6 = \sqrt{R^2 + p_2{}^2 - 2Rp_2Cos[\theta/2]}\ w_6 = 45.7742 \text{ mm}$$

$$\text{and } r+w_7 = \sqrt{R^2 + p_2{}^2 - 2Rp_2Cos[\theta/2]},\ w_7 = 69.54 \text{ mm}$$

In this configuration, three types of holes exist, which vanishes at different occasions and their burning area calculation is also in phases. First type of hole is central hole, for which an auxiliary parameter A1 is defined after web transition w_1.

$$A1 = Cos^{-1}[p_1/2(r+w)] \text{ for } w > w_1, \text{ else zero}$$

$$\text{Burning area, } A_{b1} = 6(\pi/3\text{-}A1)(r+w)L \text{ for } w<w_5, \text{ else zero}$$

Second type of hole lies at the inner pitch circle, for which additional auxiliary variables are defined. These holes interact with central hole and also with at least two nearest holes on the outer pitch circle.

$$A2= Cos^{-1}[(r+w_3)/(r+w)] \text{ for } w > w_3$$

$$\text{Web transition, } w_8 = [(r+w_3)/Cos(\theta)]\text{-}r = 63.022 \text{ mm.}$$

$$\text{Burning area, } A_{b2} = 6(\pi\text{-}A1\text{-}2xA2)(r+w)L \text{ for } w<w_5$$

$$= 6(\pi\text{-a-}A2)(r+w)L \text{ for } w_5<w<w_8$$

$$= \text{zero } w>w_8$$

Third type of holes lies at the outer pitch circle. These holes interact with holes at inner pitch circle, other holes at outer pitch circle and with outer periphery of the propellant grain. The burning area calculation needs definition of additional parameters.

$$A3 = Cos^{-1}\ [R^2\text{-}p_2{}^2\text{-}(r+w)^2]/[2p_2)(r+w)] \text{ for } w>w_2 \text{ else zero}$$

$$A4 = Cos^{-1}\,[(r+w_4)/(r+w)] \text{ for } w>w_4$$

$$A5 = sin^{-1}[p_2\,sin[(\pi/3)-(\theta/2)]/(r+w)]-(\pi/3)+(\theta/2) \text{ for } w>w_8 \text{ else zero}$$

$$\text{Burning area, } A_{b3} = 12(\pi\text{-}A3\text{-}A2\text{-}A4)(r+w)L \text{ for } w<w_6$$

$$= 6[a-(\pi/3)-(\theta/2)-A3-A2)(r+w)L \text{ for } w_6<w<w_8$$

$$= 6[\pi-A2-A5)(r+w)L \text{ for } w_8<w<w_7$$

$$= \text{zero } w>w_7$$

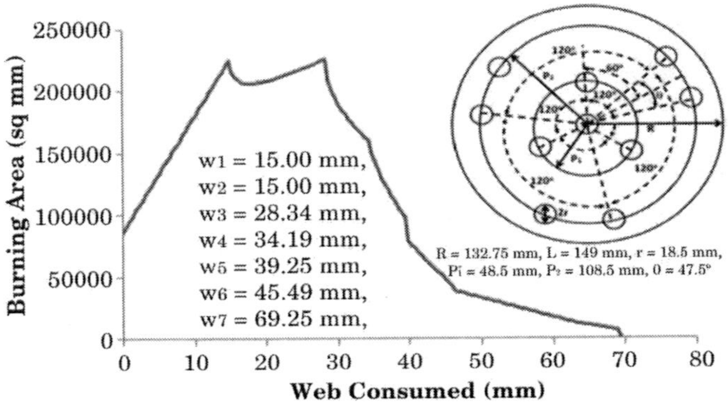

Fig. 84: Predicted burning area variation for 10-hole configuration

$$\text{Total Burning Area, } A_b = A_{b1} + A_{b2} + A_{b3}$$

Using formulations and numerical values developed earlier, burning area variation with web consumed is generated for 10-hole multi-pitch circle specific configuration and plotted as Fig. 84. It is clear that the curve is having 'M'-shape. All the web transitions are visible on the generated profile by variation of slope of the burning area profile. The pressure-time profile for the above mentioned configuration using following input parameters for global analysis is shown as Fig. 85. The nature of curve resembles burning surface area variation curve depicted in Fig. 84.

Characteristic velocity, $C^* = 1370$ m/s

Density of propellant = 1650 kg/m^3

Throat diameter of nozzle = 42 mm

Burning rate of propellant at 7 MPa = 29 mm/s

Burning rate index of propellant = 0.12

Total propellant weight is calculated as 12.95 kg. This is calculated by simply deducting hole volume from a solid cylindrical configuration. The initial burning surface area of 86598 mm^2 becomes 227027 mm^2 at first web transition of 15 mm. This gives burning surface area

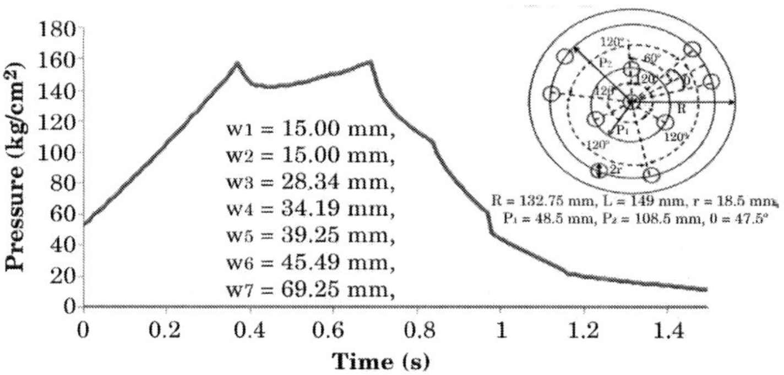

Fig. 85: Predicted pressure-time profile for 10-hole multi-perforated configuration

enhancement of 2.6 approximately. This is sometimes referred as progressivity. But in rocketry, progressivity of pressure and thrust is more important. The time corresponding to first web transition is 0.371 seconds. The initial pressure of 53.29 kg/cm^2 becomes 157.65 kg/cm^2 at the first web transition. The progressivity on pressure is around 2.96. The progressivity of burning area and that for pressure is related by the relation: $P_1/P_2 = [A_{b1}/A_{b2}]^{1/(1-n)}$. So, this progressivity variation depends on the value of burning rate index. Higher value of burning rate index (n) gives higher value of pressure progressivity for a given burning surface area progressivity. If burning area progressivity is maintained as 2.6, then for burning rate index of 0.3, pressure progressivity becomes 3.91.

The next web transition for the propellant configuration occurs at 28.34 mm of web consumed. At this point, the burning surface area regains value, almost equal to the burning surface area obtained at first web transition of 15 mm. This corresponds to time elapsed of around 0.691 seconds. The burning area at this point is 226420 mm^2 and corresponding pressure is 154.56 kg/cm^2.

This type of configuration, if used as gas generator application, requires assessment of mass flow rate variation. The equation for mass flow rate variation is a function of pressure and is derived in last chapter of section 1. The expression for mass flow rate through the nozzle of the rocket is P x A_t/C^*. Since throat area and characteristic velocity is invariant for a curve, mass flow rate variation has same value as pressure variation. The initiation mass flow rate for the configuration is calculated as 5.286 kg/s. At first web transition (15 mm web consumed and 0.317 seconds time elapsed), this becomes 15.64 kg/s. The rate of change of mass flow rate is calculated as 32.66 kg/s^2 for the initial progressive portion. A peculiar requirement of high initial progressivity ratio followed by retention of high pressure is fulfilled by this configuration. However, a large tail-off is not desirable for such configurations.

Similarly, for any multi-hole configuration, the mathematical formulations can be developed using suitable web transitions and auxiliary parameters (mostly angles) for the generation of burning area variation. This chapter mostly discusses the multi-perforated configuration in inhibited situations, where burning is permitted at the periphery of holes, only. In the beginning of the chapter, only, some insight into the uninhibited configuration is put forth. Overall, for specific purposes, such configurations are preferred.

REVIEW QUESTIONS

1. What are features of multi-perforated configuration?
2. Compare uninhibited tubular and multi-perforated configurations.
3. What are limitations of multi-perforated configurations?
4. Comment on sliver formation in multi-perforated configurations.
5. Why multi-perforated configurations have many web transitions?
6. Derive expression for Web transitions in uninhibited 7-perforated configurations.
7. How does mass of propellant during combustion of 7-perforated uninhibited configuration varies?
8. How propellant mass varies with web consumed in uninhibited multi-perforated configuration?
9. How burning surface area varies for uninhibited multi-perforated configuration?
10. How progressivity is achieved in uninhibited multi-perforated configurations?
11. What is effect of length of propellant grain on the burning profile of the 7-perforated uninhibited propellant configuration?
12. Calculate first web transitions for various multi-perforated configurations?
13. Correlate grain diameter and hole diameter in 2-hole multi-perforated propellant configuration with minimum number of web transitions.
14. What are various 2-hole multi-perforated configurations possible?

15. Explain various web transitions in 2-hole multi-perforated configurations?
16. What is maximum number of web transitions possible in 2-hole mult-perforated configuration?
17. How are 4 web transitions introduced in 2-hole multi-perforated propellant configurations?
18. Explain features of 6-hole non-central hole multi-perforated configuration.
19. What advantages multi-perforated propellant configuration with a central hole has over configuration without a central hole?
20. What are advantages of multi-perforated configuration with holes at multi-pitch circles?
21. What are limitations of multi-perforated configuration?
22. Find initial burning surface area for 7-hole multi-perforated uninhibited propellant configuration with following dimensions – (i) Grain diameter = 200 mm (ii) Grain length = 500 mm (iii) Hole diameter = 30 mm. [Ans: 499041.99 mm^2].
23. Solve problem 22 with external inhibition on all surfaces. [Ans: 131946.89 mm^2].
24. Find maximum burning surface area for 2-hole multi-perforated externally-inhibited propellant configuration with following dimensions – (i) Grain diameter = 200 mm (ii) Grain length = 500 mm (iii) Hole diameter = 30 mm. The configuration has only 2 web transitions. [Ans: 314159.27 mm^2].
25. Find the 2 web transitions in problem 24. [Ans: 35 mm and 96.8 mm].
26. Find initial burning surface area for problem 24. [Ans: 94247.78 mm^2].

9

Internal Star and Fin Port Configurations

CHAPTER SUMMARY

This chapter discusses another important solid propellant configuration, where internal perforation in right circular cylinder is made with non-circular geometry with curved surfaces. The internal perforation is concentric to propellant external diameter and is in the form of star, where fin or nodular protrusions are placed over a cylindrical perforation. This configuration has been widely researched and used in operational rockets in 20th century due to high level of attained neutrality with high volumetric loading and radial-outward burning configurations. In this chapter, this important configuration is deliberated on the following aspects:

- Features of star-shaped configurations
- Governing geometrical parameters of star configurations
- Numerical calculations for Lobed shape star configuration
- Mathematical formulations for fin-shaped lobed star configuration
- Top land star configurations and their features

INTRODUCTION

Internal star port configuration is the *most researched configuration* and has been *used in various operational rockets*. The star shaped configurations are conceived in internal outward burning mode with end inhibition such that burning takes place in radial outward direction. These restrictions make the configuration completely two-dimensional. The configuration is stated to give *neutral burning profile with high level of volumetric loading* (70-80%). However, they generally give *higher tail-off*. These propellant configurations are obtained by both casting and extrusion routes. This chapter is devoted to various types of star configurations, which are prevalent. Star configuration has been described in earlier publications as circular port propellant configuration with discontinuities. The number of discontinuities is mentioned as number of star tips or petals. The configuration has rotational symmetry depicted by number of star petals (n). For analysis of star configuration, half of a star is considered and accordingly central angle is equal to ð/n. Geometrically, star shaped configurations also have propellant diameter/radius and length as their salient parameter. Star diameter is the highest diameter/opening of the propellant port, while lobe diameter is the minimum diameter of the propellant port. For the definition of star-port configuration, certain specific geometrical parameters are also defined. Star angle is the angle between sides of adjacent star-petals. Angular fraction is one such parameter, which indicates angle subtended by slant section of the star profile in 2-D plane. Propellant webs and web transitions are also derived by several researchers. In the current chapter, mathematical formulation for the star configuration from the simplest to the complex configuration is developed.

Lobed Star Configuration

The lobed star configuration has circular arcs on the profiles of star. These configurations are created by rotation symmetry of two arcs. When curvatures of these arcs are one same side, it is called concave star and when they are on opposite sides, it is called convex-concave star. Concave star configuration develops into a true fin shaped configuration, while other results in true star configuration.

A typical convex-concave configuration is shown in Fig. 86. The configuration depicted in Fig. 86, has four number of star petals (n = 4), but the mathematical formulation developed later in this section is valid for any number of star petals. The configuration is defined in terms of following parameters (6).

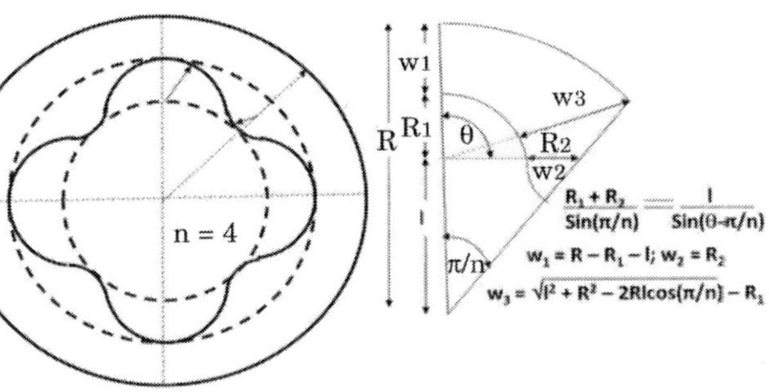

Fig. 86: Convex-concave configuration

Propellant radius = R

Propellant Length = L

Number of star petals/points = n

Star tip radius = R_1

Valley radius = R_2

Ray length = l

The curvature of propellant configuration at the maximum opening point of the propellant port (R_1) is more than its separation from the center of the propellant grain. Accordingly, radius of curvature of maximum opening point is less than radius of maximum opening point with center at the propellant center. The separation of center of propellant and center of maximum tip opening is defined as ray length. Valley is generally introduced with a radius to avoid sharp corners in propellant, which may otherwise get damaged during handling, transportation or storage. Although calculation starts with half angle subtended at the center by the star configuration, which is depicted by π/n, another angle 'q' is also of importance for calculation. The value of radii of curvatures at star and lobe surfaces is not independent but they are related to ray length and half angle at star center. For a valid lobed star configuration, $l \times sin(\pi/n) < (R_1+R_2)$. Some of the calculated parameters for evolution of burning surface area are as follows:

$$q = (\pi/n) + \sin^{-1} [l \times \sin(\pi/n)/(R_1+R_2)]$$

$$\text{Star radius} = R_1 + l$$

$$\text{Lobe radius} = [(R_1+R_2) \times \sin q / \sin(\pi/n)] - R_2$$

Propellant burns in parallel layers and star radius continuously increases with time and web consumed. The concave curve at the highest radius of the star reaches outer radius in course of time, which is point of one web transition (w_1). The radius of convex surface of propellant port reduces in course of time and vanishes when web consumed is equal to R_2. This depicts another web transition (w_2). Till this web transition, central angles for star as well as lobe portion remain constant. Maximum web consumed is attained, when star portion of propellant burning surface reaches corner of the half star portion. This is depicted by web transition w_3 and is a point of complete consumption of propellant. These web transitions are calculated and subsequent calculation of burning area is attempted through auxiliary variables. The propellant volume is also calculated for verification of burning surface area calculation.

Web Transition, $w_1 = R - R_1 - l$
Web Transition, $w_2 = R_2$
Web Transition, $w_3 = \sqrt{l^2 + R^2 - 2RlCos(\pi/n)} - R_1$

***Auxiliary parameters*:**

$A1 = Cos^{-1}\{[R^2-l^2-(R_1+w)^2]/[2l(R_1+w)]\}$ for $w_1<w<w_3$ else zero
$A2 = \theta-(\pi/n)-sin^{-1}[l \times Sin(\pi/n)/(R_1+w)]$ for $w_2<w<w_3$ else zero
$A3 = Sin^{-1}[l \times Sin(\pi/n)/(R_1+R_2)]$ for $w_2<w<w_3$ else zero
Burning area, $A_b/[2nL] = (\theta\text{-}A1\text{-}A2)(R_1+w)+[q-(\pi/n)-A3](R_2-w)$
Volume of propellant, $V = L[\pi(R^2-R_2^2)-n(R_1+R_2)\{lSinq+(R_1-R_2)\theta\}]$

For numerical calculation, a propellant configuration with following geometrical parameters is considered. Calculations are made for generation of burning area variation profile, also, which is depicted in Fig. 87.

R=98 mm, L=1000 mm, n=6, R_1=20 mm, R_2=10 mm, l=30 mm.
$\pi/n = 30^o$ or 0.523598 radians
$\theta = (\pi/n) + sin^{-1}[l \times sin(\pi/n)/(R_1+R_2)] = 60^o$ or 1.047198 radians
Star radius = $R_1 + l$ = 50 mm
Lobe Radius = $[(R_1+R_2) \times sin\theta / sin(\pi/n)] - R_2$ = 41.96 mm
Web Transition, $w_1 = R - R_1 - l$ = 48 mm
Web Transition, $w_2 = R_2$ = 10 mm
Web Transition, $w_3 = \sqrt{l^2 + R^2 - 2RlCos\left(\frac{\pi}{n}\right) - R_1}$ = 53.56 mm
Volume of propellant = 23296204 mm^3

Although one web transition occurs at 10 mm, but the same is not clearly visible as a change of slope (kink) in the Fig. 87. However, other

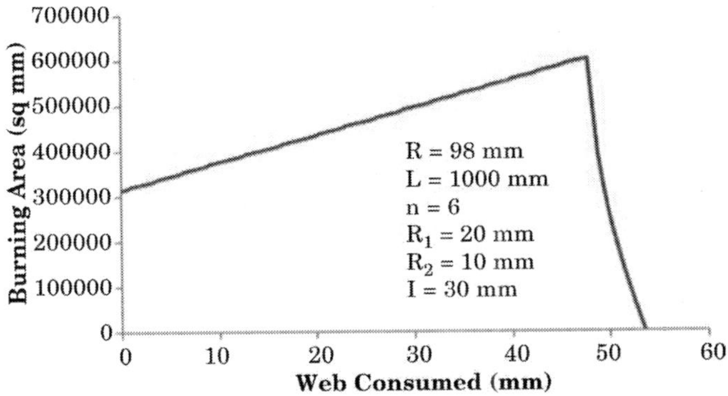

Fig. 87: Burning area profile for convex-concave lobed star

two web transitions (at 48 mm and 53.56 mm) are clearly visible in the figure. Initial burning area is 314159.3 mm^2, while burning surface area at around 48 mm is 608749.3 mm^2. Assuming a linear variation, area under burning area profile is calculated approximately as follows:

$$\text{Area} = (1/2)\text{x}(48\text{x}314159.3+53.56\text{x}608749.3)\ \text{mm}^3=23832998\ \text{mm}^3$$

This calculated area is very near to volume of propellant calculated. This validates the developed mathematical formulation. This approximate calculation introduces an error of around 2.3%, which is permissible for the crude adopted mathematical simplification. However, availability of numerical values of burning surface area at intermediate web consumed requires calculation of area under burning area profile by direct numerical integration. The calculated value of area under burning area profile is 23286030 mm^3, which is off from exact calculation (of volume) by 0.04367%. This validates the developed mathematical formulation. For this given configuration, changes are made in certain geometrical parameters to observe the effect on burning area profile.

First of all, *ray length is increased* from 30 mm to 60 mm, keeping all other parameters same. This *reduced web transitions drastically*. As valley radius, depicting at the lobe of configuration is not changed, w_2 is unaltered. However, both other web transitions are drastically reduced. The values of w_1 and w_2 are 18 mm and 34.95 mm. As ray length is increased, *more empty space is created in same cross-sectional area, reducing the propellant loading*. The volume of propellant is reduced to 16734711 mm^3. Another major change comes in the value of q, which increases from 60° to 120°. The burning area profile is superimposed over earlier profile (Fig. 87) as depicted as Fig. 88.

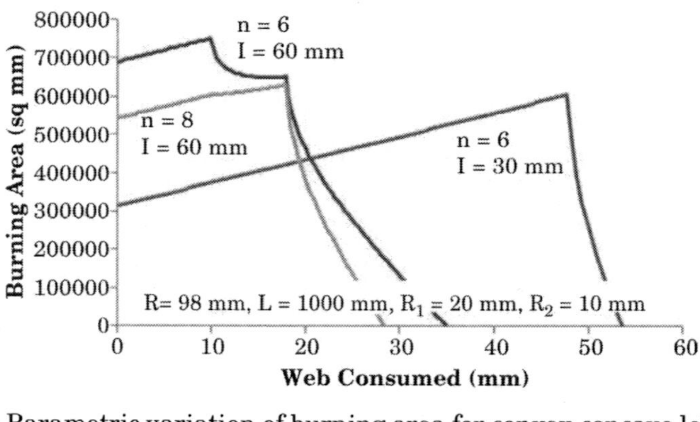

Fig. 88: Parametric variation of burning area for convex-concave lobed star

It is clear that web consumed axis is greatly compressed due to reduction in value of w_3. Area under burning area profile is also reduced as it depicts volume of propellant. The value of initial burning area is increased because of rise in value of the value of θ. The length of circular section at star and lobe ends increases, thus increasing the burning area. One peculiarity of the burning area profile is clear visibility of all three web transitions. For further parametric study, number of star petals (n) is increased to 8 (Fig. 88). The web transitions are refined slightly and web for complete consumption of propellant is reduced, further. Propellant volume is also reduced and burning area is also on lower side compared to corresponding 6-petal configuration. The prominence of web transitions is reduced in this configuration.

From parametric study, it is clear that increasing 'l', 'R_1' or 'R_2' reduces propellant volume, whereas increasing 'R' and 'L' increases propellant loading. Almost similar effects are observed on maximum available web (for complete consumption of propellant). Increasing number of star-petals (n) reduces maximum web available for complete consumption of propellant. Propellant volume is also slightly reduced.

Concave lobed star configuration has inversion of lobe-curvature. *Star tip remains same as earlier configuration, but center of curvature of lobe section is the center of propellant grain.* A typical configuration is shown as Fig. 89. As compared to previous configuration, role of ray length is eliminated, as center of star curvature lies at the lobe periphery. Ray length for this configuration is equal to lobe radius. Number of geometrical parameters for complete definition of propellant is reduced by 1. Propellant radius (R), propellant length (L), number of star points (n), Star tip radius (R_1) and lobe radius (R_2) are relevant geometrical

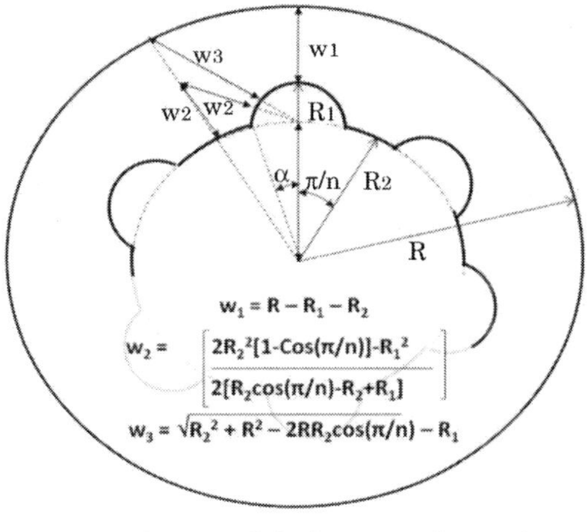

Fig. 89: Concave lobed star configuration

parameters, only. Total number of variables reduces to five only. The web transitions are also depicted in Fig. 89. Star tip touches outer periphery at web transition w_1, while maximum web available for complete consumption of propellant is given by w_3. The contribution of lobe periphery is eliminated at another web transition, depicted by w_2.

One of the angular parameter, a changes with web consumed. This continues to change numerical value till web consumed is equal to w_2. After this point, this attains a constant value of π/n till complete consumption of propellant. Expression for initial value and general value of this angular parameter can be calculated using following relation. Calculation steps for burning area, using auxiliary parameters are also depicted.

$$a = \mathrm{Cos}^{-1}\,[\{R_2^{\,2}+(R_2+w)^2-(R_1+w)^2\}/\{2R_2(R_2+w)\}] \text{ for } w<w_2$$
$$= \pi/n \text{ for } w_2<w<w_3$$
$$a_{initial} = 2 \times \sin^{-1}\,[R_1/2R_2]$$

Auxiliary parameters:

$$A1 = \pi\text{-}a\text{-}\sin^{-1}[R_2 \sin a/(R_1+w)]$$
$$A2 = \mathrm{Cos}^{-1}\,[\{R^2-R_2^{\,2}-(R_1+w)^2\}/\{2R_2(R_1+w)\}] \text{ for } w_1<w<w_3 \text{ else zero}$$
$$\text{Burning area, } A_b = 2nL[(\pi\text{-}A1\text{-}A2)(R_1+w)+(\pi/n - a)(R_2+w)]$$
$$\text{Volume of propellant, } V = L[\pi(R^2-R_2^{\,2})+nR_2^{\,2}(a\text{-}\sin a)\text{-}nR_1^{\,2}(\pi\text{-}A1)]$$

For validation through numerical example, 5 geometrical parameters are assigned numerical values and calculation of various parameters is carried out. Variation of burning area with web consumed is shown as Fig. 90.

$R = 98$ mm, $L = 1000$ mm, $n = 6$, $R_1 = 20$ mm, $R_2 = 40$ mm.
$\pi/n = 30^\circ$ or 0.523598 radians
Star radius = $R_1+R_2 = 60$ mm
Web Transition, $w_1 = R - R_1 - R_2 = 38$ mm
Web Transition, $w_2 = 0.98$ mm
Web Transition, $w_3 = \sqrt{R_2^{\,2} + R^2 - 2RR_2\cos(\pi/n)} - R_1 = 46.44$ mm
$a_{initial} = 2 \times \sin^{-1}\,[R_1/2R_2] = 28.95^\circ$ or 0.505631 radians
Volume of propellant = 20972845 mm^3

All three web transitions are clearly visible in the burning area profile. Main portion of burning area profile is progressive in nature. Area under burning area profile is calculated as 20967516 mm^3. Compared to earlier calculated volume of propellant, this is 0.025% off, only. Thus the

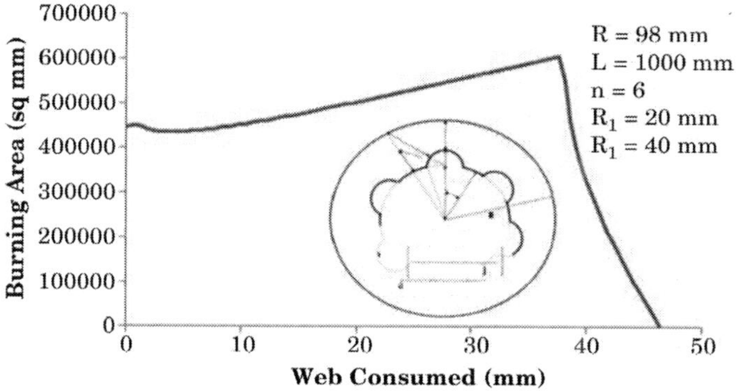

Fig. 90: Burning area variation for concave lobed star

mathematical formulation is validated. With similar geometrical parameters, Convex-concave and concave propellant configuration is compared. The burning area variation profile is shown in Fig. 91.

For same dimensional parameters, concave configuration has higher propellant volume and same is depicted by higher area under burning area profile. For both the configuration two of the web-transitions namely w_1 and w_3 have same values. However, w_2 varies. For convex-concave configuration, it is equal to radius of curvature at lobed portion, while for concave lobed star, it is equal to web for which lobed section disappears. The concave lobed star has better neutrality.

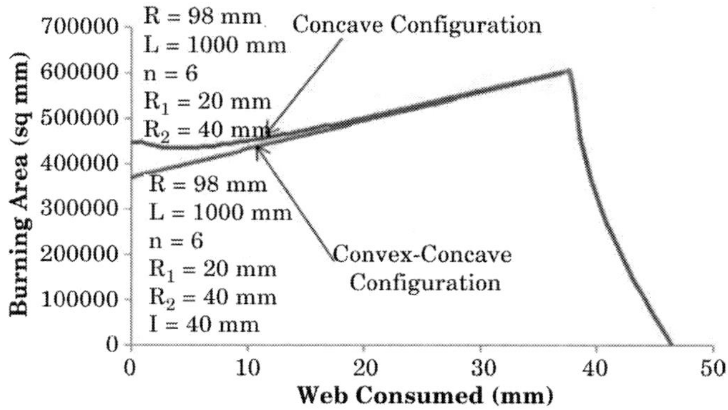

Fig. 91: Comparison of lobed star configurations

Fin-shaped Lobed Star Configuration

Fin shaped lobed star have one more straight section introduced between star tip and lobed sections. This makes calculation step a little more complicated, but better tailoring of burning area profile is possible for such configurations. These improvements are possible for both lobed classes of star port configurations, discussed in the previous section of this chapter. For both classes, mathematical formulation is developed for generation of burning area profile. Lobed star construction is not feasible for many geometrical parameters. Many a times, it is not possible to generate lobed construction. For example, in the example of concave configuration discussed in previous section, if R_2 is taken as 30 mm, the configuration becomes impossible to create. However, if a straight section is added in-between, the configuration becomes feasible. Fin-shaped concave lobed star configuration is shown in Fig. 92.

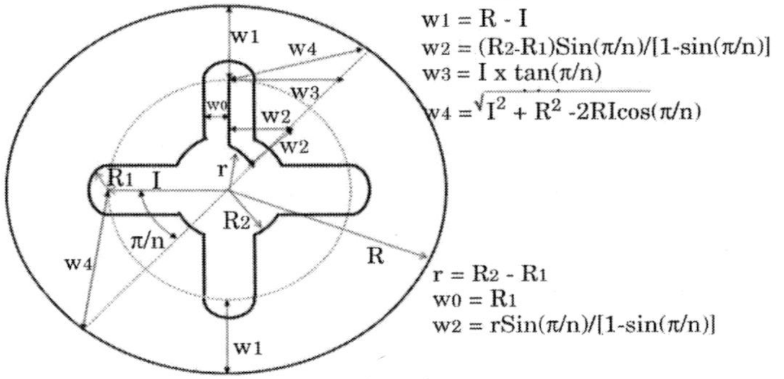

Fig. 92: Fin shaped concave lobed star configuration

This configuration has *6 geometrical parameters* denoted by R, L, n, R_1, R_2, and l. Although this configuration can be analyzed in similar way as discussed in previous section, but to explore a new approach to develop mathematical formulation, the periphery of port configuration can be assumed to be made from two primitives - a straight line and an arc. The curved section has a radius of $R_2 - R_1$, which is denoted by 'r' in the figure. The straight portion may be assumed to be of length (l-r). The initial configuration of the propellant port is attained when web consumed is equal to R_1. Initial web of the propellant is taken as w_0 (= R_1) and all calculations starts after attaining this web. With this modified assumptions, the web transitions are now completely altered.

First web transition is w_0 (=R_1) after which all calculations of burning areas start. Another web transition occurs, when so called star tip touches outer periphery. This is denoted by w_1. Another web transition occurs, when contribution of lobed section is over and this is denoted by w_2. One more web transition occurs, when straight section is also consumed and only star tip section remains. This is denoted by w_3. Complete consumption of propellant is attained at web transition w_4. Expressions for these web transitions are also shown in Fig. 92. Calculation of burning area and initial volume of the propellant can be done using mathematical formulation derived below.

$\alpha = \sin^{-1}[w_0/(r+w_0)]$

Volume of propellant= $[\pi(R^2\text{-}nw_0^2/2)\text{-}(2nlw_0)+\{nw_0^2/\tan(\alpha)\}\text{-}(\pi\text{-}na)(r+w_0)^2]L$

***Auxiliary parameters*:**

$A1 = \sin^{-1}[w/(r+w)]$ for $w_0<w<w_2$ else π/n

$A2 = A1$ for $w<w_3$ else $\tan^{-1}(w/l)$

$A3 = \pi - \text{Cos}^{-1}$ $[(l2+w2\text{-}R2)/(2lw)]$ for $w>w_1$ else zero

$A4 = \pi/n + \text{Sin}^{-1}$ $[l\sin(\pi/n)/w]$ for $w>w_3$ else $\pi/2$

Burning area, $A_b=2nL[(\pi/n–A1)(r+w)+l\text{-}\{w/\tan(A2)\}+(A4\text{-}A3)w]$

For analysis, numerical values are assigned to various geometrical parameters of the fin-shaped star configuration. The input parameters are as follows:

R = 98 mm, L = 1000 mm, n = 6, l = 55 mm, r = 20 mm, w_0 = 10 mm.

Volume of propellant = 23334120 mm^3

Various web transitions are calculated for the configuration using mathematical formulation developed earlier and with the help of auxiliary parameters, burning area variation is plotted as Fig. 93. Since mathematical formulation is changed for this configuration, burning area calculation starts with web transition of 10 mm. However, shifting the curve by 10 mm (w_0) can result in same type of curve as derived in earlier section. The mathematical formulation with this modified web consumption is relatively simpler.

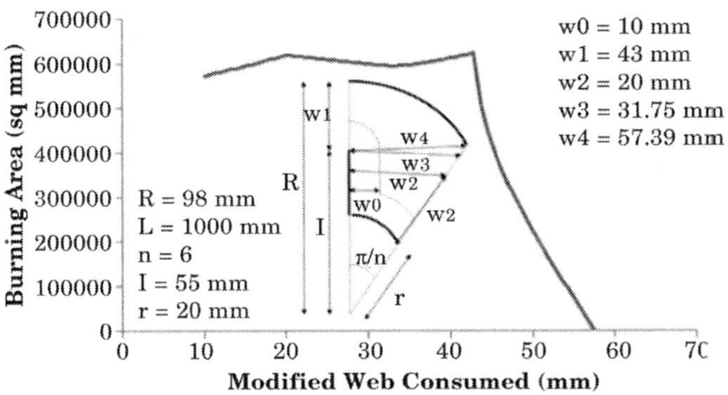

Fig. 93: Burning area variation for fin-shaped star

Area under burning area profile is obtained as 23329196 mm^3, which is only 0.02% off from the calculated volume of the propellant. Thus, derived mathematical formulation and its implementation for the given configuration are validated.

This configuration has 6 geometrical parameters for complete definition. Effect of various geometrical parameters can be derived from the nature of the problem. If propellant grain radius (R) is increased, the value of w_1 and w_4 increases. This results in increase of progressive section at the end of burning. So, neutrality will be compromised. Further, the tail-off section will also increase. So, net effect of increasing propellant radius affects later portion of the burning profile. It increases progressivity and tail-off. When length of propellant is increased, the burning area profile shifts in vertically upward direction and there is no change in any of the web transitions.

Similar to lobed star configuration, another variation of this fin-shaped star is possible, which may be called convex-concave fin configuration. In this case, the lobe diameter has center of curvature on the opposite side of the center and the figure may resemble Fig. 86 with a straight stem separating the two curved sections of the surface. Since all the configurations discussed in this section are two-dimensional in nature, analysis of only half of star-petal is sufficient to get complete burning surface area. So, the configuration of only half star-petal, with salient parameters, is shown in Fig. 94.

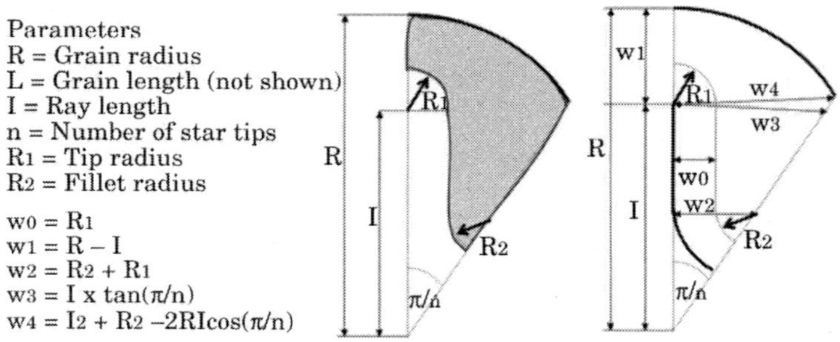

Fig. 94: Convex-concave fin shaped propellant configuration

There are 6 relevant parameters needed for complete definition of the configuration. The propellant section is shown with filled area in Fig. 94 and the dimensional parameters are superimposed over the figure also. Similar to earlier fin-shaped configuration, the propellant profile of half-star-petal can be offset by tip radius and the actual fin-configuration can be reduced to combination of a straight line and a curved section. From this initial configuration, various web transitions can be defined. The first web transition (w_0) is derivation of actual propellant configuration from ideal (dark and thick) propellant shape and this occurs at web burnt of R_2. All calculation starts from this web transition. The subscript '0' indicates that propellant burning surface area calculation starts after this only. The second web transition (w_1) occurs, when propellant burns and fin tip touches grain radius. This is given by (R-l). Third web transition (w_2) denotes the point, when fillet radius section of the propellant is completely consumed and it is given by (R_1+R_2). After this transition, propellant tip becomes sharp and fillet is lost. But this does not cause any problem, as it occurs during combustion of propellant and a transition state. The fourth web transition (w_3) depicts consumption of lower diameter straight section completely and after this propellant will have only curved section extracted from fin portion. This is given by l x Tan (π/n). The fifth web transition (w_4) indicates complete consumption of the propellant and the expression for the same is given in Fig. 94. The physical location of these web transitions are also depicted in Fig. 94. For physical existence of the configuration, check can be imposed on the inner radius of the cross-section. This dimension is not depicted in the current figure, but has been shown for the previous configurations. The inner radius of the configuration must have a positive value and there is no circumstance, when inner radius is negative for this configuration. However, for configurations to be discussed later in this section, this may be a governing parameter.

$$\text{Inner radius, } r = [(R_1+R_2)/\text{Sin}(\pi/n)] - R_2$$

Using these web transitions, mathematical expression for the burning surface area can be evolved for the different web-burnt of the propellant configuration.

Volume of the Propellant=$[\pi(R^2-R_2^{\,2})+n(R_2^{\,2}-R_1^{\,2})\text{x}(\pi/2-\text{Cot}(\pi/n))-2nR_1 l]L$

A1 = 0 for $w_0<w<w_1$ else $\pi - \text{Cos}^{-1}[\{l^2+w^2-R^2\}/\{2lw\}]$ for $w_1<w<w_4$

A2 = 0 for $w_0<w<w_3$ else $\text{Cos}^{-1}[l\text{Sin}(\pi/n)/w] - \pi/n$ for $w_3<w<w_4$

S1 = $[\pi/2\text{-A1-A2}]w$ for $w_0<w<w_4$ else 0

S2 = $[l - w_2/\text{Tan}(\pi/n)]$ for $w_0<w<w_2$, else $[l - w/\text{Tan}(\pi/n)]$ for $w_2<w<w_3$, else 0

S3 = $[(n\text{-}2/2n)\pi(R_1+R_2\text{-}w)]$ for $w_0<w<w_2$, else 0

Burning Surface Area, A_b = 2nL(S1+S2+S3)

For illustration propellant configuration with following dimensions are considered: R = 98 mm, L = 1000 mm, n = 6, l = 55 mm, R_1 = 10 mm, R_2 = 5 mm.

Inner radius, r = 25 mm.

Volume of propellant = 23565880.55 mm^3

Variation of burning surface area with web burnt is plotted as Fig. 95. The numerical values of the parameters and the web transitions are also depicted in the figure. The transitions or discontinuities of slope of burning surface area variation are clearly visible at all the web transitions. Burning profile is offset by 10 mm and actual web burnt comes out to be 47.4 mm only (not 57.4 mm as depicted by w_4).

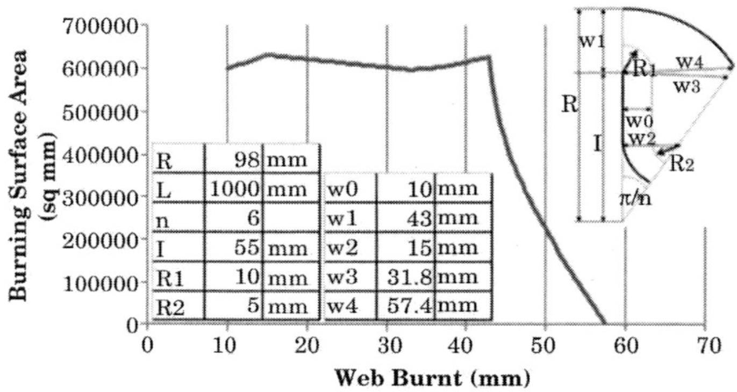

Fig. 95: Variation of burning surface area for fin-2 configuration

The burning surface area profile starts from 10 mm and first transition is encountered at 15 mm (w_2). During this period, the burning profile is slightly progressive in nature. Actually, fin tip is providing progressive portion and fillet part imparts regressive burning profile. The straight section does not show any variation with web change during initial phases. Since, progressive nature of fin is dominant, the profile looks progressive. After 15 mm, the fin tip continues to give progressive profile, but fillet is consumed and straight section starts receding till 31.8 mm (w_3). This gives slightly regressive profile. After this the burning profile is entirely made of fin tip section. It remains slightly progressive. At 43 mm (w_1), the top portion of fin-tip is consumed and although radius of fin-tip increases, but angle formed is reduced and tail-off section is obtained till complete consumption of the propellant. The shape of curve is typical for the star-port configuration.

In fact the curve can be derived from the burning circumferences of the half-star. This is divided into three parts. They are represented by S1, S2 and S3. Here S1 represents tip section, S2 represents straight section and S3, the fillet section. S3 is valid till fillet section exists and has a life of 5 mm (= R_2), only. S2 is valid till web transition w_3. The

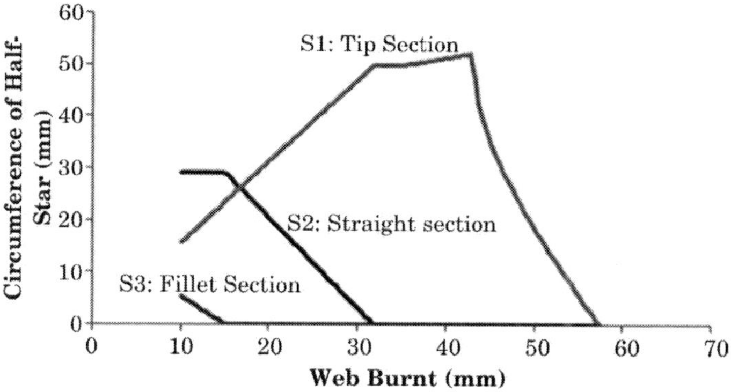

Fig. 96: Variation of burning circumferences of fin-configuration

Circumference S1 continues till complete propellant consumption. The variations of these sections are shown as Fig. 96. Initial progressivity in burning surface area is because of nullification of variation of S1 and S3, as during this period, S2 remains constant. Slight regressivity, after web transition w_2 till w_3 is due to balance of regressivity of S2 and progressivity of S1. After web transition w_3, the burning surface area profile is similar to variation profile of circumference S1. So, later section of burning profile is a function of star-tip always.

Area under the burning area versus web burnt profile is calculated as 23560782.54 mm^3, which is very near to the calculated volume of the propellant configuration. The error is only 0.02163%. So, the developed mathematical formulation is correct. Pressure-time and thrust-time profile can be generated out of the developed mathematical formulations.

For the effect of various geometrical parameters, if length of propellant grain (L) is increased, propellant volume (weight) will increase, but burning profile will not change, web transitions and other parameters on the curve remain same. When radius of propellant grain is increased, web transition w_3 and w_4 are affected and tail-off section is attained later. The effect of ray length (l) is to change web transition w_1 and point of tail-off is delayed, if ray length is reduced. The effect of fillet and tip radius is to affect location of end of progressivity. One parameter is number of star point, which cannot be directly correlated with the burning surface variation curve. So, number of star point is changed from 6 to 5 for the same configuration and burning surface area variation profile is shown as Fig. 97.

When number of star point is reduced, the vacant space in port is reduced (less number of sections unfilled with propellant in the port)

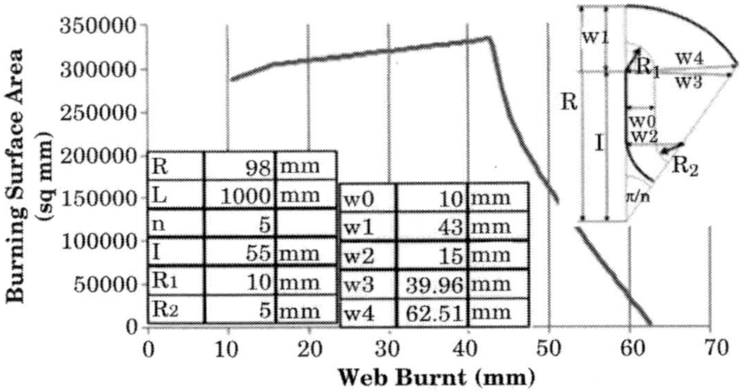

Fig. 97: Effect of number of stars on profile

and more quantity of propellant filled in the given cross-section. This results in higher volumetric loading. The volume for the 5-petal star port convex-concave fin configuration is 24520410.63 mm^3, which is more than earlier configuration. The area under burning area profile is calculated as 24516588.61 mm^3 and error in calculation is only 0.016%. The effect of reducing number of star points is reduction in inner radius to 20.52 mm. this may result in higher volumetric loading of the propellant. The web transitions affected by changing number of star point are w_3 and w_4. The web transition w_3 shifts nearer to w_1. As number of star point is reduced, the angle subtended at center by half-fin is increased and as a result, the web transition for complete consumption of propellant (w_4) is increased. So, reducing number of star-points result in (i) increase in propellant volumetric loading (ii) increase in web transition for consumption of straight section of configuration (iii) increase in web for complete consumption of propellant (iv) increase in tail-off portion of curve.

As far as nature of the curve is concerned, the intermediate regressive nature of the curve is completely lost. The configuration is entirely containing progressive burning profile till tail-off starts. So, reduction in number of fin makes fin-tip a dominant part in the burning profile. As number of star-points is reduced, progressivity increases.

Top Land Star Configuration

The most general form of star-port configuration has a top land section. This indirectly indicates that radius of curvature at the highest radius portion of the port is equal to radius at that section. In all the previous configurations, the tip radius is lower than radius of the highest empty space. If in the fin section a sector is added with radius equal to ray length plus tip radius, the star configuration changes and more number of geometrical control parameters are available for tailoring the configuration. Although analysis of this type of star configuration can be made using skeletal configuration, as discussed in previous sections, but to introduce another approach, this configuration is analyzed with web burnt starting from zero and not from a finite value (as depicted earlier). This approach does not change numerical values, but web-offset is already captured in this mathematical formulation.

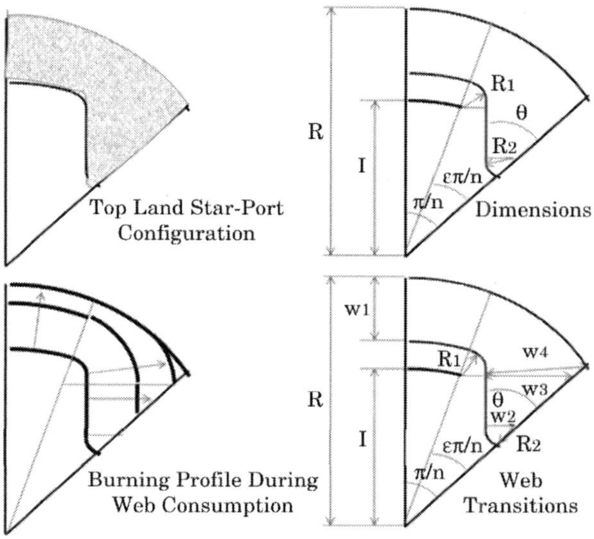

Fig. 98: Top-land star port configuration

A typical top land star port configuration is shown as Fig. 98. The first sketch in the figure depicts half star configuration and propellant portion is shown by filled portion. This type of configuration has the following dimensional parameters needed for complete definition of the configuration (shown as second sketch in the figure).

Outer radius of the propellant : R

Length of the propellant grain : L

Number of star points : n

Ray length of the configuration : l

Angular fraction : e

Tip radius of star : R_1

Fillet radius of star : R_2

Half-angle of the star opening : θ

With the help of these *8 parameters, the geometry of the configuration is completely defined*. Since the configuration is purely two-dimensional, length of propellant is generally not depicted and it is only used as multiplication factor to get propellant volume (weight, if density is known) from the end surface area. So, the two-dimensional configuration has 7 valid parameters for complete definition. Out of the 7 parameters, 5 parameters (R, n, l, R_1 and R_2) are introduced for the earlier star-configurations, also.

The angular fraction (ε) is introduced as a parameter, which represents fraction of central angle covered by the slant section of the star profile. The star-port configurations, discussed in previous sections (lobed star and fin-shaped star) represents angular fraction equal to 1 and ray length containing center of tip radius lies at the half star point. Angular fraction zero represents a circular-port (tubular) configuration. Angular fraction indirectly gives idea about the deviations that star port configuration have, with respect to the circular port. Higher angular fraction means transition from tubular towards fin-port configuration.

Another angular parameter is half angle of star opening (θ). In the given star-geometry, there are four curves – 3 are arc and 1 is straight line. This parameter affects nature of straight line of the star port configuration. This represents inclination of straight section of the star with respect to the reference half-star line. If parallel port or fin configurations are conceived, the value of this parameters is equal to π/n. If the value of θ is less than π/n, the configuration has less circumferential recess at lower radius. The hanging part of propellant in port area has bulk weight and it is not preferred for horizontal storage of the propellant. In fact, the configuration with θ more than π/n, is generally adopted for ease of manufacturing, storage, processing (decoring) and safety. If θ is more than π/n, the configuration has gas flow circumference reducing as radius increases. So, this parameter represents variation of straight section of star-profile from fin-shape.

The propellant burns in parallel layers and the two-dimensional burning profile in course of burning is shown as sketch 3 in the figure. It is clear that top land section of star proceeds towards the grain outer radius and merges with this in course of time. This portion always gives progressive burning profile. The fillet section gets consumed when web burnt is equal to fillet radius and it always gives a regressive burning profile. The straight section also gives a regressive burning profile as depicted. The star tip portion is progressive in the beginning and continues till top land section is consumed. After this tail-off section starts and burning profile gives dropping burning surface area variation. The sketch 4 of the figure gives different web transitions.

Top land consumption: w_1
Fillet consumption: w_2
Straight section consumption: w_3
Propellant consumption: w_4

The values of these web transitions can be derived from the geometry of the figure and the same is depicted below.

$w_1 = R - l - R_1$
$w_2 = R_2$
$w_3 = [l\,\mathrm{Sin}(\varepsilon\pi/n)/\mathrm{Cos}\theta] - R_1$

$$w_4 = \sqrt{l^2 + R^2 - 2lRCos\left(\frac{\varepsilon\pi}{n}\right)} - R_1$$

The burning perimeter of half star configuration is divided into three parts, first part is called tip radius portion (S1). This part determines nature of burning surface area profile at the later part of propellant (web) consumption. The second part (S2) is combination of straight section and fillet section and it is regressive in nature and gets consumed at web transition w_3. In fact fillet gets consumed at w_2 ($w_2 < w_3$, in general). The third portion of burning circumference (S3) is top land section, which is progressive till web transition w_1, after which it becomes zero. If these burning circumferences are known, the burning surface area can be obtained.

$$\text{Burning Surface Area, } A_b = 2nL(S1+S2+S3)$$

The mathematical expressions for the circumference of half-star are derived below. Two auxiliary parameters A1 and A2 are also derived for ease of calculation. Expression for the calculation of initial propellant volume is also derived.

***Auxiliary parameters*:**

$A1=0$ for $w<w_1$ else $A1=\pi-\mathrm{Cos}^{-1}[\{l^2+(R_1+w)^2-R_2\}/\{2l(R_1+w)\}]$ for $w_1<w<w_4$

$A2 = 0$ for $w<w_3$ else $A2=\mathrm{Cos}^{-1}[l\mathrm{Sin}(\varepsilon\pi/n)/(R_1+w)] - \theta$ for $w_3<w<w_4$

Burning Circumference

$S1 = [\pi/2+\varepsilon\pi/n-\theta-A1-A2](R_1+w)$ for $w<w_4$ else 0

$S2 = [\{l\mathrm{Sin}(\varepsilon\pi/n)/\mathrm{Sin}\theta\} - \{(R_1+R_2)/\mathrm{Tan}\theta\}+\{(R_2-w)(\pi/2-\theta)\}]$ for $w<w_2$,

$S2 = [\{l\mathrm{Sin}(\varepsilon\pi/n)/\mathrm{Sin}\theta\} - \{(R_1+w)/\mathrm{Tan}\theta\}]$ for $w_2<w<w_3$, else 0

$S3 = [(l+R_1+w)(1-\varepsilon)\pi/n]$ for $w<w_1$, else 0

$\text{Volume} = nL \times [\{(\pi/n)(R^2-R_1^{\,2})\} + \{(R_1^{\,2}-R_2^{\,2})(q+\mathrm{Cot}\theta-\pi/2)\} - \{(\pi l^2/n)(1-\varepsilon+ \varepsilon\mathrm{Sin}(\theta-\varepsilon\pi/n)/\mathrm{Sin}\theta)\} - \{(2lR_1)\,\{(1-\varepsilon)\pi/n+ \mathrm{Sin}(\varepsilon\pi/n)/\mathrm{sin}\theta\}\}]$

For appraisal of the developed mathematical formulations, above, a configuration is selected and salient dimensions are as follows: R = 97 mm, L = 960 mm, n = 6, l = 53 mm, e = 0.8, R_1 = 3 mm, R_2 = 3 mm, q = 33^0. For this configuration, web transitions are calculated and the values are w_1 = 41 mm, w_2 = 3 mm, w_3 = 22.7 mm, w_4 = 50.15 mm. Using these web transitions, burning perimeter for different sections of the star is calculated. This is shown in Fig. 99.

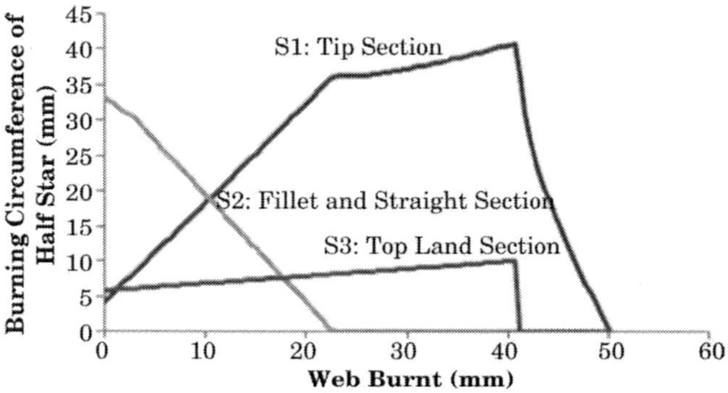

Fig. 99: Top land star configuration

From the figure, it is clear, that top land section (S3) is progressive for the entire available web. This is because the radius of top land continuously rises giving a rising trend of perimeter. The perimeter abruptly becomes zero at web transition w_1 (= 41 mm), when top land touches outer radius. Fillet and straight section of the configuration is combined and perimeter is shown by S2. The variation is regressive. As usual, the tip section (S1) is responsible for the burning surface area profile.

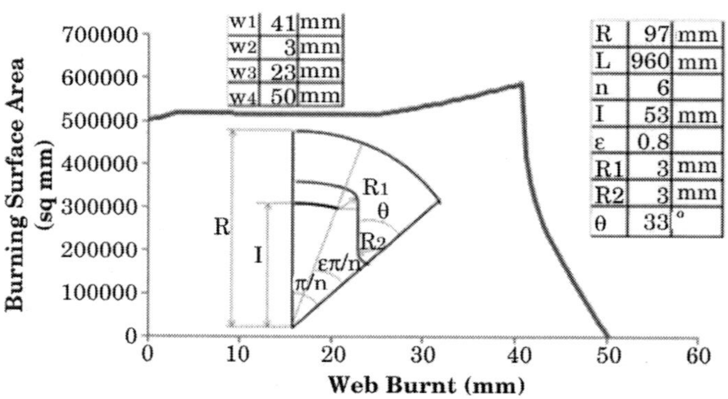

Fig. 100: Burning surface area variation for top land star configuration

The combination of these perimeters gives burning surface area and the variation is shown as Fig. 100. The profile has neutral burning profile from (w_2) 3 mm to (w_3) 23 mm of web burnt, indicating that progressive and regressive variations matches exactly in this portion. At w_3, regressive variation end and progressive burning of top land and tip section continues. At w_1 (= 41 mm), top land is consumed and highly regressive variation of tip section starts giving a tail-off.

Volume of propellant from the above derived mathematical formulation is obtained as 23149026.93 mm^3. Area under burning surface area profile is calculated as 23185196.46 mm^3 and this is volume of the propellant. The error in calculation of volume is around 0.16%. The closeness of propellant volume from the derived formulation ensures corrected of the derived mathematical formulation. For propellant with density of 1750 kg/m^3, this amount to propellant weight of 40.574 kg.

Expressions can also be derived for the smallest radius of the propellant port. This is useful for the assessment of the feasibility of the propellant configuration.

$$\text{Smallest radius, } r = [\{l\,\mathrm{Sin}(\theta - \varepsilon\pi/n) + R_1 + R_2\}/\mathrm{Sin}\theta] - R_2$$

For the example considered, the value of the smallest radius is calculated as 23.24 mm. In some cases, this smallest radius (r) is given as one of the known geometrical parameters and angular fraction (*e*) or fillet radius (R_2) is not available. So, out of three parameters, namely angular fraction, fillet radius and smallest radius, at least two should be known for the complete and definite top-land star-port geometry. The expressions for calculation of one unknown parameters, of the three is derived below.

$$\varepsilon = (n/\pi)\,[\theta - \mathrm{Sin}^{-1}(\{(r+R_2)\mathrm{Sin}\theta - (R_1+R_2)\}/l)]$$

$$R_2 = [r\,\mathrm{Sin}\theta - l\,\mathrm{Sin}(\theta - \varepsilon\pi/n) - R_1]/(1 - \mathrm{Sin}\theta)$$

As mathematical formulation for the star port configuration is developed, parametric study can be carried out for the given configuration, by changing one parameter at a time and observing effect of various geometrical parameters on the burning surface area variation.

First of all, angular fraction is changed and burning profile is generated, keeping all other parameters same. The increment in angular fraction makes top land section smaller and lower diameter of propellant-port is reduced. So, more propellant will be filled and volumetric loading will increase with increase in angular fraction. If propellant density is taken as 1750 kg/m^3, then for angular fractions of 0.7, 0.8 and 0.9,

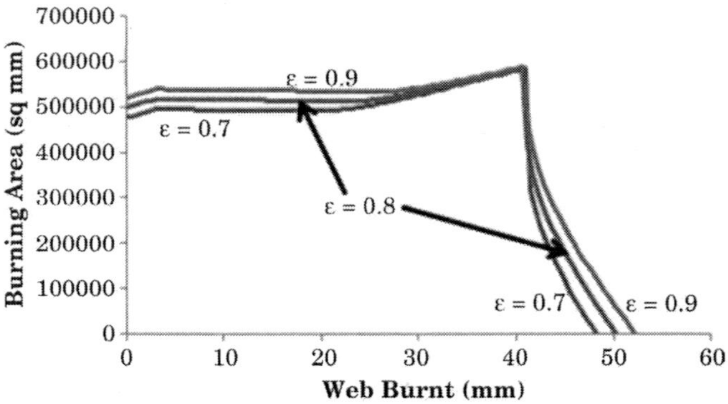

Fig. 101: Parametric study for change of angular fraction

propellant weight is calculated as 38.75 kg, 40.51 kg and 42.82 kg, respectively. Straight section is also enhanced. This results in increase in initial burning surface area with increase in angular fraction. The web transition w_3 is also delayed by increase in angular fraction. Since top land and outer radius is unaltered, web transition w_4 remains unchanged. However, tail-off section is higher for higher angular fraction. These effects are shown as Fig. 101. Major changes are tabulated below for change of angular fraction.

e	*r (mm)*	*w_3 (mm)*	*w_4 (mm)*	*Weight (kg)*	*Volume (mm^3)*
0.7	28.25	19.65	48.18	38.75	22140960.09
0.8	23.24	22.70	50.15	40.57	23185196.46
0.9	18.19	25.69	52.29	42.82	24470201.19

For the next parametric study, half angle of the star opening (q) is changed. Three values, 28^0, 30^0 and 33^0 are considered. The variation in parameters is tabulated below. As half angle for star opening affects only inclination of straight section, web transitions w_1, w_2 and w_4 are unaffected. Only w_3 changes and for higher value of q (or inclination), higher value of w_3 is observed. For higher value of q, the propellant volume and propellant weight reduces, because higher inclination of straight portion of star increases empty section in the cross-section. Overall, the variation in half-angle of star opening affects initial portion of the burning surface area. The variation is depicted as Fig. 102. High value of q gives lower initial burning area and neutrality of central section is also affected. However, web transition of w_2 is obtained in all the configurations at same web burnt.

Although tail-off section is always available in star port configuration, it is attempted to get neutral (no change in burning surface area with

q (^{0})	*r (mm)*	w_3 *(mm)*	*Weight (kg)*	*Volume* (mm^3)
28	17.65	21.42	41.79	23878617.8
30	20.08	21.89	41.26	23577525.17
33	23.24	22.70	40.57	23185196.46

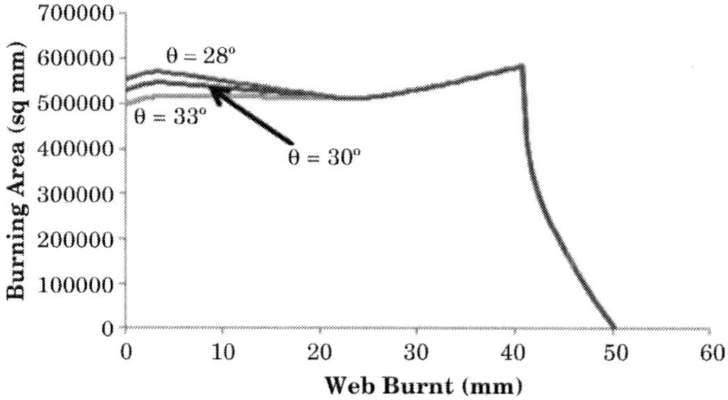

Fig. 102: Parametric study for change of half angle for star opening

web burnt) burning profile. However, tail-off section is compromised in this case and very long tail-off is observed in those cases. In fact, if web transition w_3 is increased to bring it near w_1, then the rising trend of burning profile just before tail-off can be controlled. A typical burning profile, where this condition is attained is shown as Fig. 103. The burning profile is neutral with a long tail-off. The propellant weight in this case for 1750 kg/m^3 propellant density, is 23.64 kg. Propellant burnt till beginning of the tail-off is 20.33 kg and in tail-off section 3.31 kg propellant burns, which is around 14% of the initial propellant weight. For the reference configuration with 40.57 kg propellant, propellant burning in tail-off portion is 3.06 kg, which is only 7.54% of initial

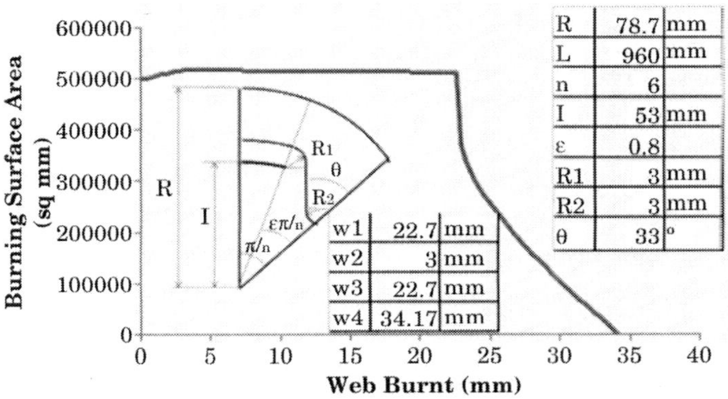

Fig. 103: Neutral burning star-port propellant

propellant weight. So, neutrality of burning can be achieved in star port configuration at the cost of increased tail-off.

The other side of the configuration tailoring may be for zero tail-off portion. Such configuration will demand equalization of web transitions w_1 and w_4. Such configuration cannot be conceived with star port configuration and is possible for tubular burning profiles, only. In fact, zero tail-off means zero sliver propellant, which is difficult to achieve in complicated geometries. Even for tubular or solid sustainer propellant configurations, where zero tail-off is ideally expected, the practical firing curves give tail-off.

CONCLUSION

Star-port configuration is slowly developed from lobed star to fin-shaped star to top-land star with convex-concave and concave configurations. Compete mathematical formulation is developed for each of the configurations and illustrated numerical examples are also given. Parametric study is also presented to understand the effect of various parameters on the burning area profiles. Such treatment of star-port configuration is not available elsewhere.

REVIEW QUESTIONS

1. What is effect of angular fraction on the burning area profile for star-port configurations?
2. What is effect of half-angle of star opening on the burning area profiles?
3. Draw a neat sketch to describe various geometric parameters for star-port configuration?
4. Why tail-off is obtained in star-port configurations?
5. Which part of burning curve profile for star-port configuration depicts contribution of the fillet portion?
6. How many essential geometrical parameters are needed for lobed star-port rocket propellant configurations?
7. How many web transitions are there for the lobed star-port propellant configuration?
8. What is significance of web transitions in burning surface area profile generation?
9. What is difference between convex-concave and concave lobed star configuration?

10. How many geometrical parameters are needed to describe fin-shaped star-port configuration?
11. Why is star-port propellant configuration versatile?
12. How are tailoring of geometry of fin-shaped star port configuration affect burning area profile?
13. What are conditions for which progressive followed by tail-off burning profile is obtained for the star-port configurations?
14. What is role of burning perimeter in calculation of burning surface area for star-port configurations?
15. How is inner radius of the port calculated for top-land star-port configuration?
16. What is significance of area under burning surface area versus web burnt curve?
17. How can propellant volume be calculated for star-port configurations?
18. How is developed mathematical formulation validated for the star-port configurations?
19. Can star-port configuration be obtained without any tail-off?
20. If tail-off is permitted, can neutral burning profile be obtained from star-port configurations?

10

Simple Configurations and Numerical

CHAPTER SUMMARY

This is summarizing chapter for predictive rocket propellant ballistics and it covers different non-two-dimensional features of solid propellant configurations. Propellant with step or ramp in the port profile, ridges or slots on the external surfaces and combination of inhibition on non-circular sections are discussed in this chapter with certain numerical examples. However, all the new configurations discussed in this chapter are simpler and have known geometrical features. Following points are illustrated in this chapter:

- Mathematical formulations for longitudinal irregularities
- Mathematical formulations for circumferential irregularities
- Mathematical formulations for non-circular sections
- Derivation of Mathematical Correlations
- Limitations of predictive ballistics of solid propellants

INTRODUCTION

As discussed in earlier 4 chapters of this part of the book, the mathematical formulations are derived and final expressions are given for the mostly two-dimensional configuration with circular or sectorial symmetry. In this section, some designed longitudinal irregularities are introduced in the propellant configuration and they are analyzed for derivation mathematical formulation. The mathematical formulation for the burning surface area is converted into pressure-time profile and thrust-time profile. In this section digital data for the mathematical formulation is also given for practice, validation and cross-checking the understanding of mathematical formulations derived.

Tube with Ramped Enlargement of Port

Simple tubular configuration with inhibition at lateral and end surfaces is considered. The propellant burning is permitted in radial outward direction, only and there is no change in the propellant length in course of burning. The port geometry has two separated diameters, separated by a ramp section. A cross-section of the proposed profile is shown in Fig. 104. Although there is possibility of introducing sudden change in geometry, but step enlargement of pot results in jump of flow, flow reversal, eddy formation, erosive burning at edges and subsequent conversion to ramped geometry. Additionally, the sudden step enlargement has chances of blow-hole formation in the transition zone during propellant casting.

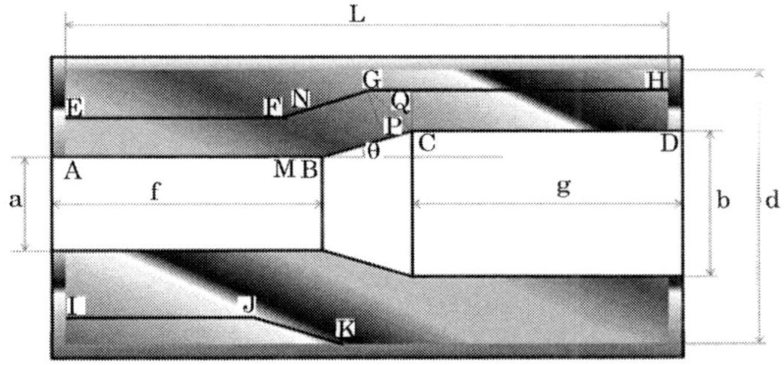

Fig. 104: Propellant configuration with longitudinal variations

The initial propellant profile is shown by ABCD. The propellant length and diameter are shown as 'L' and 'd', respectively. The diameters at smaller end and bigger end are 'a' and 'b', respectively. Lengths of these sections are 'f' and 'g', respectively. The length of the ramp section becomes (L-f-g). The ramp angle with respect to the longitudinal axis of the propellant configuration is shown as q.

$$\text{Tan } \theta = (b - a) / \{ 2 (L - f - g)\}$$

After certain web burnt, the cross-section of central port is shown as EFGH. It is clear from the configuration that length of smaller diameter portion reduces and the same of larger diameter portion increases. Perpendiculars are drawn at the transition points (B and C) on straight and slant sections and web burnt are equal, *i.e.*, BN=CQ=FM=GP. However, the length change can be denoted by product of web burnt and *Tan (θ/2)*. The first transition for this propellant configuration occurs, when larger port diameter (b) reaches outer diameter of the propellant grain. Till this web transition, contribution of all three sections exists and the length of ramp section remains constant. However, after this web transition, contribution of larger diameter portion of the configuration ends and length of ramp section is also lowered. Such situation is shown by curve IJK. The web transitions and corresponding burning surface area can be derived as given below.

***Web Transitions*:**

Larger diameter consumed; $w_1 = (d-b)/2$

Propellant consumed, $w_2 = (d-a)/2$

***Burning surface area*:**

For $w < w_1$,

$$A_b = \pi[2w(f+g)+(af+bg)+w(b-a)Tan(\theta/2)+(b-a)(a+b+4w)/(4Sin\theta)]$$

For $w_1<w<w_2$,

$$A_b = \pi[(a+2w)\{f-wTan(\theta/2)\}+\{d^2-(a+2w)^2\}/(4Sin\theta)]$$

Although the curve is similar to funnel shaped propellant configuration discussed in chapter 6, where near-neutral burning profile is obtained, there are more complexities in this case. The funnel is attached with a larger diameter section, which gets consumed. The ramped web transition remains intact till complete consumption of the propellant. The angle of ramp remains constant during propellant combustion. The volume of propellant in this configuration is calculated by the following formula.

$$\text{Volume} = (\pi/12)[3d^2L - 3a^2f - 3b^2g - (b^2+ab+a^2)(L-f-g)]$$

Calculations are made for the following geometrical parameters: a = 100 mm, b = 150 mm, d = 300 mm, L = 500 mm, f = 250 mm, g = 200 mm.

This gives, q = 0.46365 radians = 26.565°, w_1 = 75 mm, w_2 = 100 mm, volume = 29223356.66 mm^3. Burning area variation is shown as Fig. 105. The burning surface area variation is progressive because burning is taking place in radial outward direction. Higher diameter always results in higher burning area. The progressivity continues till first web transition, when larger diameter side of the propellant is consumed. At this point, propellant on one side is consumed and contribution of only ramped and smaller port-opening side of the configuration remains. The variation of burning surface area is slightly regressive because length of propellant and slant section length, both reduces after this point. This reduction compensates for increase in port diameter.

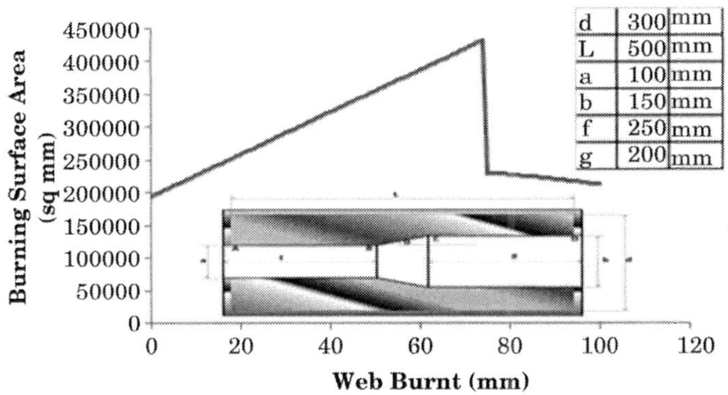

Fig. 105: Behavior of ramped tubular inhibited propellant configuration

If smaller port diameter (a) is increased, the initial burning surface area increases. The peak value of total burning surface area also increases. Increase in smaller diameter brings it closer to larger diameter (b) and difference between both the web transitions reduces. The small regressive portion after a transition becomes smaller and more regressive.

One change that can be introduced in the configuration is making it sleeve cast configuration of tubular propellant grain by removing the end inhibition. This change leads to reduction in the length of the propellant from both the sides. This adds two extra end portions in the burning surface area and initial burning surface area increases for the same configuration. However, these additional surfaces reduce in numerical values, as burning progresses. The progressivity is reduced and mathematical expressions for various parameters are shown below. Volume and web transitions are not changed. The burning profile is shown for same geometrical parameters as Fig. 106.

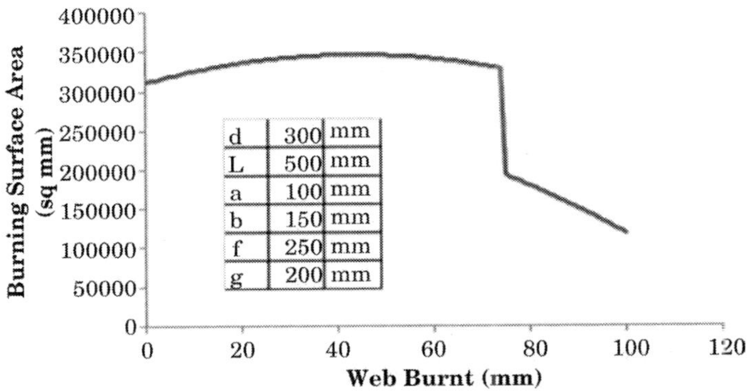

Fig. 106: Burning profile for end uninhibited ramped tubular configuration

$$\text{Volume} = (\pi/12)[3d^2L - 3a^2f - 3b^2g - (b^2+ab+a^2)(L\text{-}f\text{-}g)]$$

***Web Transitions*:**

Larger diameter consumed; $w_1 = (d\text{-}b)/2$

Propellant consumed, $w_2 = (d\text{-}a)/2$

***Burning surface area*:**

For $w < w_1$,

$A_b = \pi[2w(f+g)+(af+bg)+w(b\text{-}a)\text{Tan}(\theta/2)\text{-}w(a+b+4w)+(b\text{-}a)(a+b+4w)/(4\text{Sin}\theta)] + (\pi/4)[2d^2\text{-}b^2\text{-}a^2\text{-}4bw\text{-}4aw\text{-}8w^2]$

For $w_1<w<w_2$,

$A_b = \pi[(a+2w)\{f\text{-}w\text{-}w\text{Tan}(\theta/2)\}+\{d^2\text{-}(a+2w)^2\}/(4\text{Sin}\theta)]+(\pi/4)[2d^2\text{-}a^2\text{-}4aw\text{-}4w^2]$

The nature of burning profile for the said configuration is derived, but no useful purpose is served by such design or configuration of propellant. This configuration is relatively simpler and can be used as starting point for performance prediction for configurations with longitudinal variations in port dimensions. The portion of burning surface profile after first web transition is virtually useless for propulsion of rocket and thus this configuration is not utilized in general. However, a longer propellant grain with only end inhibition can be helpful in getting a near-neutral burning profile. This propellant configuration has academic interest only.

Dendrite Configuration

One of the peculiar configurations mentioned in the rocket propellant literature is dendrite, where multiple fins emanating from a central stem is observed in a circular cross-section of propellant. The configuration has over-all circular geometry and is designed with external and end inhibition generally such that burning of propellant must take place in two-dimensional fashion in radial outward direction from the central perforation. The burning pattern may be treated as same as simple fin design, except for introduced complexities due to more burning orientation and steps. The central perforation has many stems emanating from center, on which at least two branches are located. Although, the configuration can be generated with many branches and stems, but equi-spaced identical stems and branches are preferred. For ease of calculation and structure, 3 stems and two branches per stem is analyzed in this section. However, it is reiterated that many varieties are possible with this configuration and many number of variables can be introduced for this configuration. The configuration is depicted in Fig. 107 (sketch X).

The configuration has the following salient features:

- The configuration can be manufactured by both casting and extrusion routes.
- The configuration has very few control geometry parameters.
- The configuration has very good volumetric loading of propellant.
- The configuration does not very small sliver fraction.
- The propellant configuration may have erosive burning for longer grains.
- The case-bonded and cartridge-loaded, both configurations are possible.
- The configuration has been reported in books, but operational rockets are not reported.

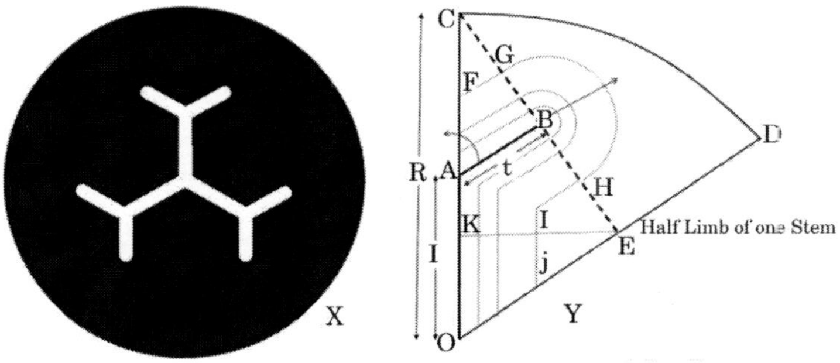

Fig. 107: Dendrite configuration

The salient geometrical parameters of this configuration include grain radius (R), grain length (L), stem length (l), branch length (t), branch angle from stem (q), and number of stems (n). The stem and branch width may be same or different. Another parameter '*d*' can be assigned for this parameter, if both stem and branch width are identical. Since, many variables for different configurations are introduced and analyzed in this section of predictive ballistics, some idealizations for ease of calculation is made in this section for this configuration. For analysis of this configuration, half-limb of a stem is considered (Fig. 107: sketch Y). All parameters, except grain length and stem/ branch width are shown in the figure and value of π/n (<DOC) may be taken as a.

In the figure, 'O' depicts centre of the propellant configuration. The curve 'CD' bounds or makes external boundary of the geometry. Thus

OC-OD = R. The stem length, depicted by *OA (length l)* is shown and branch *AB (length t)* comes out of this at an angle of 'θ'. The burning of propellant can be divided into four zones clearly, as depicted by profile observed in course of burning (typical profile is FGHIJ). Upper part of stem is depicted by FG, while lower part is shown by HI. The stem of propellant is shown to move in course of burning to IJ. The end of branch AB (*i.e.,* point B) acts as center of generated circular profile and profile GH is created as a circular profile in course of burning. The perimeter can be calculated by taking length of various line segments and then multiplying it by 2n to get completer perimeter of burning surface area. Multiplying this number by length of the propellant grain, burning surface area of the propellant can be obtained.

The configuration can be designed in such a way that perpendicular from tip of the branch (point B) passes through point C. This situation indicates that in course of propellant burning, the upper part of branch with straight section is consumed at the same time when curved section is touching outer diameter of the propellant grain. This situation gives a correlation, $Cos\theta = t / (R\text{-}l)$.

For getting another two important web transitions (BE and EK) desire many intermediate relations. Although not created in Fig. 107, but there exists a triangle OAB, where $<ABO = \chi$ and $<AOB = \eta$. Some of the mathematical expressions valid for this triangle are given below.

$(OB)^2 = l^2 + t^2 + 2.l.t.Cos\theta. \rightarrow$ OB is calculated.

$OB/Sin\theta = l/Sin\chi = t/Sin\eta, \rightarrow \chi$ and η are calculated.

Next triangle OBE is considered, where sine-law of triangle is applied to give the relation: $OB / Cos(\alpha\text{-}\theta) = OE / Cos\chi = BE / Sin(\alpha\text{-}\eta)$. This helps in the calculation of lengths BE and OE in the triangle. Now, triangle OEK is considered and value of *EK* is calculated as $OExSin(\alpha)$. Although not required in calculation, but $AK = l - OExCos(\alpha)$. The burning surface area is given by $2nL[(2t+l)+w\{\pi\text{-}(1 / tan\theta)\text{-}2tan(\theta/ 2)\text{-}\{1 / tan(\alpha)\}]$ till first web transition, which is deliberately not specified here, but can be calculated easily, based on logic and exercises given throughout this section of predictive ballistics. The burning surface area seems to have linear variation with web consumed and form of burning surface area can be reduced to A+Bw form.

This profile is not analyzed for generation of full burning surface area, but it analyzed, only for some critical condition. The method of analysis and equations used are depicted more lucidly for this profile and similar approach has been used in this entire section. The burning area for this profile is not derived completely, but can be derived based on concepts discussed in previous chapters.

For calculation sake, following parameters are considered: R = 100 mm, L = 500 mm, n=3, l = 60 mm, t = 20 mm, d = 10 mm. Clearly, $\alpha = \pi/n = \pi/3$ = 1.0472 rad (60°). The value of q [$=Cos^{-1}\{t/(R-l)\}$] is also 1.0472 rad (60°). The value of BC ($=tTan\theta$) is 34.641 mm. The value of OB is calculated as 72.111 mm. The values of angles χ and η are found to be 0.8046 rad (46.102^o) and 0.2425 rad (13.898^o), respectively. Lengths of BE and OE are calculated as 51.962 mm and 50.0 mm, respectively. From this value of EK *{=OExSin(α)}* is obtained as 43.301 mm. Here BC, BE and EK may be assumed to acting as web transitions. The minimum value is that for BC. So, burning surface are calculation is repeated from half width of the branches and stems (=5 mm) to the length of BC (=34.641 mm). The variation of burning surface area is depicted as Fig. 108. Clearly the variation is linear.

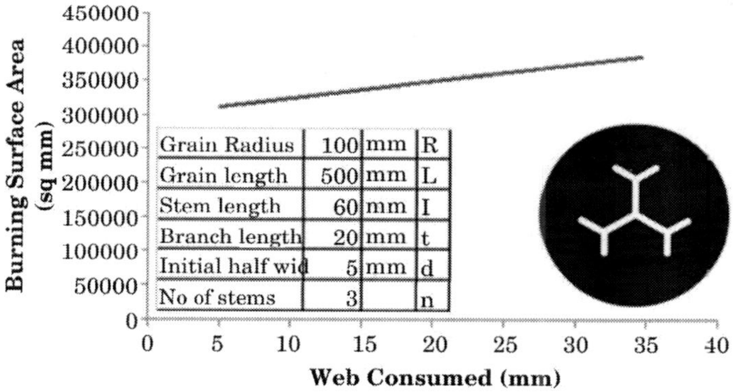

Fig. 108: Burning surface area for dendrite configuration

For a typical variation, it is assumed that all straight sections of the burning profile are consumed almost at the same time and later on only curved section (circular arc with center at B) remains. Additional constructions are also depicted in the figure and for fulfilling the mentioned criteria, BC = BE = EK. For this situation, the configuration has only one planar parameters → R. The value of α for three stem case is $\pi/3$ (60°). The equality of web transitions leads to following values of remaining planar parameters → $\theta = \pi/3$ *(60°)*, $l = R/2$, $t = R/4$. For propellant grain radius of 100 mm, 'l' becomes 50 mm and 't' becomes 25 mm. The constant web comes out to be 43.3 mm. The web is higher than earlier calculated minimum value (=34.641 mm) of web transitions of the three. The variation of burning surface area is linear as depicted in Fig. 109. However, the most surprising part is that till web transition of 34.641 mm, the value of burning surface area is identical to that depicted in Fig. 108. Because of calculation till higher web transition, the linear variation of burning surface are attains higher value.

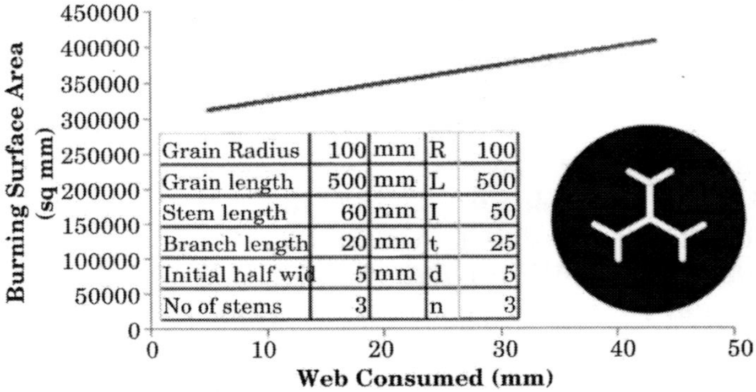

Fig. 109: Special case of dendrite configuration

Cross-shaped Propellant Configuration

Contrary to earlier discussed configurations, where radial-outward burning occur cross-shaped propellant configuration is radial-inward burning. The propellant configuration has symmetry and has simpler configuration. This depicts circumferential irregularities, as well as non-circular cross-section in single shape. This configuration has simple mathematical formulations and main features of the configuration are depicted below.

- Propellant configuration can be conceived in cartridge-loaded as well as case-bonded configuration.
- Propellant manufacturing through both, casting or extrusion is possible.
- Passage of combustion gases from head end to nozzle end exists.
- Volumetric loading of the propellant in rocket motor is high.
- The burning profile has very good neutrality.
- The configuration has no web transition and direct consumption of propellant leads to negligible tail-off.
- Mathematical formulation of the configuration for ballistic prediction is very simple.
- Simultaneous Initiation in all the flow channels is difficult to achieve.
- Chances of differential pressure exist due to presence of multi-channel for flow of combustion gases.
- Propellant may rupture at the end of burning due to low web thickness.
- Propellant pieces may be thrown out in course of combustion inside rocket motor chamber.

- Application of inhibition requires specialized arrangements, if thread/tape winding is adopted.
- Rocket motor is exposed to combustion gases during rocket operation. High thickness of rocket motor casing and inhibition is desired.
- The legs of the cross may rupture in transportation and handling.
- The propellant grain is stressed more due to pressure of the combustion gases.
- If propellant holding grids are not designed properly, there are chances of propellant grain closing the nozzle of the rocket motor, leading to catastrophic failures.

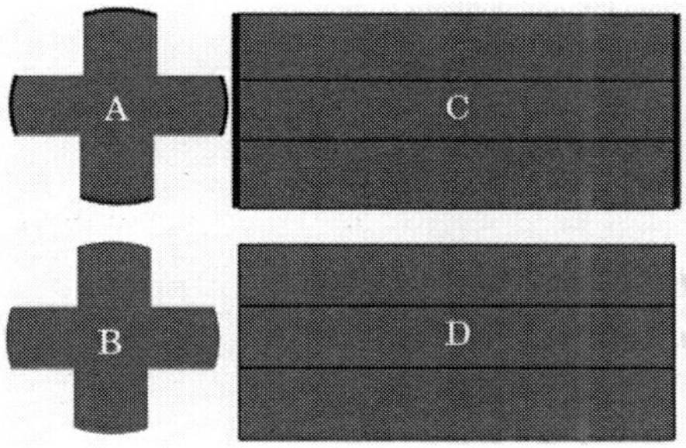

Fig. 110: Cross-shaped propellant configuration

Cross-section and longitudinal profile of cross-shaped propellant configuration is shown as Fig. 110. The cross-section of the configuration has two varieties. The outer radius (arc) of the cross-section may be inhibited along the length as shown in 'A' of the figure or the longitudinal section may be completely uninhibited ('B' of the figure). Similarly the ends of the configuration may be inhibited ('C') or uninhibited ('D'). In total there are 4 different combinations possible for this shape of propellant grains.

End inhibited and longitudinal inhibition on arc of cross-section (A+C)

End inhibited and no longitudinal inhibition (B+C)

End uninhibited and no longitudinal inhibition (B+D)

End uninhibited and longitudinal inhibition on arc of cross-section (B+D)

Although, arc of cross-section is shown inhibited (A), this inhibition has tendency to open-up in course of burning (for penetration of flames) and due precautions must be taken to avoid such incidences. End inhibition prevents length of grain to reduce in course of burning and proper holding of propellant grain by grids is ensured till end. However, if end-inhibition is not present, propellant length reduces and burning occurs at the end surfaces also. This may result in poor holding of propellant grains. So, overall, end inhibited arc uninhibited configuration is best suited for safe and reliable operation.

This propellant configuration has only four controlling geometrical parameters for ballistic analysis.

Outer radius of the configuration: R

Length of the configuration: L

Width of legs of the cross: 2t

No of legs of the cross: n

As depicted for other propellant configurations with sector-symmetry, one important angle for the calculation is π/n *(defined as θ)*. Maximum web available for the propellant configuration is 't'. The calculation of burning surface area, for different inhibition conditions, at progressively changing web consumed (w), for half of one leg, are given below.

Auxiliary Angle, a = $\sin^{-1}$ [(t-w)/(R-w)] for Arc Uninhibited
= $\sin^{-1}$ [(t-w)/R] for Arc Inhibited
Arc Periphery in half limb of a leg, P1 = (R-w). α for Arc Uninhibited
= 0 for Arc Inhibited
Periphery of Straight Section in half limb of a leg,
$P_2 = (R-w).[\alpha + \sin(\theta-\alpha)/\sin\theta]$ for Arc Uninhibited
$= R.[\alpha + \sin(\theta-\alpha)/\sin\theta]$ for Arc Inhibited
Longitudinal Burning Surface Area,
$A_1 = 2.n.L.(P_1+P_2)$ for End Inhibited
$= 2.n.(L-2w).(P_1+P_2)$ for End Uninhibited
End burning Surface Area, $A_2 = 0$ for end inhibited
For End Uninhibited and Arc Uninhibited
$A_2 = 2.n.(R-w)^2.[\alpha(1+\sin\alpha) + \sin(\theta-\alpha).\sin\alpha/\sin\theta]$
For End Uninhibited and Arc Inhibited
$A_2 = 2.n.R^2.[\alpha(1+\sin\alpha) + \sin(\theta-\alpha).\sin\alpha/\sin\theta]$
Total Burning surface area, $A_t = A_1 + A_2$
Volume of the Configuration,
$V = n.L.R^2.[\beta(1+\sin\beta) + \sin(\theta-\beta).\sin\beta/\sin\theta]$, where $\beta = \sin^{-1}[t/R]$

The auxiliary angle (a) changes with web consumed (w) and is an instantaneous variable. If arc is inhibited, outer radius remain constant through-out burning of propellant configuration. If arc is uninhibited, radius of grain reduces in course of burning. Same is reflected in the equation for arc periphery, periphery of the straight section and end burning surface area. If end face of the propellant configuration is inhibited, contribution of end face is absent and length of propellant grain remains constant. Variation of length is accounted in the equation for longitudinal burning surface area. The volume of propellant configuration can be obtained by multiplying initial end burning surface area with initial propellant length.

For calculation, following parameters are considered: R = 100 mm, t = 25 mm, n = 4, L = 300 mm. The value of θ is calculated as (π/n=) 0.7854 radians or 45°. The value of $\beta = \text{Sin}^{-1}[t/R] = 0.25268$ radians (14.47°). Hence volume of the configuration can be calculated.

$$V = n.L.R^2.[\beta(1+\text{Sin}\beta) + \text{Sin}(\theta-\beta).\text{Sin}\beta/\sin\theta] = 5944941.3 \text{ mm}^3.$$

First of all, condition for no longitudinal inhibition is considered and burning is assumed to take place on the entire longitudinal surface. The arc is reducing and straight section is also accordingly getting modified in course of burning. Out of two cases, first case with end inhibited condition (depicted by B+C) is considered. In this case, end area is not accounted and length of propellant grain is not reducing. The length of periphery of arc and straight portion of the configuration in half limb of a leg is depicted in Fig. 111. If is clear that both the portions are continuously reducing with consumption of propellant web. Arc section has smaller numerical value of the two and it reduces to zero at the end of web consumption. In course of propellant consumption, radius of top arc is not reducing to zero but reduction in the angle at profile center reduces to zero, thus making it zero at the end. The straight portion of the profile does not reduce to zero length, even at the end of consumption and it is forced to become zero at the end of web consumption.

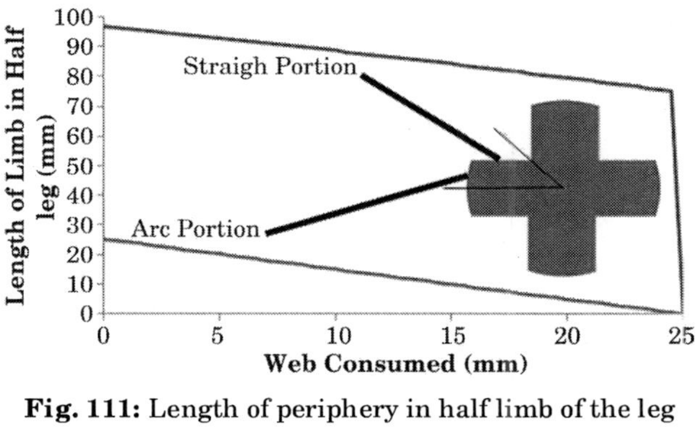

Fig. 111: Length of periphery in half limb of the leg

When length in Fig. 111 is multiplied by 2.n.L (=2x4x300), it gives the variation of burning surface area with the propellant consumption. In the Fig. 111, initial lengths of arc and straight section are 25.268 mm and 97.0926 mm, respectively. This gives burning surface area as 293665.5 mm^2. Since both length are reducing in value, the burning surface area will depict a regressive burning profile.

The variation of burning surface area with web consumed is depicted as Fig. 112. The initial value is same as obtained by manual calculation. The trend is regressive as predicted. Area under burning surface area versus web consumed is calculated as 5896334 mm^3, which is only 0.8176% off from the earlier calculated value. This indicates correctness of calculation strategy and correct implementation of devised mathematical formulation.

In the same figure, variation of burning surface area for the end uninhibited propellant configuration is also shown. Compared to inhibited configuration, end surface area is included, which increases initial burning surface area. However, burning at end faces, reduces length and adds more to regressivity. So for uninhibited end face, the slope of burning surface area variation is more and at the end of propellant consumption, uninhibited end face condition gives lower burning surface area than uninhibited end face condition. So compared to inhibited end face, uninhibited end face shows higher initial burning surface area and steeper reduction in burning surface area with web consumed. The volume is calculated by area under the burning surface area versus web consumed profile as 5904015 mm^3, which is deviating by -0.6884%, only.

These two cases are analyzed for case, where longitudinal arc portion is inhibited. If arc portion is inhibited, the straight portion is extended

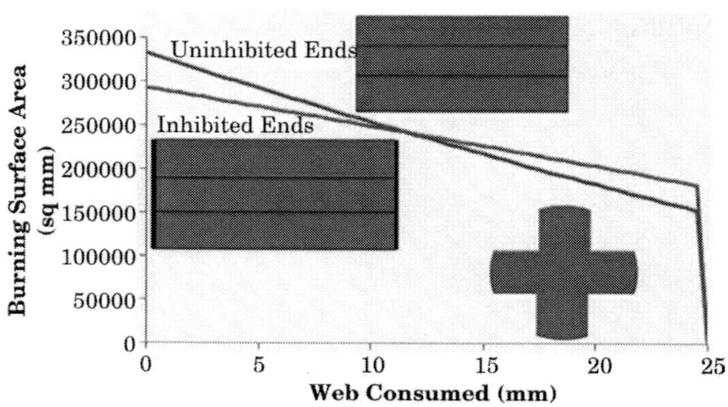

Fig. 112: Burning surface area of cross-shaped grain

to outer radius till complete consumption of web. The regressive nature of arc portion is unavailable. The straight portion is slightly progressive in nature. If end face is available for burning, the initial burning area is higher, and the rate of change of burning surface area is also faster.

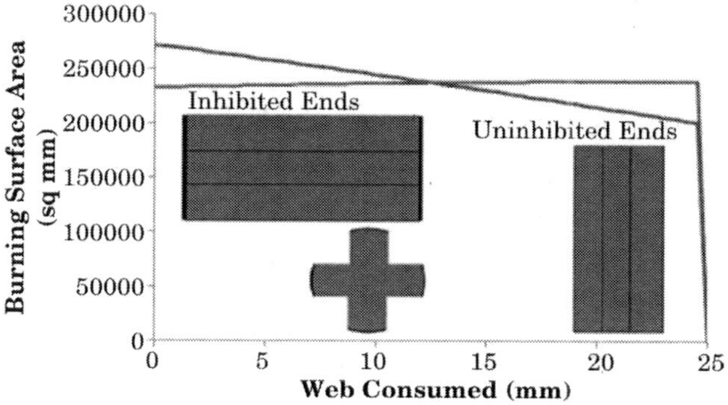

Fig. 113: Burning surface area for longitudinally inhibited cross-shaped propellant

So, for end inhibited condition, the burning profile may be slightly progressive, but for end uninhibited condition, the burning profile is regressive in nature due to regressive burning of end face and introduction of regressive nature by reduction in length. For both the cases, burning surface area profile is generated using mathematical formulation depicted in the section, above as Fig. 113.

The propellant volume is matched to the area under the depicted profile to ensure correctness of calculation. This propellant configuration is very useful for small burning time calculation. This can also be manufactured from solid sustainer configuration by machining. The configuration is predominantly regressive in nature, but slight progressivity can be introduced by selecting suitable inhibition conditions.

Mathematical Equations Used for Prediction

In the entire section of predictive ballistic, development of mathematical equation for calculation of burning surface area for each step of web consumed is based on simple trigonometrical relations and the concept of parallel burning of solid rocket propellant. However, for clear understanding mathematical relations and conceptual burning surface area evolution is expressed in this section. Some of the precautions and common mistakes in developing mathematical expressions of burning surface areas are also highlighted in this part.

Before entering into intricacies of geometry, some mathematical relations and derivations, useful for the calculation of burning surface profile is given below. Most of the time burning surface area is calculated by multiplication of length of propellant grains with burning profile/ perimeter and calculation of perimeter becomes an important parameter.

Triangle is obtained in the burning profile too often. Some correlations for triangle ABC is depicted below. As standard representation, capital letters indicates angles of the triangle and small letter represents side opposite to the angles.

Area of the triangle = ½ x base x height = ½ ab SinC

Sine law of triangle → a/SinA = b/SinB = c/SinC

Angular sum in triangle A+B+C = π

Cosine law, $c^2 = a^2 + b^2 = 2.a.b.CosC$

Quadrilaterals are not handled for the derivation of mathematical formulations directly, but they are observed in course of burning of propellant configurations. They are generally converted into triangles and analyzed. Hence for completeness these surfaces are also included for their geometrical derivations.

Area of quadrilateral = base x height.

For the curved portions, arc is the most important configuration. The arc is characterized by only one parameter 'D' (Diameter).

Length of an arc = $\alpha D/2$

Area of sector = $\alpha D^2/8$

For circular cross-sections many correlation are possible. Diameter (D) and axial length (L) are two parameters for various mathematical correlations.

End face of circular cross-section = $\pi D^2/4$

End face of tubular cross-section = $\pi[D^2 - d^2]4$

Lateral surface of the cylinder = πDL

So, circular end face behaves as lateral surface of cylinder with length D/4.

Conical sections are obtained because of central slant portion. In addition to the propellant manufacturing by casting require certain draft taper, which ultimately converts the simple cylindrical section into frustum of a cone. Correlations for these configurations are also worth consideration.

Slant area of cone = πDs, where s = slant length.

Sometimes calculation of mass of propellant is also carried out for the initial configuration. This requires calculation of volume of the configuration. Important volume calculation formulations are given below.

Volume of pyramid = area of base x height

Volume of prism = 1/3 x area of the base x height

Volume of the Cylinder = $\pi D^2L/4$

Volume of cone = $\pi D^2L/12$

Volume of frustum of cone = $\pi[D^2+Dd+d^2]L/12$

Since burning surface may be straight line or curved with some radius of curvature, the variation in burning surface area is depicted for both the situations separately. Any straight surface remains straight in course of burning and this condition is valid for end faces, central perforations, concentric circles etc. Handling of straight sections for some typical situations is depicted in Fig. 114.

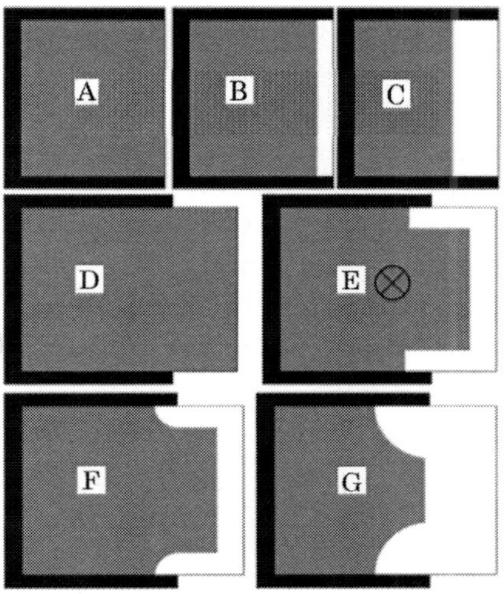

Fig. 114: Burning surface area with straight sections

Sketch 'A' of Fig. 114 depicts end burning cylindrical solid propellant configuration with inhibition, restricting the burning at all lateral and one end face. Sketch 'B' depicts situation after certain web consumption. The inhibition will have a slower burning/erosion rate than propellant and naturally, some portion of inhibition will be projected out beyond the burning surface. Sketch 'C' depicts much more inhibition surface exposed to combustion gases and under certain situations, these inhibition may disengage from the main part and passes through rocket nozzle, resulting in formation of sudden kink in pressure time profile due to obstruction in flow passage. This is associated with a dip in the thrust time profile. In all three sketches, the burning surface area of the propellant will be constant ($=\pi D^2/4$).

A peculiar situation exists, when propellant is projecting beyond the inhibition surface (Sketch 'D'). It seems that propellant here also burns in parallel layers and will have burning surface made of only straight section (Sketch 'E'). The projected portion continues to behave as made of straight profiles, but a curved profile is generated in this configuration in course of burning. The junction, where inhibition ends on lateral surface, acts as center of fresh curved profile all along the external periphery (Sketch 'F'). The lateral uninhibited burning surface of the propellant disappears in course of time (Sketch 'G') and later on the end face is also consumed (not depicted in the figure) by the curved surface emanating in course of burning with center at the end of inhibition on lateral surface. So, if longitudinal projection of the uninhibited propellant is 'L' and end face diameter is 'D', initial burning surface area is given by $\pi D(D+4L)/4$. For Sketch 'F', which depicts profile at web consumption of 'w', the burning area from straight sections is $\pi(D-2w)(D+4L-6w)/4$. However, this is not the burning surface area at that instance. For actual burning surface area, the curved portion at the periphery must be considered. So long as lateral burning surface is present, the generated curved section constitutes an angle of $\pi/2$ at the center and the burning surface area is given by $\pi w(4w+\pi D)/2$. This configuration is discussed in chapter 6 under blind hole section (Fig. 39). It must be kept in mind that not considering the created curved section, which is absent in initial configuration, may result in error in calculation.

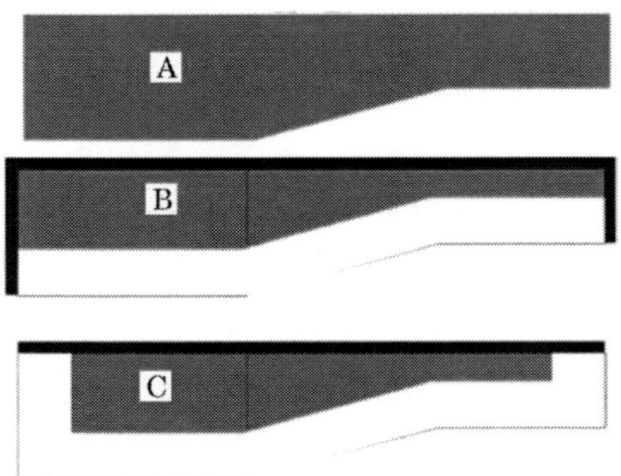

Fig. 115: Variation in longitudinal geometry

Another straight section calculation is present for the longitudinal variation is profile, where central perforation has two different diameters separated by a slant portion (slant angle q). This configuration is depicted as Fig. 115. In this case slant angle of the configuration does not change in course of burning. Two situations may exist. In first case, the burning is permitted at the central perforation alone and both ends as well as external lateral surface are inhibited. For the second case, the inhibition is placed at the lateral surfaces only and ends are free to burn, resulting in reduction in the length of the propellant. This configuration is handled in this chapter earlier and in the initial phases, *the axial length of slant section does not change.*

For Sketch 'B' of Fig. 115, overall length of the propellant does not change. But due to slant section, the length of portion of propellant with higher internal diameter (right hand side of the sketch) increase. The axial length of the slant section remains constant. The enhancement in length of the portion with larger diameter for a given web consumed 'w' is given by w x tan(θ/2). Similarly, the length of larger diameter (left side of the sketch 'B') reduces by the same amount. For the End uninhibited situation (Sketch 'C'), the overall length of the propellant reduces from either side. As axial length of the slant section does not change, the length of the smaller diameter side (left side) of the propellant reduces by more amount than the length of larger diameter side. Reduction in the length of smaller diameter side at any web consumed is given by w+w x tan(θ/2). Reduction in the length of smaller diameter side for any web consumed is given by w-w x tan(θ/2). No curved portion is generated in course of burning in these configurations.

Many a times, propellant configuration has curved burning surface in the beginning of combustion. The curves portion may be in the form

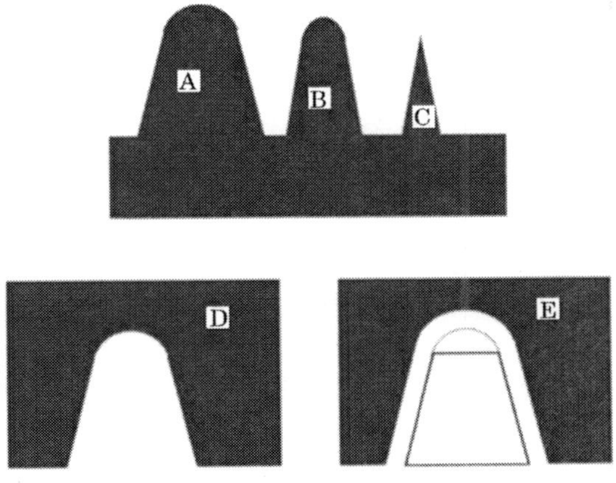

Fig. 116: Curved projections and cavities in configurations

of projections (Sketch 'A' of the Fig. 116) or cavities (Sketch 'D' of the Fig. 116). Both types of the curved profiles are available in star-shaped configurations. Both types of curved surface has certain initial radius of curvature and center of curvature. The center of curvature does not change in course of burning for these configurations. For the projected portion, the radius of curvature continuously decreases in course of burning (Sketch 'B' of Fig. 116) and finally curved section disappears, when web consumed exceeds the radius of curvature (Sketch 'C' of Fig. 116). For the curved cavities, the radius continuously increases in course of burning with fixed center of curvature of the curved section (Sketch 'E' of Fig. 116). Burning in parallel layer is exhibited by curved section also.

Limitations of Predictive Ballistics

Although, derivation of mathematical formulation in this part of the book is based on purely geometrical understanding with simple logic of offsetting initial burning profile by known distances, the burning in parallel layer may not be exhibited by solid propellant during rocket operations. The conversion of burning surface area profile into pressure time profile needs several idealizations and that also may lead to several variations. This part of the chapter summarizes major limitations of the predictive ballistics with solid propellants for rocket applications.

- ***Burning in parallel layers***: Solid propellants have initial configurations and initial burning surface area can be derived from geometrical treatment. However, subsequent burning surface area requires an assumption of burning in parallel layers. This assumption may not be valid for the entire propellant geometry. Sudden change of cross-section, sharp corners, longitudinal variations, erosive burning and similar other factors may produce a pressure-time profile, which may not be possible with the derived burning surface area versus web burnt profiles. A protrusion in the flow field may level off and projections may exhibit higher burning rates than otherwise flat surface. Similarly a depression on the surface may give a local lower burning rate and burning in parallel layer may not be valid. So, mostly a progressive profile may become neutral and a neutral profile may give regressive burning profile on static evaluation. The differential burning rate due to propellant geometry, flow field parameters (pressure, velocity, temperature, particles), flow obstructions and secondary combustion may result in non-parallel burning and deviated firing profile than prediction.

- ***Dimensional accuracy of propellants***: Solid Rocket propellants are manufactured by casting, followed by curing. As curing of solid rocket propellants take place at elevated temperatures, it is not possible to maintain the dimensional tolerances as per the mould, core, mandrel or metallic guiding components. The differential shrinkages are also observed in many cases. Another method of manufacturing propellants is extrusion and by this route also dimensional tolerances matching to die or hole size could not be achieved. So initial propellant configuration assumed, ideally is seldom achieved in reality. This introduces variations from prediction in the pressure and thrust parameters.
- ***Faults in geometrical treatments***: Geometry of propellants at the burning surfaces is offset by equal amount during propellant consumption. However, in course of burning, this is difficult to achieve. Localized higher web consumption is observed in those areas, where combustion gases are in dynamic mode. Stagnant gases give lower web consumption rates. However, if combustion gases have more metalized or condensed hot particles, higher web consumption is possible. So, overall uniform web reduction is practically difficult to achieve during propellant burning.
- ***Variations in propellant properties***: Propellant properties are assumed to be uniform and homogenous throughout the propellant mass, but differential mechanical properties due to uneven curing, settling of metal or high density particles, presence of local blow holes, cracks etc may lead to variation in properties of propellants, which cannot be predicted by any simplified assumptions. It is reported that propellant obtained by casting has more solid particles near mandrel and have higher burning rate there. Additionally, hump effect at mid-web is also pronounced in certain specialized cases.
- ***Unknown properties of combustion gases***: The theoretical means of calculating ratio of specific heats are available, but there is no practical means readily available for the determination of this term for the combustion gases. Although attempts are made, but this term is generally not available for the combustion gases readily and thumb rule is prevalent in these theoretical predictive calculations. Similarly adiabatic flame temperature and species in combustion exhaust products are also unknown. This clearly affects the value of characteristic velocity calculated from such assumed parameters.
- ***Flow field anomalies***: The flow reversal, stagnation of gases, flow obstructions, secondary nozzle throat creation, propellant port taper behaving as nozzle, blast tube parameters, eddy formation

etc are some of the incidences, which may lead to variation in combustion behavior from ideal. Prediction under such anomaly is not possible.

- ***Governing equation***: The global analysis of performance prediction assumes that mass generation rate by combustion matches mass discharge rate through the nozzle. This assumption may not be true, because this becomes a lumped parameter model, where contribution of local variations in geometries and flow field are not considered. The contribution of flame temperature and 2-phase flow is assumed to be incorporated in burning rate and characteristic velocity. So, the mass flow rate balance may not be sufficient to predict transient pressures during ignition and tail-off phase. If ignition phase is changing any of the geometrical parameters, even stable pressure cannot be predicted correctly by this global analysis scheme.
- ***Variations in nozzle throat***: Nozzle erosion is one effect that has to be accounted and incorporated in the global analysis model for matching of prediction with practical firing curves. In fact, throat diameter of nozzle is used for the derivation of pressure from burning surface area, any change in throat diameter during firing may lead to deviation in the prediction from actual curves. At nozzle flow velocities of high temperature combustion gases are maximum and nozzle erosion is evident. Generally graphite nozzles are used, but nozzle erosion may result in introduction of regressivity in the burning profile. Exactly opposite to this is deposition of metal (Aluminium) particles on the nozzle throat resulting in reduction in nozzle throat. For highly aluminized propellants, this phenomena is also prevalent. The exact value of nozzle throat diameter is required for prediction, which may not be available with designers. Another effect in nozzle throat diameter is observed during firing of nozzle-less boosters, where nozzle throat diameter is continuously varying during firing and prediction for pressure-time profile from burning surface area needs specialized treatment.
- ***Acceleration dependent flow field variation***: Rocket are accelerated in course of burning and this accelerating motion may affect the burning rate of propellant by enhancing mass discharge rate. Additionally, many rocket are launched with spin for better stability. It is observed that spin enhances burning rate of propellant on those surfaces, which are perpendicular to the spinning vector. Additionally, spin creates a boundary-layer separation at the rocket nozzle, resulting in reduction in throat diameter. So, effective throat diameter is lower than physically

available throat diameter of the rocket motor. These factors affect propellant combustion and prediction for such situation needs extra variables to be introduced in the calculations than discussed in chapter 5.

- ***Non-ideal special rocket configurations***: The whole predictive ballistics, discussed till this part of the book considers calculation of burning surface area at various time steps during combustion as the starting point of calculation. Calculation of pressure from burning surface area of propellant assumes constant value of propellant density, characteristic velocity, throat diameter, burning rate index and burning rate coefficients. However, many a times, these parameters are not constant during a rocket motor operation. So the pressure calculation becomes erroneous. From pressure to thrust again require calculation of instantaneous value of thrust coefficient and throat diameter, which introduces some error in calculation.
- ***Incremental analysis***: Although role of mass flow rate balance is used in the ballistic calculation as a lumped parameter model, but for certain specific situations, the calculations demand use of energy equation, momentum equation, erosive burning factors, gas properties for detailed calculations. This is accomplished through incremental analysis, where propellant configuration divided into many small increments along the axial length. Calculation is started from head end side and mass released in one small section is accounted and flow for a given pressure drop is calculated. The pressure drop is converted into velocity gain by the combustion gases. Energy equation is also accounted. The outgoing gas from first section from head end side is input for the next section. The calculation is repeated for each section till nozzle and mass discharged is equated to mass flowing out from the nozzle. This is an iterative procedure but this calculation gives additional parameters, like variation pressure inside rocket chamber, flow velocity at various instants in rocket, differential web consumption, erosive burning parameters, more realistic pressure time profile. However, these parameters are meant for very specific cases of propellant combustion and configurations.

REVIEW QUESTIONS

1. What are features of axially ramped tubular propellant geometry?
2. Why the axial length of the ramped section is constant in course of burning?
3. Compare funnel shaped and ramped tubular configuration.

4. What are the advantages of ramped tubular configuration over step-varying tubular configuration?
5. What are the effects of removing end inhibition in ramped configuration?
6. What are salient features of dendritic configuration?
7. Comment on burning surface area variation for dendritic configuration.
8. What are various web transitions for dendritic configurations?
9. What are advantages of cross-shaped propellant configuration?
10. What are limitations with cross-shaped propellant configuration?
11. What are different variants of cross-shaped propellant configuration?
12. What are effects of longitudinal inhibition on cross-shaped configuration?
13. Derive expression for volume of frustum of a cone.
14. Whether end face of a cylinder be treated as lateral surface of cylinder with length equal to one-fourth the diameter?
15. How is curved projection handled for calculation of burning surface area?
16. How does burning surface area evolve for curved cavities?
17. What are causes of propellant not burning in parallel payers?
18. What are variations in propellant geometry and properties?
19. Which properties of combustion gas of the propellant ids not practically determined?
20. What are various flow field variations responsible for the failure of predictive ballistic approaches?
21. How nozzle affects predictions for rockets?
22. Whether flow field affects predictive capabilities for rockets?

Section III:

Practical Rocket Propellant Ballistics

Before flight trials, rockets are evaluated in static mode for ascertaining and confirming various performance parameters. During instrumented static evaluation of rocket, variation of pressure, thrust, temperature, strain, vibration, sound, etc against time are recorded and performance of rocket as integrated system of hardware, igniter, propellant and nozzle for specified operating conditions are obtained. In this section of the book, output from static evaluation of rockets in the form of pressure and thrust versus time diagram is considered as input for conformance to designed parameters. This section discusses practical internal ballistics of rockets under five chapters. One chapter is added for other parameters affecting internal ballistics of rockets and future trends.

- Static Evaluation of Rockets
- Ballistics from Pressure-time Curves
- Ballistics from Thrust-time Curves
- Advanced Evaluation of Firing Curves
- Combustion Instability in Solid Rocket Propellants

11

Static Evaluation of Rockets

CHAPTER SUMMARY

Practical ballistic covers static evaluation of rockets, where ballistic data and performance parameters are generated by practically firing the integrated rocket motor in instrumented set-up. In the first chapter of practical ballistics, static evaluation requirements, importance, methods, set-up and analysis of acquired curves are enlisted. This chapter covers the following aspects.

- Components of Static evaluation of rockets.
- Parameters measured during static evaluation of rockets.
- Instrumentation for static evaluation of rockets.
- Static test-bed for rocket testing.
- Interpretation of static testing results

INTRODUCTION

Rockets are the quickest means to deliver payload at target. However after designing a rocket and realizing its various components, flight trials are never attempted directly. A cheap, reliable and confirmatory way to assess performance of rockets is by static firing of rockets. During static firing, instrumented rocket is held against a wall, called thrust wall, to prevent its forward motion. Provision is made for pressure measurement in the rocket chamber through pressure gauges. A load cell is placed between rocket and thrust wall to measure thrust generated by rocket. On firing, hot exhaust gases, generated from the combustion of rocket propellant generates pressure inside rocket motor chamber and their outward flow from the rocket nozzle pushes rocket in forward direction. Both pressure and thrust are measured against time of burning. Skin temperature is also measured during static firing to assess adequacy of thermal insulation and possibility of any hot-spot formation ion the rocket motor casing. Measurement of strain at various locations is also carried out to confirm sufficiency of rocket casing thicknesses. In addition to this measurement and recording of displacement, acceleration, sound and visual (videography) data is also planned during static firing, to clear the system for dynamic or flight trials. Although static firing is not a substitute of dynamic firing and performance of rockets deviates from ballistic data obtained during static evaluation, but static firings represents behavior of rockets in flight with greater accuracy than results from any other component level tests. In addition to this characterization of propellants during design phases need frequent static firings. Finalization of composition, tailoring of compositions and ballistic properties, optimization of igniter quantities, assessment of burning rate behavior etc are some of the requirements, which frequently demand static evaluation of rockets. In addition to this, cost-effectiveness, confirmatory results and repeatability of performance can be very well ascertained by static firings. A typical static firing set-up for rockets is given as Fig. 117. The purpose and advantages of static firing of rockets are summarized below.

- In static firing to test all the components of rockets are tested simultaneously and interplay of various components are ascertained.
- The static firings of rockets are economical in terms of materials and investments.
- The static firings of rockets can be conducted in relatively small space than flight trials requirements of space.
- The static firings of rockets need less logistic supports than flight trials.
- The static firings of rockets are instrumented to get quantitative results on performance.
- The static firings of rockets are conducted to verify predictions.
- All components are tested in almost actual loading conditions, except for aerodynamic and kinetic loadings.
- Various transient phenomena like pressure build-up, tail-off, pressure transitions etc are captured in the static firings of rockets.

In the set-up displayed as Fig. 117, rocket is mounted horizontally on the trolley. Trolley is mounted on rails and is free to move along longitudinal axis of the rocket. The trolley has V-blocks or saddles (minimum 2 in numbers) on which motor is mounted. Alignment in horizontal and vertical plane is essential for rocket evaluation in static mode and provisions are made in the mounting of V-blocks or saddles to move them in both the planes. Load cell is attached to rocket motor through thrust block and adapter. The part of load cell in contact with the thrust wall is rounded to ensure point contact during firing and proper noise-free recording of thrust. The trolley must have smooth movement so as to avoid any oscillations in thrust data. Additionally, rocket motor is secured through minimum overhang of nozzle. To suppress oscillations in thrust data, pre-loading of around 10% of the expected maximum thrust is also incorporated in the static evaluation set-up.

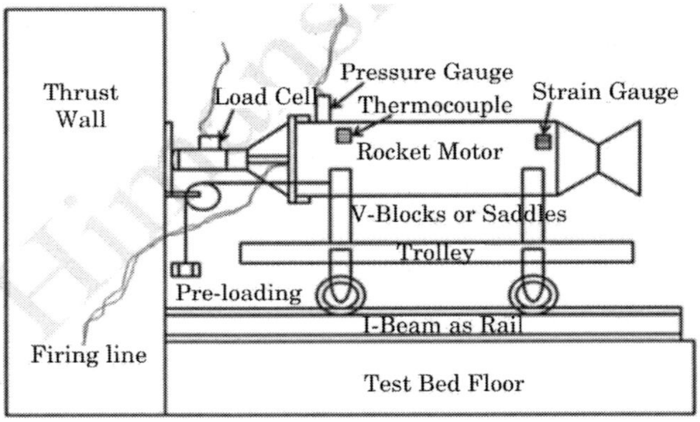

Fig. 117: A typical set-up for horizontal static firing of rockets

Pressure gauge is generally mounted on the rocket motor directly, either at head end or nozzle end. Sometimes, they are mounted at any in between location also. In yet another case, mounting of pressure gauge through igniter port is also planned. Thermocouples and strain gauges are generally patched with the help of bonding cements at the outer surface of rocket motor after due cleaning. Joints and nozzle areas are prone to show high temperatures. However, variation of temperature during burning time of rockets is found to be almost negligible, but after firing, temperature in the nozzle region increases by conduction. The upper limit of skin temperature, for rocket motor is 100-150°C. Strain gauges are placed near joints and the head end dome section of the rocket motors. Dome section is subjected to tri-axial state of strain, while strains in the motor casing section are bi-axial. Values of strains in rocket motors are generally less than 5000 micro-strains. The firing line gives initiation pulse to firing circuit, which initiates igniter. Igniter generates large amount of hot combustion gases. It increases temperature and pressure in the empty space of the rocket and brings surface of propellants to auto-ignition temperature. Once propellant is ignited, pressure time and thrust time profiles are recorded.

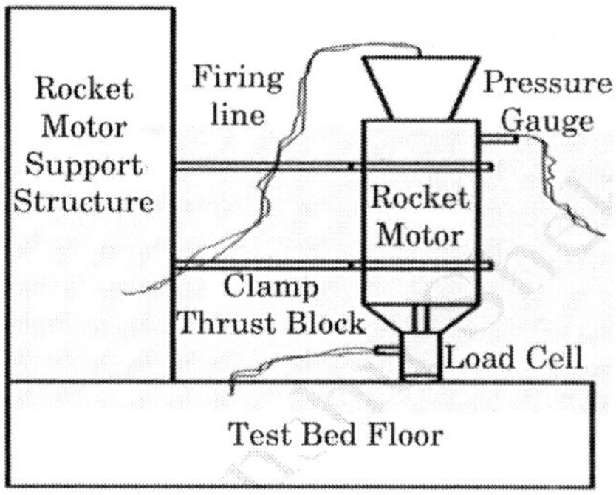

Fig. 118: Vertical rocket firing set-up

Similar to horizontal firing, vertical firing of rockets is also attempted. For vertical firing, rockets are mounted in nozzle up arrangement. Rocket motor is mounted on load cell and initial reading of load cell has contribution from weight of the assembled rocket motor. As propellant is consumed in course of time, the weight of rocket reduces with time. So, during firing, if initial weight of rocket is taken as tare weight, then thrust data at the end of firing is underestimated. This clearly indicates that during vertical firing of rocket in static mode, thrust calculation needs further analysis of data. However, if propellant weight is insignificant relative to the rocket hardware weight, firing in vertical mode can give thrust data with reasonable accuracy. The vertical firing set-up for static evaluation of rocket motors is shown in Fig. 118. All arrangements are similar to horizontal firing. To show alternative arrangements, firing line is shown to be coming out through the nozzle and pressure gauge is mounted at the nozzle end of the rocket motor. Comparison of static firings of rockets in horizontal and vertical modes is given below.

- In horizontal firing of rockets, gravity assists in placement of rocket on trolley and clamping provisions can be dispensed with. In vertical firings, the clamping provisions must be in place.
- In horizontal firing, handling of rockets is easier with the help of tackling fixtures, belts etc. In vertical firings, the rocket motor needs special provisions for rocket motor case handling and they are generally handled in a cage or frame.
- Horizontal firing is carried out for all sizes and weight classes of rockets. Even space-shuttle solid rocket booster motor (SRBM) is fired in horizontal mode. Vertical firings are used for smaller rockets, generally.
- In vertical firing thrust measurement needs correction for weight of the rocket motor, but in horizontal firing correction for pre-load or trolley friction are to be accounted.

- Thrust measurement in horizontal firing may have oscillations in absence of pre-load, but rocket motor itself act as preload in vertical firings.
- In vertical firings, exhaust gases are discharged and given motion in upward direction, but in horizontal firings, exhaust gases move in horizontal space.
- Area requirements for horizontal firing is more, else exhaust gases are reflected from obstructions.

Complete static firing exercise can be classified under following heads

- Test article details for static firing
- Test bed for static evaluation
- Instrumentation for static firing
- Data analysis and interpretation
- Each is defined in different sections.

Test Articles for Static Firing

Although, rocket propellants are evaluated during static firing, mainly, but complete rocket motor is considered as the test article. The rocket motor may be a flight worthy motor, consisting of thin rocket motor casing, in-situ thermal insulation, propellant grains with proper inhibition, pyrotechnic igniter for initiation, electrical connectivity for initiation and convergent-divergent nozzle. However, for evaluation purpose, proof motors are also fabricated and tested in static mode. The proof motor resembles flight motor in internal and nozzle configuration, but thickness of proof motor is generally very high for its repetitive use and high burst pressure. In fact, need of static evaluation arises right from preliminary design of rocket motors. Once rocket motor is designed, ballistic parameters of propellants need to be evaluated. Burning rate of propellants, characteristic velocity, specific impulse needs static evaluation of rocket motor. After propellant composition passes these ballistic characterizations, propellant configuration is tested in proof rocket motors. For evaluation of ballistic parameters of rockets, ballistic evaluation motors are considered as test articles, in which rocket propellants are tested in end-burning solid sustainer mode, progressive tubular, regressive tubular or near neutral externally inhibited tubular propellant configurations. These propellant grains are fired in cartridge loaded modes to assess ballistic parameters, mainly burning rates of propellants. Simultaneously, characteristic velocity and specific impulse are also obtained from area under pressure-time and thrust-time profiles respectively. In such firings propellant quantity for evaluation is less than 2-3 kg. However, provision of 40 kg BEM configurations, for the assessment of ballistic parameters of propellants also exists.

Propellant characterization, during any development begins with laboratory evaluation of density and burning rate. Burning rate is laboratories are determined by strand burner equipments. This is a closed chamber in which propellant strands (5 mm x 5 mm x 100 mm) is suspended. Provision to pressurize the closed chamber also exists. Two probes at marked distances are placed for assessment of arrival time on oscilloscopes. The propellant strand is initiated at one end and time for travel of combustion surface from one point to the other is recorded. Burning rate is evaluated through division of pre-measured distance by arrival time. This gives burning rate at different pressures. From pressure and burning rate data, burning rate law is computed. Although for preliminary estimates, this calculation is satisfactory, but there are certain issues, which makes this test a symbolic test only. First is propellant is burning in closed chamber, where combustion of propellant adds certain combustion gases and increases pressure. So burning rate evaluated may be at some different pressure than the pre-set pressure of the closed chamber. Second, Measurement of distance by vernier and time by oscilloscope is also not acceptable because of independent system of measurement and different measurement accuracies. Third, the strand burner does not simulate conditions prevalent in the rocket motors. In rocket motors, propellant is subjected to flow of hot gases and pressure simultaneously. In strand burners, pressure is simulated to certain extent, but temperature and above all erosive burning due to flowing gases is not correctly simulated. Fourth, measurement of length of propellant strand is taken in un-pressurized condition. However, during strand burner evaluation, propellant under pressure will deform and their length may reduce or increase depending on dimension and Poisson's ratio of the propellant. Whatever may be the situation, measurement of distance between two points has some discrepancies. Fifth, except burning rate, no other parameter is extracted from strand burner about the propellants. Sixth, since dimension of propellant grains are small, assuming them to represent bulk of propellant charge is also an extraneous assumption. Except time, all other parameters, namely distance and pressure, are estimated inaccurately in strand burner. So strand burner tests for burning rate evaluation are not considered sufficient for characterization of propellants. However, in practical scenario, the burning rate index of power law is assumed to be accurate and repeatable from strand burners.

So, static evaluation of rocket motors in ballistic evaluation motors is attempted. One of the simplest forms of propellant configuration utilized for burning rate evaluation is firing propellants in end burning solid sustainer mode. Here again, propellant length is measured by independent vernier calipers and burning time is obtained from the firing curves. Burning rate is computed as length divided by burning time. Although some of the drawbacks of strand burners are addressed, but measurement of length by vernier and no simulation of propellant facing combustion gases, may not favour usage of such configuration for burning rate evaluation. Additionally, during assessment, in addition to propellants, igniters and hardware are also included as test articles. Final performance is a combination of behavior of all sub-components of the test articles. The ideal burning profile is neutral and burning rate of propellant at a single pressure is obtained by this firing. Practically, when propellant is static evaluated in this configuration, the lateral inhibition may not sustain for prolonged duration and firing curve becomes progressive at the end of firing. Extra burning surface is made available by the failure of inhibition and rather than giving a neutral burning profile, the profile becomes progressive. Although, this may not occur in all the firings and the progressive nature of propellant-burning is dependent of burning rate of propellant, type of propellants, mechanical properties of propellant, propellant composition, pressure levels used for firing and length of propellant. Slight progressive nature of the burning curve is generally obtained in these firings due to enhanced burning rate of propellant by heat feedback from hot combustion gases. This leads to certain reduction in burning time leading to enhancement in burning rate than normal, because of underestimated burning time. Suitable factor for such variations must be considered for any calculation from deviated firing data. This simulates propellant combustion in actual configurations for a missile. Overall this gives characteristic velocity of propellant combustion and specific impulse of rocket motor, as additional data.

An improvement over end burning solid sustainer propellant grain is tubular propellant grain, without any inhibition. This leads to regressive burning profile and a wide range of pressure is covered during such firing in the rocket motor. This helps in evaluation of burning rate over a wide range of pressure. As propellant grain is uninhibited, processing is quicker. Burning rate at different pressures are evaluated and burning rate law can be fitted to the data. However, this type of configuration gives very small web for complete consumption of propellant. Another problem is combustion at both internal and external surface of the tube. This results in problem of pressure non-equilibrium between inner (inner diameter of propellants) and external port (between propellant and rocket motor casing). Additionally, chances of erosive burning make prediction difficult for such situations. Another variant for the tubular propellant grain is neutral burning end inhibited propellant grain. This gives neutral profile and is utilized for confirmation of burning rate at a single pressure. One of the popular tubular propellant grains, used for burning rate evaluation is progressive burning tubular configuration. This type of propellants is inhibited at external lateral surface and also at both the end faces. Burning is allowed at the inner surface only in radial outward direction. This covers much wider range of pressure than regressive tubular propellant grains. Another variation is obtained by inhibiting tubular propellant grain at the external lateral surface only. During combustion, such propellant grains are consumed from inner surface as well as from both the ends. This type of configuration is capable of giving near neutral burning profiles for the propellant grains and thus burning rate at a constant pressure is ascertained. For the regressive and progressive mode of propellant configuration in tubular form, burning rate at varied pressures, realized during firing, can be assessed by ballistic calculation. The calculation strategy is described in next chapter. Main test article during static evaluation is propellant.

For initiation of propellant, suitable ignition system must be employed. Igniter should be initiated by firing squib through electrical supply. The quantity of igniter should be sufficient to ignite around 80% of the end burning solid sustainer and around 20% of tubular propellant configuration. Another practical consideration for igniters is creation of sufficient pressure inside propellant port to establish smooth combustion surface at the propellant, which is self-sustaining. To achieve, minimum pressure level before any flow of combustion gases is achieved, the rocket motor is attached with a nozzle closure. This makes rocket operation a closed vessel in the beginning of firing. The nozzle closure bursts at certain defined pressure and rocket starts behaving as vented vessel. By that time, propellant surfaces are already ignited and smooth combustion will start at the surface. Igniter must simultaneously fulfill the pressure as well as thermal requirements. Igniter quantity is generally finalized as the higher quantity obtained by the two controlling criteria. Suppose, igniter quantity for the igniting given percentage of burning surface of propellant is higher than igniter quantity needed for generation of sufficient pressure in the port. In such situation, higher quantity of igniter is selected, which may lead to very high pressure during ignition transient. This initial pressure peak may lead to motor burst in adverse situations. Additionally, the initial pressure peak may alter the burning surface of propellant and predicted combustion behavior is not realized in the static evaluation. So, quantity of igniter is coupled with igniter configuration. The igniter composition may be in powder, granules or pellet form. The combination of these forms, for satisfying all the requirements of rocket ignition, makes igniter design, realization and use an art. Igniter composition in small rockets is generally kept in cloth bags, aluminium containers or celluloid containers. During ignition of rocket propellants, the container should not create problem of port erosion, nozzle erosion, nozzle blockage or incomplete combustion of igniters. The containers has small holes to distribute combustion gases over the surface of propellants and they also acts as retainers for the igniter pellets, which may be otherwise ejected out of the rocket motor in unburnt state.

Hardware is visible part of the test articles. It is made from metals or from composite materials. It retains combustion gases inside rocket motor chamber to generate high pressure. The main requirement of rocket motor is to withstand pressure and temperature of combustion gases. Main rocket motor casing is a pressure vessel which is subjected to thermal loads also. For proof rocket motors, thickness of casing is on higher side. For design of rocket motor, consideration of reduction in strength of material at elevated temperature must be considered. The rocket motor has nozzle attached at the rear end. Nozzle has convergent and divergent sections for smooth flow of gases without any shock. The throat diameter (minimum diameter) is selected to generate given average pressure in the rocket motor. If throat diameter is increased, pressure inside rocket motor is reduced. Throat area faces maximum temperature and it is made of graphite. Graphite has very high thermal conductivity, which takes heat away. Molybdenum and tungsten alloys are also utilized for the fabrication of throat section of the nozzle. Hardware is to be qualified separately, but during static evaluation, hardware is also inadvertently tested. Both rocket motor casing and nozzle is lined at the inner surface by suitable thermal insulation material. This restricts temperature rise in both the sub-components of the test article.

Static evaluation of rocket motors is invariable a test for rocket propellant performance. Igniter and hardware are usually tested inadvertently. Failure of these sub-components also leads to failure of the test and data from static firing becomes unreliable. Failure modes, from test article point of view, during static firing are enlisted below.

- Igniter does not initiate due to insufficient electric current.
- Igniter does not initiate due to electrical discontinuity.
- Ignited gives very high pressure peak.
- Flow passage for combustion gases is blocked.
- Propellant does not pick up due to nozzle closure opening at lower pressure.

- Propellant does not pick-up due to insufficient igniter quantity.
- Propellant does not pick up due to less exposure (time and quantity) with igniter gases.
- Propellant burning behavior is erratic.
- Propellant crumbles during static evaluation.
- Nozzle blockage occurs by propellant chunks or igniter container.
- Nozzle melts by excessive heating during firing.
- Motor casing burst opens at joints.
- Motor casing overheated at localized places (hot spots).
- Motor casing leaks at joints.
- Thermal insulation erodes at certain places.
- Bonding between thermal insulation and motor casing breaks

Overall, propellant, igniter and hardware constitute the test articles for static evaluation. The propellant is mainly tested for energetic, while hardware is tested for structural and thermal compliance. If testing is carried out in simulated proof motor, then obtained parameters can be realized in actual flight trials also.

Instrumentation for Static Evaluation

Test bed for static evaluation is one of the most sophisticated infrastructures needed for the development of any new rocket, missile or propellant. It is simultaneously subjected to mechanical, electrical, explosive and electromagnetic hazards. Test bed is a facility and it is safety concern, which governs design of a static test bed. The test bed comes under hazard division 1.1 of UN classification of dangerous goods. The classification is compiled in orange book and all explosive materials come under hazard division (HD) 1. This division is classified in 6 segments. HD1.1 depicts materials with bulk explosion hazards. HD1.2 covers materials with projection hazards. HD1.3 has materials with mass burning hazards. Propellants in general come under HD1.3 classification, but test bed comes under HD1.1. This makes this place prone to explosion or detonation hazards. Since rocket motors, during operation acts as pressure vessel, any eventualities may lead to detonation. The safety distances, as per capacity of the test bed is observed, while citing the facility. Propellant has inherent combustion hazards and due to mass burning capabilities, propellants are treated as HD 1.3. The rocket motor with nozzle and without ignition system has propulsion hazard and it may be treated as HD 1.2. If both nozzle and igniters are attached then the complete system is classified as HD 1.1 and utmost precautions must be taken to handle them.

Test bed received rocket motor for static evaluation in assembled condition. The transportation of assembled rocket motor is generally on a battery operated vehicle with proper earthing provisions. If petrol or diesel vehicles are used, the exhaust should be in the front and the exhaust pipe should have spark arrester. The motor must be mounted on saddle with proper vibration isolation provisions. During movement of vehicles, rocket motor must be tightly held in position by lugs, strings, belts and other suitable devices. The speed of vehicle should be below 20 km per hour and all norms valid for transportation of hazardous materials must be followed.

The test bed itself has many facilities, which needs special attention. The thrust wall is main civil structure to support the rocket motor during evaluation in horizontal state. The civil structure is embedded below ground and is full RCC structure. It supports thrust frame, which acts as a rigid member during static evaluation. The rocket motor is mounted on a wheeled trolley, with provision to prevent sideways movement of the trolley. Trolley is free to move in longitudinal direction. The trolley and other support structures at static test bed are designed with very high margin of safety of the order of 6. The trolley on which rocket is mounted has four wheels and the saddles for supporting rocket motors are such that overhang of rocket on both the sides is not beyond ¼ th of full length of the rocket motor. High overhang result in high oscillations in the recorded pressure time and thrust time profiles. This trolley contains V-block with clamping arrangements for securing rocket motor tightly, where rocket can be made integral part of the trolley. Movement of trolley independent of trolley is not expected during static evaluation. If any such movement occurs, failure of rocket motor on test bed is expected.

After rocket is mounted over the trolley and secured tightly with the help of suitable clamps, one of the key arrangements is pre-loading. Although rocket motor is touching thrust wall before the static evaluation and thrust pushes rocket in forward direction, there are some oscillations, which moves rocket in back-and-forth. This causes oscillating thrust-time profile. A usual means for prevention of such unwanted longitudinal oscillation is application of preloading. For preloading some hanging weight are used to pull the rocket towards thrust wall. Generally 2-3% of maximum expected thrust is used as preload. This ensures positive touches and prevention of any jerk at the initiation of firing. For all large rocket motors preloading is a must.

As far as electrical connections are concerned, the firing line is secured last on the rocket motor and is the first line, to be activated in static firing of rockets. The purpose of firing line is to provide adequate current to the igniters for initiations. The firing current is fixed for a given rocket and is the main feature of pyro-cartridge utilized in the rocket firing. The firing current is of the order of 4 mA to 20 mA. Firing pulse is said to be given, once 75% of the stated firing current is passed to the igniter of the rocket. Once igniter has functioned, the firing line becomes an open circuit. The time of passage of firing current is taken as reference time and all time-related calculations for the rocket start, considering this point as zero time. In many data-recorders, data acquisition is initiated at this zero time and in others, data acquisition is started beforehand with proper marking of time in data for giving firing pulse.

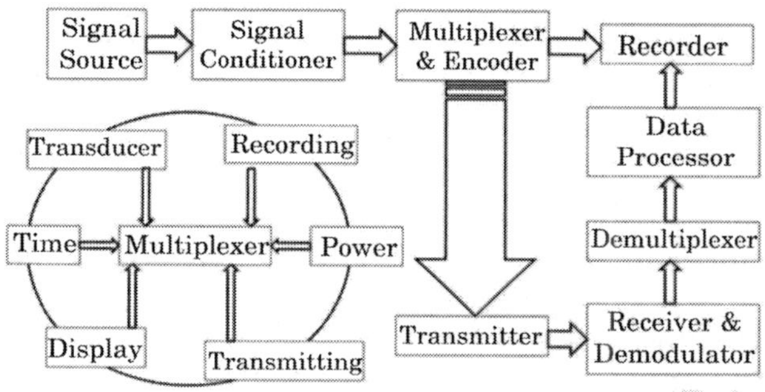

Fig. 119: A general instrumentation schematic

Once rocket is initiated, data acquisition should start. For data acquisition of rockets a tentative block diagram of the instrumentation is shown in Fig. 119. Once rocket is initiated, data for pressure, thrust, temperature, strain, vibration, sound etc are generated. This is to be converted to electrical signals for transmission through cables. Conversion of these engineering parameters into proportional electrical signal is executed by transducers. The transducer output is only an electrical replica of the rocket parameters and for converting them into standard output forms, various signals conditioning equipments are used. They may include amplifier, filter, sampling, digitization, data compression, digital formatting, modulation, correction factor, conversion etc. since simultaneously multiple channels will be acquiring data for the same rocket firing, for correct correlation between them, multiplexing unit is needed to systematically arrange the data. The data is encoded and have two options. First, it may be send to recorder directly for display and storage. Second, data may be transmitted through wires or otherwise to distance locations. Transmitted data is received, demodulated and demultiplexed in separate unit. It is then processed for storage and display. In the complete system multiplexing unit is responsible for efficiency of the rocket instrumentation. It needs power and takes input in terms of transducer output with respect to time. Data is then modified (or encrypted) for display and transmission.

Transducer converts physical energy into electrical energy, which may be charge, voltage, frequency etc. Selection of sensing element of transducer needs system and signal study and many factors govern selection of sensors:

- ***Type***: The sensor selection starts with comparison and matching of data to be measured and capabilities of the sensor. Data type includes rate of variation of signals, sampling rates, response, transience, acquisition time etc. Capabilities of sensor include, nominal capacity, overload capabilities, hysteresis etc.
- ***Range***: The maximum and minimum values expected to be measured must be known beforehand for any type of data. This is helpful in identification of sensor capabilities.
- ***Sensitivity***: This indicated capability of sensor to discriminate and identify any small change in the input measured quantity. A highly sensitive sensor may be able to show any small variations in input signals and at times, it may show hunting behavior by capturing spurious noise. For strain gauge transducer, it is written as 0.08 mV/MPa.
- ***Resolution***: This is capability of the sensor to measure smallest signal increment. For example, resolution of a transducer is expressed as 2 mV/V.
- ***Accuracy***: Although accuracy is indicated as a governing factor, but correct reading is a relative term, which is measured by another calibrated more accurate instrument. Since reference of correct reading is always evasive, accuracy of sensor is a relative, which measures difference of measured value with respect to standard sensor.
- ***Repeatability***: If same input is given again and again and sensor is giving same output, the sensor readings are said to be repeatable. However, it depends on sensitivity and resolution of sensors. A low sensitivity sensor may show high repeatability of results.
- ***Response time***: Signals from any event is transient and has fast variation in numerical values. However, sensor has to match the variation and quickness with which the variation is recorded and displayed by a sensor, also governs the selection criteria.
- ***Others***: Cost of sensor, compatibility with other sub-systems, usefulness in hazardous (corrosive, high temperature, vacuum, high pressure) environment are some other criteria for the selection of sensors.

For the measurement of pressure and thrust in a rocket firing, strain gauge based pressure gauge and load cells are used. In strain gauge, strain gauges in the form of wheat-stone bridge (4 numbers) are placed on a pressure sensing diaphragm. The diaphragm deforms proportionally under pressure, along with the sensing strain gauge. The change in length of wires in strain gauge due to deformation unbalances the bridge. Corresponding to given input voltage, the unbalanced bridge gives certain proportional variation in output voltage. The variation in output is recorded and converted to pressure by suitable calibration. As far as load cells are concerned, instead of diaphragm, a solid rod of metal is placed, which gets compressed under thrust of rocket. The attached strain gauge show proportional variation and output is recorded after applying suitable gauge factor as thrust variation data. The placement of strain gauges in bridge format is to amplify the signal, as deformation of four strain gauges add-up to give higher voltage variation than a single strain gauge.

Signal conditioning can be applied to both, analog or digital signals. Once data is suitably formatted, it is bound to go for recording, display or storage. The signal from rocket instrumentation needs prior definition of acquisition time for each rocket motor, sampling rate of data acquisition, data compression format etc. one of the important issue, which need proper attention is calibration. Each transducer is associated with a transfer function, which acts as combining thing for electrical and measured parameters. Since transfer function of transducers is not clearly known, it is generally obtained by calibration. This is explained by using a numerical example.

Let us take a pressure transducer with gauge calibration factor of 151 bars. Assume a general 12 bit analog to digital converter for $\pm$ 10 Volt input. If the signal conditioner has binary gain step, the counts can be expressed as 0, 1, 2, 4, 8, 16, 32, ... 4095 for offset binary output format. The zero of such signals lies at 2048 count, ideally. If sensitivity of the strain gauge transducer for pressure measurement is stated as 2 mV/V, then for full scale reading (10 V), the output is 20 mV. If gain is adjusted in such a way that maximum voltage of 8.5 V (corresponding to gauge capacity) is obtained, then gain is automatically set. Gain in this case is (8500/20 =) 425. In 2's complement output format, there are total 2048 counts, which is divided into 10 volts. Count per volt is given by (2048/10 =) 204.8 counts per volt. For 8.5 V input, the count displayed will be 1741. Since voltage need not be seen later, the unit scale factor is calculated as 151 bar/1741 counts = 0.08673 bars per count. Now, counts are generally recorded. For a count of 3000, the value of pressure is (3000-2048) x 0.08673 bars = 82.57 bars. This is theoretical analog to digital resolution in terms of engineering units for this channel.

In practical situation, major problem comes with zero setting, which changes due to noise, hysteresis and other spurious inputs. It is possible that at zero pressure, zero voltage is not observed and some counts are recorded. So, the reference of 2048 is disturbed. To overcome this, balancing of bridge is must before each firing and also zero calibration must be done before each firing. Zero readings (of pressure in terms of counts) just before firing are always recorded and used for calculation.

However, there are several deviations in this calculation. First the 204.8 counts per volt may not be maintained by the A/D converter. The gain may not be exactly 425. To carry out calibration practically, a resistance of known capacity (in terms of ohms and in terms of generating known pressure signals) is placed as one of the arms of the bridge, as fictitious load. This is called shunt resistor. Let this shunt resistor is supposed to give 8.5 V (corresponding to capacity of the pressure gauge), as per manual. If only 7.5 V is recorded by the system, then gain or scaling factor is to be adjusted to make display of 8.5 V. The count of raw data is observed, after adjustment. This is called shunt calibration. Let this be represented on count as 1636. The unit scale factor is calculated as 151 bar/1636 counts = 0.0923 bars per count. For a count of 3000, the value of pressure is (3000-2048) x 0.0923 bars = 87.87 bars. If count corresponding to shunt calibration is smaller, for same count higher pressures are calculated (provided zero count is not altered). Similar calculations are valid for thrust records also.

Data Analysis and Interpretation

During static firing of rockets, weight of propellant and throat diameter is fed as input parameter. Firing pulse is given and performance parameters are practically measured by dedicated instrumentation arrangement. The pressure-time and/or thrust time profile is the routine outcome of any static firing of rocket. The combustion profile from a rocket evaluation is shown as Fig. 120.

In the combustion profile, the progress of time is recorded on x-axis (horizontal axis). The vertical axis (y-axis) is for the rocket parameter and it may be pressure, thrust, temperature, strain or displacement. For most of the rocket firing pressure and thrust data are considered to be the ballistic parameters. Maximum values of these parameters are important numerical value for design of rocket motor. It has very little ballistic significance but has very high structural significance for flight motors. For proof motors, this value is not critical. The maximum or peak value of thrust or pressure may occur in the beginning, at the end or in-between anywhere during firing. There are certain unwanted

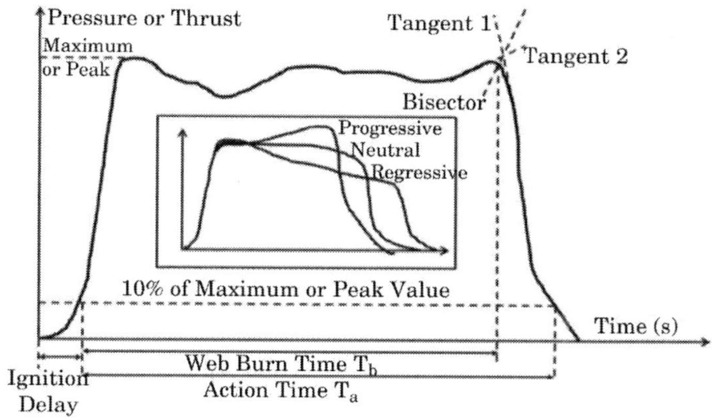

Fig. 120: Curve from static evaluation of rockets

situations, which lead to peak values, at locations other than designed conditions. Sometimes, when higher quantity of igniter is used, maximum value of the parameter is realized in the beginning. This may be in the form of spikes (sudden rise and fall to normal pressures in microseconds). This indicates optimization of ignition system parameters. When peak pressure occurs at the end in the form of relatively flat spike, this indicates propellant fracture at the end. By design, peak values can be designed at any point during operation of motors. Once correct peak pressure without any spike or kink is realized, the parameter is considered for further evaluation of ballistic parameters. The variation of ballistic parameters may be regular or irregular and it must follow the predictions. There may be variations in their numerical values. However, except peak value, no other values are directly obtained from the firing curve without some inherent calculations involved (as explained in next chapters).

Most of the direct measurement parameters are obtained on time axis. When firing pulse is given, electric current reaches ignition system of the rocket motor. The squib in the ignition system fires and ignites flammable composition surrounding squib (electric bridge). The generated combustion products ignites igniter pellet, powder or composition, generating large quantity of combustion gases, which generates pressure inside empty space of the rocket and heats adequate surface of propellant to auto-ignition temperature. Stable flame front is created at the propellant surface after certain gap in time. Overall, there remain certain delay between pressing firing switch and build-up of pressure in the rocket motor chamber. As a convention, the time lapsed from pressing ignition switch to attainment of 10% of peak value of the parameter is called ignition delay for the firing. The value is generally

20-100 milli-seconds. At the end of firing, time for 10% of peak value is again recorded. The time taken to achieve 10% of peak value during rising trend to falling trend is called, action time (T_a).

It is observed during practical evaluation of rocket motors in static firing, that at the end of firing, falling duration of the pressure or thrust value is longer than prediction. This may be due to certain differential combustion of propellants due to various reasons. The falling period may be release of accumulated gases in the rocket motor after firing. This period should not be treated as propellant performance. The actual web burn time calculation is slightly tedious from practical firing curves. Tangents are drawn at the falling as well as continuing part at the end of firing. The angle formed by tangent is bisected and the point, where the bisector touches the firing curve is called web burn time (T_b). The difference between action time and web burn time (T_a - T_b) is called tail-off time. Drawing tangent on firing curve is very extraneous and difficult. More or less, it depends on selection by operator. However, consistency and repeatability of web burn time is an essential requirement to be fulfilled from static firing.

Although firing profile may have non-uniform variation of parameters. But in many cases, monotonous variation of parameters is recorded. In the figure, progressive (increasing), regressive (decreasing) and neutral (constant) variation of performance parameter is also depicted. Locating web burn time for regressive firing is slightly cumbersome.

The firing profile for each case is typically assessed for ballistic parameters (peak, ignition delay, web burn time, action time) and area under pressure-time and thrust-time profiles have special significance from performance point of view. Area under pressure time profile is used for calculation of Characteristic velocity (C*), while that for thrust-time profile is used for calculation of specific impulse (Isp). These calculations are also used for the calculation of average values of ballistic parameters. Invariably, the portion of pressure-time or thrust-time profile before ignition delay is ignored for the calculation of area. This area is matches with igniter performance and neglecting that area does not affect propellant performance evaluation. The significance of web burn time and action time is established for calculation of area under the curves.

For calculation of action time based calculations, area under the curve as depicted in Fig. 121 (top) is considered. Entire area, when pressure or thrust is above 10% of their respective maximum value is considered. Average pressure or thrust is calculated as area under the curve divided

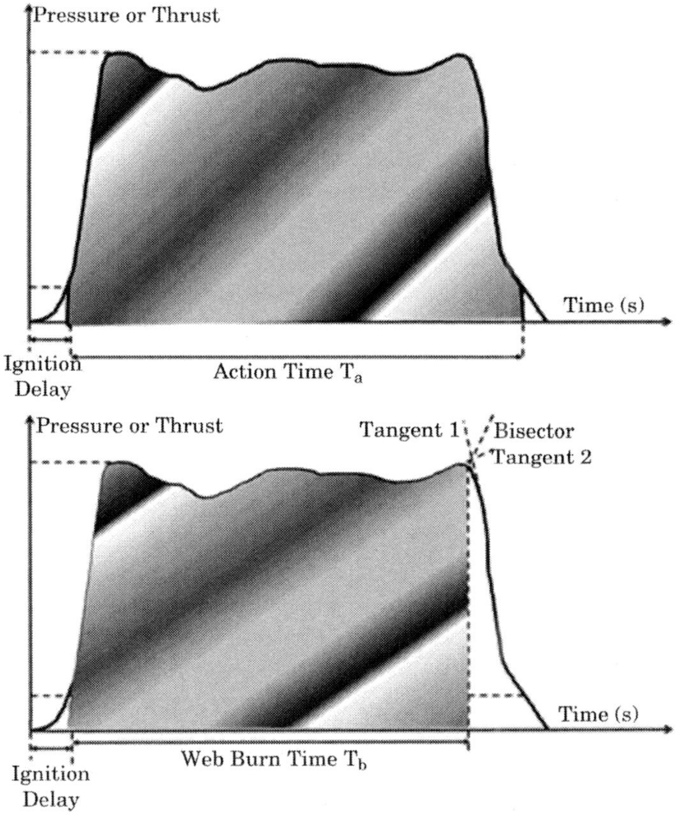

Fig. 121: Area under rocket firing curves

by action time. The value of characteristic velocity (C*) and specific impulse (Isp) is calculated by the formula given below:

C* = Throat area x Area under the pressure-time curve/Mass of the propellant

Isp = Area under the thrust-time curve/Mass of the propellant

For calculation based on web burn time, area under the curve as depicted in Fig. 121 (bottom) is considered. It is clear that this area is relatively smaller, as some portion of the tail-off portion is not included in this area. For average pressure or thrust during firing, this area is divided by web burn time (T_b). This average value is more than that calculated earlier using action time. This is because, in web burn time based calculation, area removed has relatively lower values of ballistic parameters. The value of C* is calculated as shown earlier and is on lower side than earlier calculated C* by action time based calculations. Same is the situation of specific impulse also. Thus various ballistic parameters are practically evaluated by static evaluation of rocket motors.

Auxiliary Rocket Parameters

Pressure and thrust are ballistic parameters, which are responsible for ballistic calculation and performance of a rocket. However, for system design validation, there are several other parameters, which are recorded. These are as follows:- Temperature, Strain, Sound, Vibration, Acceleration, Displacement. Each one has separate significance and use.

Temperature measurement of rocket is generally conducted to assess adequacy of thermal insulation system. Creation of hot spot is one of the important dangerous symptoms in any rocket firing. Since thermal insulation is an inert component, this must be kept bare minimum in a rocket to contain the payload. Additionally, lower insulation give extra space for the propellants inside rocket chamber for extra range and payload capabilities. Overall, temperature data are very useful for development of rockets. For cartridge loaded propellant configuration, the hot gases come in direct contact with thermal insulation layer of the rocket motor casing. This may result in more sever condition. In case-bonded propellant configurations, rocket motor casing is exposed to hot combustion gases at the far end of combustion and situation is not that severe. However, in both the cases, the nozzle throat is subjected to high temperature due to favorable condition of high heat transfer, high velocity of flow and high temperature of gases. Later on, temperature rise of rocket motor casing due to conduction from heated throat takes place. Although this is a slow process, but temperature rise of rocket motor casing in nozzle area after 100-120 seconds is observed to reach around 100°C, even for the most efficiently designed rocket motor. Although measurement of flame temperature is one area of work, but in rocket instrumentation, mostly, skin temperature of rocket motor casing at various locations is attempted. The joint area, head end dome and nozzle sections are most critical from temperature measurement point of view. For skin temperature measurement, thermocouples are generally employed. They are pasted over rocket motor casing for the measurement.

The thermocouple has two electrical conductors of different materials, which are connected at ends. One junction is placed on rocket motor, which is called measurement junction. The other junction, which is connected to a measurement device is called reference junction and the temperature of reference junction should be constant during the measurement. When there is a difference in temperature at both the junctions, a potential difference (in mill volts) is generated, which is proportional to the temperature difference. This effect is SEEBECK effect. Thermocouples are of two types:

- ***Base metal thermocouples***: They are made from copper, nickel, aluminium, iron, chromium, silicon etc and are given type name as E, J, K, N, T.
- ***Noble metal thermocouple***: They are made from platinum and rhodium and given type name as R, S, B.
- For rocket recording, K-type thermocouple is generally used.

Strain is measured by strain gauges and the gauges are generally positioned in rectangular rossete or triangular rossete fashion. Rectangular rossettes are positioned on rocket body where circumferential and longitudinal strains are measured. Triangular rossettes are generally positioned at head end dome or curved surfaces of nozzle to get three dimensional strains. Strain gauges area generally patched over the rocket motor casing directly using proper cementing compounds.

During rocket evaluation, high pressures are generated inside rocket motor chamber. This leads to radial dilatation of rocket motor casing. Since head end is prevented from expansion in longitudinal, the nozzle end moves backward due to longitudinal expansion under pressure. Displacement or dilatation of rocket motor casing is measured by linear variable differential transformer (LVDT). It is a common electromechanical transducer, which converts linear motion of an object into electrical signals. LVDT commercially measures length upto few centimeters. Physically, LVDT has a movable core of magnetic material and three coils making a static transformer. One of the three coils is a primary coil and other two are secondary coils. The moving element of LVDT is a core made of magnetically permeable material, which is free to move axially through the coils. This moving element is in contact with rocket motor and LVDT is oriented so that displacement moves the core in axial direction. For measurement of radial dilatation, axis of core is kept radial and for the axial expansion of rocket, core is placed co-axial with the rocket.

REVIEW QUESTIONS

1. Why is static firing of rockets essential before flight testing?
2. Compare vertical and horizontal static firings of rockets.
3. What is a proof motor for static firing of rockets?
4. What is a ballistic evaluation motor (BEM)?

5. Why burning rate of propellant measured through strand burner is not adequate?
6. Why burning profile in static firing of rockets vary from predictions?
7. What is significant of igniters during static firing of rockets?
8. Which properties of hardware are tested during static firings of rockets?
9. What are various failure modes in the static firings of rockets?
10. What are hazard divisions?
11. What are essential requirements in fixing trolley for the static firing of rockets?
12. Explain data acquisition for static firings of rockets.
13. What is a transducer? What are factors affecting selection of a transducer for static firing of rocket?
14. What is a strain gauge? What is its utility in static firing of rockets?
15. Explain various terms in the pressure-time profile obtained from the static firings of rockets?
16. What are parameters derived from the pressure-time and thrust-time profiles of the rocket?
17. What are auxiliary parameters measured in static firings of rockets? How is skin temperature measured on the rocket, during static firings?

12

Ballistics from Pressure-time Curves

CHAPTER SUMMARY

This chapter is first in the section of practical ballistic, giving importance of pressure-time profile of rocket motor. The key input for this chapter is pressure-time profile, obtained from the static firing of rockets and further calculations strategies are evolved. No prominence is given to the propellant configuration in this section. The obtained pressure-time profile is used for calculation of various ballistic parameters and burning rate of propellants for specific situations. One of the significant portion of the chapter is depicting the situation, where effect of changing propellant and hardware parameters is predicted on pressure-time profile using formulations derived in theoretical ballistics and prediction ballistic sections. Overall, this chapter covers the following:

- Importance of pressure-time profile
- Derivation of ballistic parameters from pressure-time profile
- Calculation of burning rate from pressure-time profile
- Analysis of progressive and regressive burning tubular propellant grain for the evaluation of burning rate
- Derivation of expression for burning area versus web consumed profile
- Effect of varying propellant and hardware parameters on pressure-time profile

INTRODUCTION

Practical evaluation of pressure time profile for a rocket motor during static firing gives much information about rocket ballistics. Maximum pressure is one of the important outputs from static evaluation of rockets. This is an input to the structural design of rocket motor. Ignition delay is a function of igniter, nozzle closure and propellant characteristics. Action time and web burn time gives idea about regularity of burning of propellant. The value of average pressure has notional value and *characteristic velocity (C*) is taken as performance index for the propellant combustion.* Although values of C* should be constant for same propellant compositions ideally, but practically, it is found to vary with chamber pressure (throat diameter, area ratio) for same propellant compositions.

ANALYSIS OF PRESSURE TIME PROFILE

Pressure time profile from static evaluation of rocket motor can be analyzed to give significant information about propellant behavior. Although propellants are evaluated for its physical properties (density) and burning rate (strand burner), but complete test of propellant is ensured in static firing and recorded pressure time profile gives valuable inputs about smoothness of propellant burning, achieved pressure levels, confirmation of burning rate, finalization of flight performance etc. during pressure time profile recording, input parameters known beforehand includes throat diameter and weight of the propellant. A typical pressure time profile recorded during static evaluation of a complex shaped propellant grain is shown as Fig. 122.

The complexity of propellant configuration is not depicted in the pressure time profile, which is relatively smooth and near neutral. The profile is recorded at a sampling rate of 1000 samples per seconds and pressure is recorded at every 0.001 seconds interval. The zero of the recording is coinciding with the time of giving firing current to the igniter of the rocket motor. The propellant weight in the firing is 39.425 kg and throat diameter of the nozzle is 42.92 mm. The throat area is equal to 1446.8 mm^2.

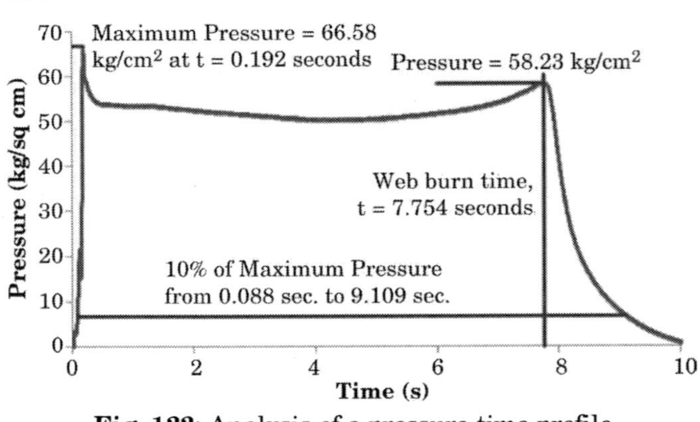

Fig. 122: Analysis of a pressure time profile

In the beginning of the firing, when pressure values are less, noise in recording is on higher side and oscillations are recorded. So, small values of pressure are generally not considered in the analysis of pressure time profile. However, the limit of small values cannot be treated as absolute. In case of rocket firing, some percentage of peak pressure is generally taken as limit for the calculation of these limiting pressures. This percentage of peak pressure decides ignition delay also. In case of firing curve depicted above, maximum pressure during firing is recorded as 66.58 kg/cm^2 and it lies at 0.192 seconds of recording.

For the calculation of performance parameter, 2% of the maximum pressure is considered as limit of valid pressure recording. The pressure limit is 1.33 kg/cm^2. In the pressure rising part of the pressure time profile this occurs at 0.03 seconds, which is taken as ignition delay for the firing. Ignition delay is a function of igniter quantity, igniter type, free volume available in the rocket motor, propellant characteristics, propellant configuration etc and is a transient function. Exact prediction of igniter delay is not possible and is practically evaluated in static firings. On the falling curve, this pressure level is achieved at 10 seconds. So, the burning time of the rocket propellant in the static firing is 9.97 seconds (10–0.03). Area under pressure time profile for this burning duration (9.97 sec) is obtained as 432.74 kg-s/cm^2. Division of area by burning time gives average pressure of the firing, which is 43.40 kg/cm^2. Characteristic velocity can also be calculated from the available data.

$$C^* = A_t g \int P dt / W = (1446.8/100) \times 9.81 \times 432.74/39.452 \text{ m/s} = 1557.86 \text{m/s}.$$

Although calculation has no flaw, but many a times, this 2% of maximum pressure gives much higher noise and 5% or 10% values of maximum pressure is considered relevant for calculation of performance parameters of propellant. Further analysis is considered with 10% of maximum pressure, as mentioned in previous chapter also 10% of maximum pressure is 6.658 kg/cm^2. It is observed on the rising curve at 0.088 seconds. So, ignition delay of the firing is 88 milliseconds. On the falling side, this occurs at 9.109 seconds. Burning time of the rocket is 9.021 seconds. It is obvious that with respect to calculation with 2% of maximum pressure, this total burning time is significantly small. Accordingly, area under pressure time profile is also small (now 429.67 kg-s/cm^2). The average pressure from this calculation is obtained as (429.67/9.021 =) 47.62 kg/cm^2. This value of average pressure is higher because, in case of this calculation that portion of pressure time profile is discarded from earlier calculation, which is operating at lower pressures (from 1.33 kg/cm^2 to 6.658 kg/cm^2). However, similar observation is not true for the calculation of C*, because lower area under pressure time profile in this case will result in lower C*.

$$C^*=A_t g\int Pdt/W=(1446.8/100)x9.81x429.67/39.452\ m/s=1546.82\ m/s.$$

Another feature in the pressure time profile is assessment of burning time for web consumption. In every pressure time profile, when main web of the propellant is consumed, fall in pressure values continues in monotonic way. In other words from end of burning when pressure time profile is traversed, the first peak gives idea about the web burn time. In case of the firing under consideration, the web burn time is 7.754 seconds and pressure at web burn time is 58.23 kg/cm^2. The portion from 7.754 seconds to 9.109 seconds is called tail-off period and is equal to (9.109s -7.754s =) 1.355 seconds. The tail-off is expressed as percentage of total burning time and in this case, it is (1.355/9.021 =) 15.0%. The burning time till web burn is (7.754 – 0.088 =) 7.666 seconds. Sometimes area under pressure time curve till web burn time is consider, which gives a higher value of average pressure, because again lower pressure part of pressure time profile is discarded. However, the value of area under pressure time profile till web burn time is always on lower side. The value of C* will be lower. For the firing curve, the value of òPdt till web burn time is 397.62 kg-s/cm^2.

$$\text{Average pressure} = 397.62/7.666\ kg/cm^2 = 51.87\ kg/cm^2$$

$$C^*=A_t g\int Pdt/W=(1446.8/100)x9.81x397.62/39.452\ m/s=1431.53m/s.$$

In the calculation given above, throat area remains constant for a rocket nozzle during firing except for some nozzle erosion resulting in nozzle enlargement at the end of firing. If this is taken constant, then another parameter is weight of the propellant. In fact, after consumption of main web of the propellant, sliver burning continues and this part of propellant is not assumed to be contributing significantly to the thrust generation. However, the weight of propellant till consumption of propellant web is to be ascertained and used for the calculation of C* and not the initial total weight of the propellant. One way to get this propellant weight is to calculate it from the geometry of the propellant. This weight will be less than total weight of the propellant used in static firing. In absence of any information about propellant configuration, this lower value of C* with reduced òPdt and original weight of propellant is not acceptable alternative. Web burn time is useful for the calculation of average pressure alone.

If weight of propellant till web burn is known beforehand as 36.5 kg, then C* can be calculated with this modified information.

$C^* = A_t g \int P dt / W = (1446.8/100) \times 9.81 \times 397.62/36.5$ m/s $= 1546.15$ m/s.

If web of the propellant configuration is known, then burning rate at average pressure may also be calculated. Suppose web of the propellant grain is measured before firing as 40 mm, burning rate of the propellant is = 40/7.666 mm/s = 5.2178 mm/s at 51.87 kg/cm^2.

For any static firing these information are derived from the static firing. As a summary, pressure time profile gives the following parameters for known throat diameter of nozzle and weight of the propellant:

- Maximum pressure
- Ignition delay
- Total burn time
- Area under pressure time profile
- Average pressure
- Characteristic velocity
- Web burn time
- Tail-off duration
- Average pressure till web burn
- Burning rate (known web)

One peculiarity for pressure time profile is observed, when pressures at head end or igniter end are compared. Pressure at the igniter end or head end (fore end or away from nozzle end or vented end) is always on higher side than pressure recorded at nozzle end side (aft end). It is also observed that difference in pressure reduces as burning progresses. In fact head end pressure represents stagnation pressure and nozzle end pressure has a significant portion of pressure energy converted into kinetic energy (flow velocity). This reduces pressure at the nozzle end side. For many rockets, nozzle end pressures are measured additionally, to ascertain longitudinal compression of propellant grains during rocket operation. These differential pressures are critical for long tubular grains, which may crumble at the end of burning due to small web left. These types of pressure differential are predominantly in those rocket motors where high volumetric loadings are applied. High volumetric loading leads to less port volume and flow passage may not be sufficient in the beginning for smooth flow of combustion gases. However, as propellant gets consumed, more and more port volume is made available and flow of gases becomes smooth. In the initial phases of burning, accumulation of combustion gases may result in enhancement in head end pressure.

Similar pressure differential is envisaged in those propellant grains, where combustion gases are generated at two ports and flow also occurs at two port. A typical situation may occur for uninhibited cartridge loaded tubular propellant, which burns at inner port as well as outer port. If pressure difference exists at inner and outer port, propellant will be subjected to compressive radial pressure. This may result in crumbling of propellant, if propellant strength is insufficient. There is no instrumentation mechanism to measure this pressure differential.

However, location of pressure sensor may be responsible for variation in calculated propellant performance. Since calculated characteristic velocity is dependent on area under pressure time profile, smaller area under pressure-time profile is obtained, if pressure sensor is mounted at nozzle of the rocket motor (than when it is mounted on head end of the rocket motor). This may result in reduced characteristic velocity. Same will be the case with average pressure.

Calculation of Burning Rate

Calculation of burning rate of propellants from static evaluation of rockets is one of the significant practical propellant characterization approaches. One of the simplest methods for such activity is firing the propellant in neutral burning configuration to get burning time. Burning rate can be achieved by division of maximum web available with the burning time. Since pressure during neutral firing is almost constant, the value of burning rate can be obtained at single pressure by single firing. So minimum two firings are needed to get burning rate law ($r = aP^n$). For completeness, since two parameter burning rate law is to be derived, minimum three firings are conducted. To get burning rate at many pressures and get burning rate law from single firing, firing of propellant compositions in simple tubular propellant configuration is needed. Burning rate law and burning rates at different covered pressure ranges are obtained directly from the firing curves. However, this is a mathematically intensive exercise. The method of calculation is explained in Fig. 123.

In the pressure time profile, digital data of pressure at different time steps are known. In step 1 (Fig. 123), the acquired data must be analyzed for smoothness and nature of the firing curve. Additionally throat diameter and propellant weight during firing must be known. It is also needed that propellant dimensions are known and geometric relations for calculation of burning area and volume can be represented in terms of propellant dimensions by simple correlations. If nature of the curve is as expected and non-anomalous, then area under pressure time profile

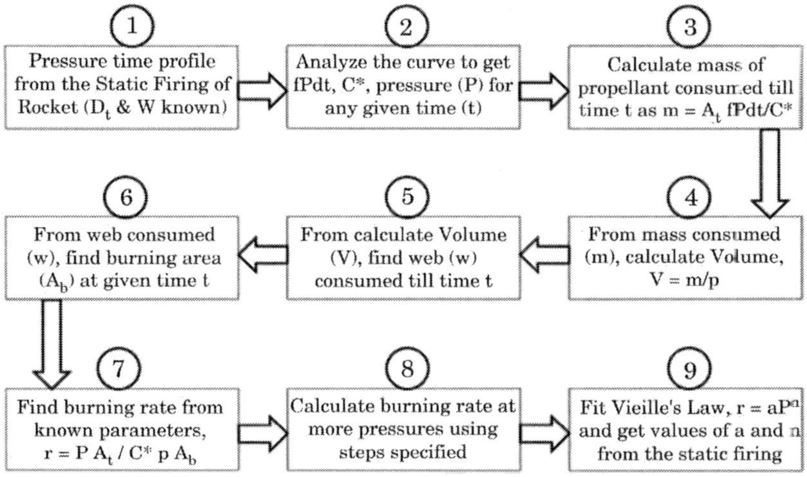

Fig. 123: Flow chart for calculation of burning rate law from pressure time profile

is calculated (step 2) and C* is estimated ($A_t\int Pdt/W$). C* is utilized again and again for calculation. Now a particular time (t) is considered and value of pressure (P) at that time is known. Calculation for finding area under pressure time profile till selected time (t) can be done using similar relation as used for the calculation of C* (step 3). Only difference remains that now the calculated value of C* is utilized and mass consumed till selected time is calculated. From mass consumed, volume of propellant consumed can be calculated, because density of propellant is known before firing (step 4). From propellant geometry, web versus volume consumed data can be generated by simple numerical means. However, reverse of this calculation is needed, because now with known volume web consumed is to be decided (step 5). This requires simple propellant geometry and this step needs help of geometry. The geometrical calculation is also needed to get burning area from the web consumed data but this is a relatively simple exercise (step 6). From burning area, burning rate can be calculated by mass flow rate balance, as discussed in theoretical ballistic section. This gives burning rate for a given pressure (step 7). A new time is selected and steps from 3 to 7 are repeated. At least 10 values of pressure and corresponding burning rate data is calculated (step 8). Power law of burning rate ($r = aP^n$) is fitted to pressure and burning rate data (step 9). This gives burning rate law for the pressure range covered by the static firing.

However, this entire calculation is valid for those firings, where burning profile is either progressive or regressive. For neutral or near neutral firing, for almost same pressures different burning rates may be obtained, which leads to erroneous results. So, it is cautioned that the calculation steps depicted in Fig. 123 should be utilized for progressive or regressive firings only. Obviously solid sustainer propellant configuration may not be suitable for such calculations as it gives neutral burning profile. However, attempt can be made with uninhibited solid sustainer, which is burning from all the sides. But again this is not a good option for calculation. Another constraint is calculation of web consumed from volume of propellant consumed. This needs simple propellant geometry. Obviously, tubular propellant configuration is very suitable. In one firing pressure range covered depicts validity of pressure range for the derived burning rate law. So number of firing may be restricted, if wide pressure levels are covered in single firing.

For ease of calculation, the simplest propellant configuration is radial outward burning tubular propellant grain characterized by outer diameter (D), inner diameter (d) and length (L). The web consumed at any time is given by w. Outer diameter and length is constant during propellant combustion in the rocket motor and only enlargement of inner diameter takes place. Length of such propellant configuration does not change during static firing. The volume consumed (V) at any web is given by the following expression. However, as depicted in step 5, for known volume, web consumed is to be calculated. The reverse calculation is carried out and web consumed is derived in terms of volume and propellant dimensions. The outer diameter has no role in such calculations. Burning area (A_b) for known web consumed can be calculated easily.

$$V = \pi[D^2 - d^2]L/4 - \pi[D^2 - (d+2w)^2]L/4$$

$$\text{Or, } w^2 + dw - V/\pi L = 0$$

$$\text{So, } w = [\{\sqrt{(d^2 + 4V/\pi L)}\} - d] / 2$$

$$\text{Burning area, } A_b = \pi[d+2w]L$$

The ease of calculation is clear from the simple derived expressions. It is clear that inhibition at outer diameter and both ends of a tubular propellant grain make it an ideal choice for the calculation of burning rate. The combustion and initiation of propellant is very smooth.

A tubular propellant grain (OD: 111 mm, ID: 71 mm and Length: 202 mm) is evaluated in progressive mode (lateral and end faces inhibited). Propellant weight is 2.096 kg. The propellant grain is evaluated with throat diameter of 15.4 mm and pressure time profile during firing is recorded. The sampling rate is 2000 samples per seconds and pressure range covered is from 55 kg/cm^2 to around 126 kg/cm^2. Initial ignition delay and transient combustion exists but has little significance for the calculation. The obtained pressure time profile and certain pressure values at isolated points are depicted in Figure 124 Corresponding time is also indicated. Area under pressure time profile is evaluated as 185.665 kg-s/cm^2.

Volume of the propellant=π (111^2–71^2)x202/4000 cm^3=1154.975 cm^3

Density of the propellant = 2096/1154.975 g/cm^3=1.815 g/cm^3

C*=A_tg ∫P dt/W=(πx15.4^2/4)x98100x185.665/2.096=1618.6m/s

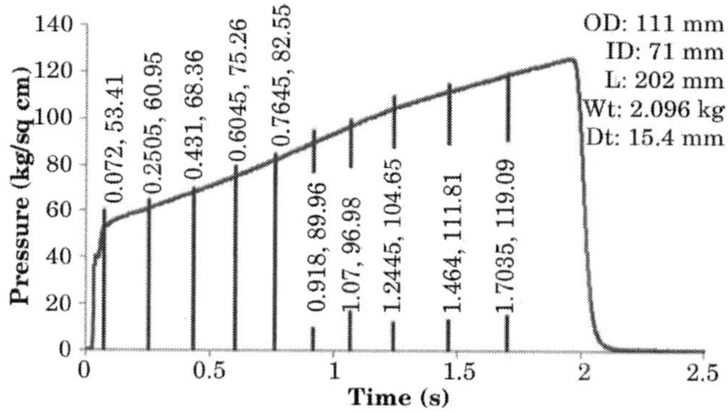

Fig. 124: Pressure time profile in progressive mode firing

For calculation of burning rate, data at 0.431 seconds is taken, where pressure is 68.36 kg/cm^2. The area under pressure time profile till this point is calculated as 23.893 kg-s/cm^2.

Mass of propellant consumed=2.096x23.893/185.665 kg=0.2697 kg.

Volume of propellant consumed = 269.7/1.815 cm^3 = 148.612 cm^3

Web of propellant consumed till 0.431 seconds, w = [{√(71^2 + {4 x 148612/ π x 202})} - 71]/2 = 3.1583 mm

Burning area of propellant at 0.431 seconds, A_b = π x (71 + 2 x 3.1583) x 202 = 49065.24 mm^2

Burning rate of propellant = P A_t g / C*r A_b = 68.36 x (p x 15.4^2/4) x 98100/(1618.6 x 1.815x49065.24) mm/s = 8.667 mm/s

Similar calculation is repeated for each of time step specified in the Fig. 123 and results are tabulated as Table 15. Now pressure and burning rate data is plotted and best fit curve is generated to obtain burning rate laws in different forms for the propellant.

Burning rate, r (mm/s) = 0.8428 $P^{0.5499}$

Burning rate, r (mm/s) = 4.4247 + 0.0612 x P

Burning rate, r (mm/s) = 5.1329 ln(P)–13.014, where P in kg/cm^2.

Table 15: Calculation of burning rate from progressive mode firing

Time (s)	*Pressure (kg/ cm^2)*	*Int Pdt (kg-s/ cm^2)*	*Weight consumed (kg)*	*Volume consumed (mm^3)*	*web (mm)*	*Burn area (mm^2)*	*BR (mm/s)*
0.072	53.41	1.9139	0.0216	11906.3	0.263	45390.87	7.320
0.2505	60.95	12.2004	0.1377	75896.1	1.646	47146.19	8.042
0.431	68.36	23.8930	0.2697	148632.4	3.158	49065.24	8.667
0.6045	75.26	36.3667	0.4105	226228.3	4.708	51033.01	9.174
0.7645	82.55	49.0011	0.5531	304823.9	6.220	52951.63	9.697
0.918	89.96	62.2399	0.7026	387178.8	7.747	54890.12	10.195
1.07	96.98	76.4541	0.8631	475601.7	9.329	56897.97	10.603
1.2445	104.65	94.0895	1.0621	585307.2	11.218	59294.65	10.979
1.464	111.81	117.889	1.3308	733362.8	13.651	62383.34	11.149
1.7035	119.09	145.562	1.6432	905510.1	16.337	65792.56	11.260

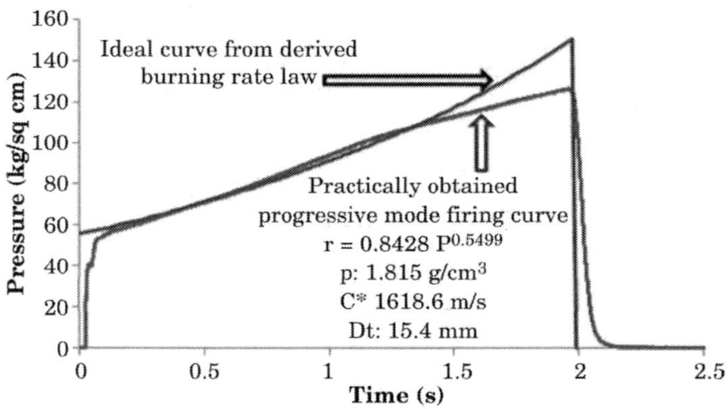

Fig. 125: Superimposed ideal and practical firing curves

With power law (first expression) of burning rate, prediction is made for the propellant configuration and superimposed over the practically obtained pressure time profile as Fig. 125. It is obvious that firing curve matches with prediction in the initial part, confirming correctness of derived burning rate law. However, after around 1.5 seconds, some deviations from ideal prediction are observed in the pressure time profile. In fact, the propellant is probably having a bi-plateau behavior and it exhibits two burning rate laws – one valid for low pressures below around 100 kg/cm^2 and another valid for high pressures. From the curve, it is observed that at high pressures, this particular propellant exhibits lower burning rate index (n). For such propellants two different burning rates can be derived from the firing curve. However, the approach for calculation of burning rate law from actual firing pressure time data is valid for any propellant formulation.

Table 16: Generation of burning rate data at higher pressures

Time (s)	*Pressure (kg/cm²)*	*Int Pdt (kg-s/cm²)*	*Weight consumed (kg)*	*Volume consumed (mm³)*	*web (mm)*	*Burn area (mm²)*	*BR (mm/s)*
1.2	102.99	89.4698	1.0100	556569.3	10.730	58676.29	10.918
1.3	106.69	99.9541	1.1283	621789.2	11.829	60070.48	11.049
1.4	110.01	110.7877	1.2506	689182.2	12.938	61477.91	11.132
1.5	113.08	121.9392	1.3766	758552.5	14.054	62893.74	11.185
1.6	116.15	133.3890	1.5058	829779.3	15.173	64315.05	11.234
1.7	119.22	145.1465	1.6386	902919.4	16.298	65742.57	11.280
1.8	122.16	157.2143	1.7748	977990.4	17.428	67176.23	11.312

For the given curve, pressure value of 104 kg/cm² corresponding to 1.2445 seconds is taken as cut-off point above which burning rate of the propellant is following different burning rate law. The calculation repeated and calculated burning rates at higher than 104 kg/cm² pressure or time more than 1.2 seconds is shown as Table 16.

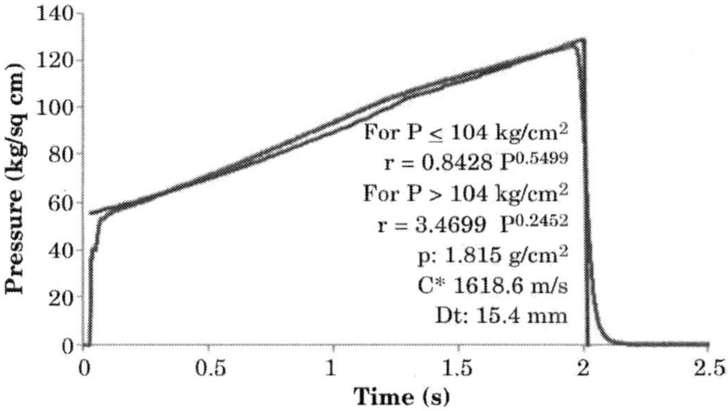

Fig. 126: Superimposition of prediction with two burning rate laws

From pressure and burning rate values, *burning rate index is obtained as 0.2452*. As variation of burning rate with pressure should not exhibit any break, burning rate at the transition by both the burning rates should be same. Using this assumption, burning rate coefficient (a) is obtained.

Burning rate at 104 kg/cm² = $0.8428 \times 104^{0.5499}$

$= 10.836$ mm/s $= a \times 104^{0.2452}$

Burning rate coefficient (a) for high pressures = 3.4699

With twin burning rate law, derived from the firing curve, one valid for low pressures (≤ 104 kg/cm²) and other valid for high pressure, the pressure time profile is predicted and superimposed over the firing curve as Fig. 126. Perfect matching of pressure time profiles indicate correctness of calculation. Progressive burning tubular propellant grains are used for practical determination of burning rate of propellants, by static evaluation.

The summary of steps is given below for assessment of burning rate law from progressive pressure time profiles (from externally inhibited tubular propellant configuration).

1. From pressure time profile get C*.
2. Consider any intermediate time (t) and corresponding pressure (p).
3. Find area under pressure time profile till time t and call it 'H'.
4. Find mass of propellant consumed as time t as W
5. Find port volume as $W_p = \pi d^2 L\rho/4$
6. Find burning rate as $r = (PA_t/\rho A_{bi}C^*)/\surd[1+(\Delta W/W_p)]$
7. Repeat step 2 to 6 for other pressures and get set of burning rate and pressure.
8. Get burning rate law from set of burning rate and pressure.

Although mathematical ease favours above mentioned configuration but the progressive burning profile depicted by radial outward burning tubular propellant configuration needs inhibition of propellant grain at the lateral surfaces and at the ends. This additional step of inhibition delays evaluation and this additional step may be avoided by taking an uninhibited tubular propellant grain, which shows regressive burning profile. The uninhibited tubular propellant grain burns at all the surfaces. Outer diameter and length of such propellant configuration reduces during firing and inner diameter enlarges. The propellant grain has three surfaces. First is inner surface, which is similar to progressive burning configuration and it increases during combustion of propellant. Second surface is external lateral surface, whose area reduces continuously. The third surface is two end faces, which reduces continuously. The pressure range covered in this configuration is on lower side than progressive burning configuration and to cover a pressure range, more number of static firings is to be conducted. This type of propellant configuration has maximum web equal to half the maximum web available for progressive burning grain. This reduces burning time for same burning rate propellant. However, for qualification purpose, this is also adopted in rocket firing. Expression for the calculation of volume and burning area for the propellant configuration at different web burnt is derived. Calculation of web burnt from volume is also derived to fulfill step 5 of Fig. 123. For regressive burning profile (D-d+2L) is present in expressions for volume and burning area. So, it is represented by another variable 'H'.

Parameter, $H = D - d + 2L$

$V = \pi\ [D^2\text{-}d^2]L/4 - \ [(D\text{-}2w)^2\text{-}(d+2w)^2](L\text{-}2W)/4$

Or, $4w^2 - (D\text{-}d+2L)w + 2V/\pi(D+d) = 0$

So, $w = [(D\text{-}d+2L)\text{-}\sqrt{\{(D\text{-}d+2L)^2 - 32V/\pi(D+d)\}}] / 8$

$= [H - \sqrt{\{H^2 - 32\ V / \pi\ (D+d)\ \}}\ ] / 8$

Burning area, A_b = outer surface + inner surface + end face area

$= \pi[D\text{-}2w](L\text{-}2w)+ \pi[d+2w](L\text{-}2w)+ 2 \times (\pi/4)[(D\text{-}2w)^2-(d+2w)^2]$

$= (\pi/2)(D+d)(D\text{-}d+2L\text{-}8w) = (\pi/2)\ (D+d)\ (H\text{-}8w)$

Since pressure is derived from burning area of the propellant configuration, an insight into the range of pressure covered in firing can be made from the expressions of burning area. In progressive burning profile, increase in burning area is because of continuous increase in inner diameter, where as for regressive configuration, the reduction in burning area is due to reduction in length and end faces. If effect of length alone is considered, the percentage change in burning area is relatively small for regressive burning tubular propellant configuration. Since length is numerically a bigger quantity than inner diameter, percentage change in burning area in case of inner diameter (numerically lower value) dependent burning area is higher. For example if maximum web is 30 mm, inner diameter is 50 mm and length is 200 mm, percentage change in burning area in progressive firing is 60% (30 x 100/50), while for regressive firing percentage change is only 25% (10 x 100/200). So regressive firing covers less pressure range for same propellant dimensions in less total time. In addition to this, maximum web available for regressive burning profile (D/4-d/4) is half the maximum web available for progressive burning profile (D/2-d/2). This also reduces not only pressure range covered, but the burning time is also on lower side for the regressive burning profile. As compared to progressive burning propellant configuration, following features of regressive burning tubular propellant grain can be enlisted.

- Regressive burning tubular propellant grain does not need additional inhibition step. This makes processing of propellant simpler.
- There is no inhibition piece to obstruct the flow of combustion gases in regressive burning propellant configuration. Inhibition pieces may cause disturbance in recorded pressure time profile and regularity of combustion is also difficult to ensure in progressive burning configuration.
- Regressive burning grain has lower web (half).
- Burning time of regressive burning grain for same propellant is less (1/2 of progressive burning grain for same average pressure).

- Regressive burning propellant configuration has higher initial burning surface area and also higher pressure in the beginning.
- For maintaining same initial pressure during firing in identical propellant dimension and burning rate, higher throat diameter is needed for regressive propellant configuration.
- Range of pressure covered in regressive burning tubular propellant configuration is less.
- Mathematical formula for calculation of web from volume consumed is more complex for regressive propellant configuration.
- Regressive nature is added due to reduction in length of propellant during combustion, where as progressive nature is because of increase in inner diameter of propellant during combustion.
- Initiation is smooth in progressive burning propellant configuration where only inner diameter is to be initiated. In regressive burning profile, initiation at end faces and two lateral surfaces may result in higher ignition delay and more deviation in pressure time profile from prediction.

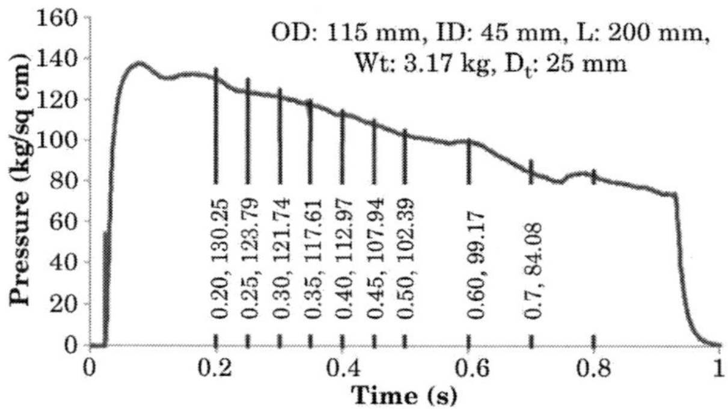

Fig. 127: Pressure time profile from static firing of regressive tubular propellant

Although calculation steps are similar as explained for progressive burning tubular propellant formulation, but the same is depicted again using appropriate inputs. For regressive mode of firing a tubular uninhibited propellant grain (OD: 115 mm, ID: 45 mm, Length: 200 mm) weighing 3.17 kg is evaluated in static mode in a rocket motor with throat diameter of 25 mm. The pressure time profile is recorded with sampling rate of 2000 samples per second giving pressure data at every 0.0005 seconds interval. The peak pressure is recorded as 137.59 kg/cm^2 and the pressure time profile is depicted as Fig. 127. The burning time is of the order of around 1 second.

For the estimation of burning rate at different pressures, realized in the static firing, constant interval of time is considered. In earlier case, arbitrarily selected time steps were considered to show strength of the calculation strategy. In this case, time step of 0.05 seconds in the beginning and later 0.1 seconds time step is considered and pressure is recorded in Fig. 127 at these time-steps. The calculation has same steps as shown earlier for progressive burning propellant configuration, but the volume and burning area expressions are different. Area under pressure time profile is calculated as 96.038 kg-s/cm^2.

Parameters, H = D-d+2L = 115-45+2x200 mm = 470 mm

Volume of the propellant=π (115^2–45^2)x200/4000 cm^3=1759.29 cm^3

Density of the propellant = 3170/1759.29 g/cm^3 = 1.802 g/cm^3

$C^* = A_t g \int P\,dt / W = (\pi x 25^2/4) \times 98100 \times 96.038/3.17 = 1458.9$ m/s

For calculation of burning rate, data at 0.20 seconds is taken, where pressure is 130.25 kg/cm^2. The area under pressure time profile till this point is calculated as 22.25 kg-s/cm^2.

Mass of propellant consumed = 3.17x22.25/96.038 kg=0.7345 kg.

Volume of propellant consumed = 734.5 / 1.802 cm^3 = 407.64 cm^3

Web of propellant consumed till 0.20 seconds, w = [470 - √{470^2 – (32 x 407640/ ∂ x | | 115+45 | |)}] / 8 = 3.558 mm

Burning area of propellant at 0.20 seconds, Ab = (π/2) x (115+45) x (470-8x3.558) = 110968.6 mm^2

Burning rate of propellant = $P A_t g / C^* \rho A_b$ = 130.25 x (π x 25^2/4) x 98100/(1458.9 x 1.802 x 110968.6) mm/s = 21.501 mm/s.

The calculation step is repeated for all the time steps depicted in Fig. 127 and the various relevant parameters are shown in Table 17. Pressure and burning rate data is compiled and fitted into various laws. From best fit curve, following burning rate laws are derived.

Burning rate, r (mm/s) = 1.746 $P^{0.5166}$

Burning rate, r (mm/s) = 9.3394 + 0.0947 x P

Burning rate, r (mm/s) = 9.842 ln(P) – 26.411, where P in kg/cm^2.

Using power law (first law) of burning rate law, prediction is made for the propellant configuration and superimposed over the regressive pressure time profile to know the validity, correctness and consistency of the result. The superimposed curves are shown as Fig. 128. For perfect matching an ignition delay of 30 milliseconds is used. This ignition delay cannot be derived unless rocket motor is fired and is a purely practical parameter. The prediction is relatively smooth and it is interesting to note that the curves matches well in spite of the fact that the pressure

Table 17: Calculation of burning rate for regressive burning tubular propellant configuration

Time (s)	*Pressure (kg/ cm²)*	*Pdt (kg-s/ cm²)*	*Weight burnt (kg)*	*Volume burnt (mm³)*	*Web burnt (mm)*	*Burning area (mm²)*	*Burning rate (mm/s)*
0.2	130.25	22.25	0.7345	407640.1	3.56	110968.6	21.50
0.25	123.80	28.57	0.9431	523433.0	4.61	108850.4	20.83
0.3	121.74	34.71	1.1458	635921.1	5.65	106752.3	20.89
0.35	117.61	40.69	1.3433	745516.4	6.69	104667.8	20.58
0.4	112.97	46.45	1.5333	850982.4	7.71	102621.9	20.16
0.45	107.94	51.99	1.7160	952367.9	8.70	100615.9	19.65
0.5	102.39	57.24	1.8895	1048669.0	9.67	98672.7	19.01
0.6	99.17	67.24	2.2194	1231731.0	11.56	94869.2	19.15
0.7	84.08	76.45	2.5234	1400480.0	13.38	91222.7	16.88
0.8	82.40	84.66	2.7943	1550816.0	15.06	87846.7	17.18

time profile during static firing is not very smooth and oscillations, intermittent nozzle obstructions or chuffing is present. However, the matching prediction indicates correctness of the derived burning rate law. There is no bi-plateau behavior of this propellant. It is cautioned here that deduction like progressive burning profile gives bi-plateau behavior or regressive burning profile is unsmooth is extraneous and should not be used. This is just a mere coincidence that selected static firing curves have these behaviours for certain type of tubular propellant configuration. Generalization may lead to unwanted results. So, any of the above mentioned two configurations (progressive or regressive) can be utilized for determination of burning rate law of propellant from practical firing data.

The summary of steps is given below for assessment of burning rate law from progressive pressure time profiles (from externally inhibited tubular propellant configuration).

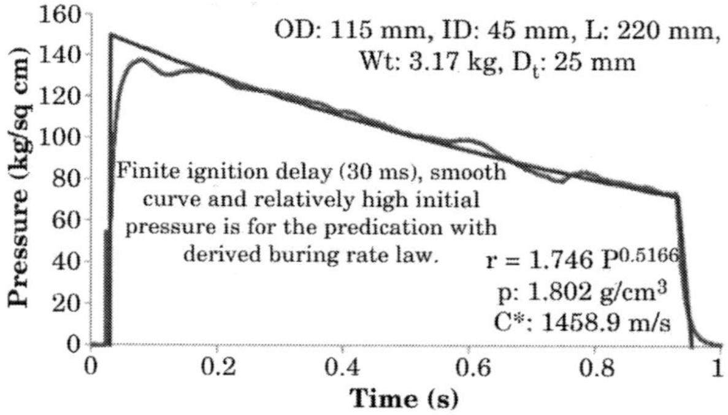

Fig. 128: Superimposed pressure time profile for regressive propellant configuration

1. From pressure time profile get C*.
2. Consider any intermediate time (t) and corresponding pressure (p).
3. Find area under pressure time profile till time t and call it 'H'.
4. Find mass of propellant consumed as time t as W
5. Define B=(ΔW/W)*[{(D-d)/4}+(L/2)] and C=(ΔW/W)*{(D-d)/4}*(L/2)
6. Find burning rate as r = (PC/H)/√(B² – 4C).
7. Repeat from step 2 to 6 for other pressures and get set of burning rate and pressure.
8. Get burning rate law for the set of burning rate and pressure.

Prediction of Burning Area

Pressure time profile obtained during static firing of rockets is a prime source of calculating burning area variation during combustion of propellants. If physical, ballistic and motor parameters of rocket during static evaluation are known, it is easy to calculate variation of burning area of propellant surface with respect to web consumed. This type of calculation is reverse of performance prediction calculation and is very helpful, if different propellants are fired in same configuration. As far as ballistic calculation is concerned, calculation identifies propellants by its parameters like burning rate (r), density (ρ), characteristic velocity (C*) and operating conditions are identified by throat area of the nozzle (A_t) and operating temperature of static evaluation. The burning area for any pressure is given by equation 1 in which burning rate is given by $r = aP^n$.

$$\text{Eqn 1: } A_b = \frac{P^{1-n} A_t}{a\rho C^*}$$

For the given input parameters, the burning rate coefficient (a) is calculated as 2.28156, when burning rate is in mm/s and pressure is expressed in kg/cm². From the pressure time profile, burning rate for each pressure value can be calculated by the equation given above. The time axis is divided into smaller time steps for ease of calculation. Multiplication of time interval and average burning rate during that time step gives web consumed. In the form of time marching solution, the plot of variation of burning area with respect to time can be obtained. For analysis a typical pressure time profile is considered as Fig. 129 (y-axis is pressure in kg/cm² and x-axis is time in seconds) for which known performance or input parameters are given in Table 18. The value of burning rate coefficient (a) is calculated from the burning rate data of Table 18.

Using equation 1 above, the burning surface area of the propellant configuration is calculated. Variation of burning area with time is shown

Table 18: Input parameters for pressure time profile

Parameter	*Symbol*	*Unit*	*Value*
Density	r	Kg/cm^3	1680
Characteristic velocity	C*	m/s	1526
Burning rate @104.8 kg/cm^2	r	mm/s	7.3
Burning rate index	n	–	0.25
Throat diameter	D_t	mm	44.5
Propellant weight	W	kg	109.4

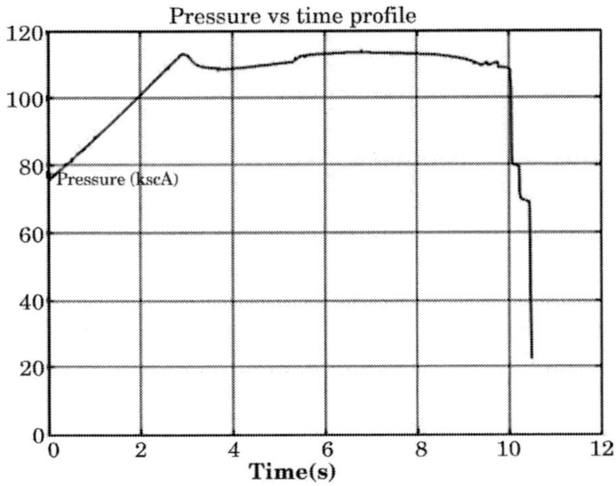

Fig. 129: Pressure time profile from static firing of rocket

in Fig. 130. The nature of curve is same as pressure-time profile, but the relative variations in values are less. As pressures are known, corresponding burning rates can be calculated and from time, web consumed can be arrived at. The burning area with respect to web consumed is plotted as Fig. 131.

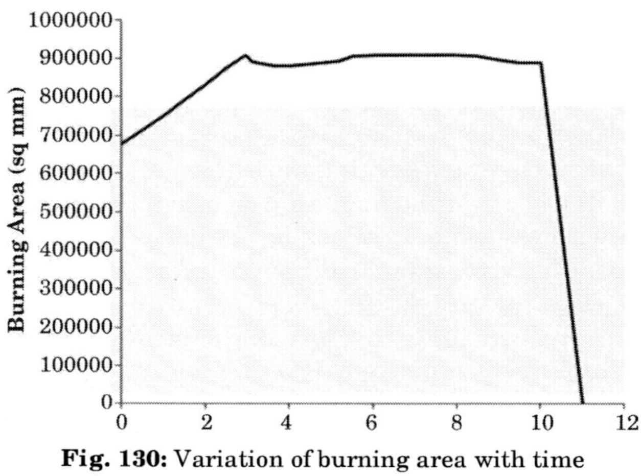

Fig. 130: Variation of burning area with time

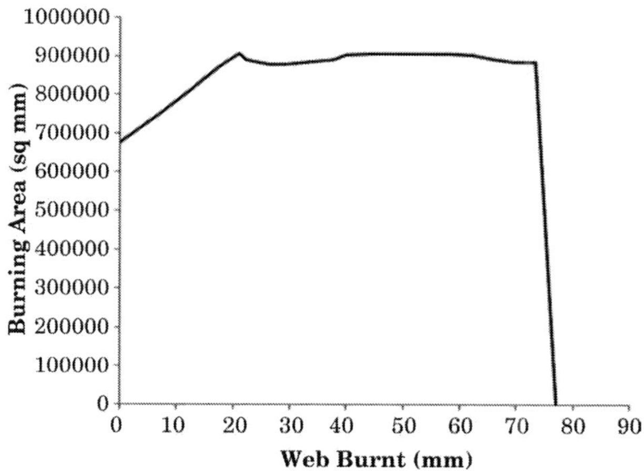

Fig. 131: Variation for burning area with web consumed for the static firing

The maximum web of the propellant is obtained as 76.9 mm. Area under burning area and web consumed profile is calculated as 65.12118628 m^3. This should represent volume of the propellant. When this is multiplied with density of the propellant (from Table 18), weight of the propellant can be calculated. The weight of propellant is calculated to be 109.4 kg, which is comparable to the input parameters given in Table 18. This matching weight of propellant indicates correctness of the calculation.

Now, each of the input parameter is changed one by one and the variation in pressure time profile is depicted in Fig. 132, keeping propellant configuration (burning area versus web consumed profile) same. First of all, density is changed from 1680 kg/m^3 to 1760 kg/m^3. This gives higher pressure and corresponding burning time is reduced.

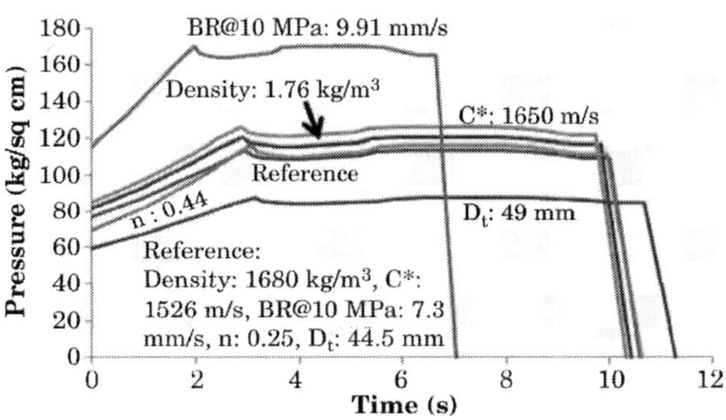

Fig. 132: Predicted pressure time profiles for same configuration with different compositions

Peak pressure is increased from 113.5 kg/cm^2 to 120.76 kg/cm^2. The burning time is reduced from 10.59 seconds to 10.43 seconds. Then the value of C* is changed from 1526 m/s to 1650 m/s and peak pressure rose to 125.96 kg/cm^2. Corresponding reduction in burning time of propellant is also clear. Similarly, values of throat diameter, burning rate and burning rate index are changed to 409 mm, 9.91 mm/s and 0.44, respectively. Except throat diameter, all other enhancements in values of parameters results in increased pressure and reduced burning time with respect to reference configuration. Increase in throat diameter lowered the peak pressure and enhanced burning time significantly.

The most peculiar curve is obtained for enhanced value of burning rate index (n). The curve with higher burning rate index started with a lower pressure. This is because higher burning rate index gives lower burning rate at less than specified pressure. This line needs explanation. In this case, burning rate is specified at 104.8 kg/cm^2. So, for all pressures less than this reference pressure (of burning rate), the calculated pressure with higher burning rate index is lower than reference. If burning rate at 104.8 kg/cm^2 is 7.3 mm/s, then with n = 0.25, burning rate at 90 kg/cm^2 is calculated as 7.027 mm/s. For same input, if 'n' is changed to 0.44, burning rate at 90 kg/cm^2 is 6.827 mm/s. However, at higher pressure pressures, reverse trend is observed. The process is repeated for 120 kg/cm^2 pressure. With n= 0.25, burning rate is 7.55 mm/s, while for n = 0.44, burning rate is 7.75 mm/s. High burning rate leads to higher rocket chamber pressure. So, for all pressures less than 104.8 kg/cm2 (pressure at which burning rate is specified), higher 'n' leads to lower pressure and vice-versa. So, pressure time profile at n = 0.44 crosses reference pressure time profile (n = 0.25) at 104.8 kg/cm^2 pressure. All curves are depicted in Fig. 132.

Suppose, new propellant composition (Density: 1760 kg/m^3, C* : 1600 m/s, BR@ 7 MPa: 8.3 mm/s, n: 0.44) is to be evaluated in same configuration (geometry), which has given pressure time profile as given in Fig. 129. It is asked to calculate throat diameter, which restricts peak pressure to same values (113.5 kg/cm^2) as the reference configuration. The burning rate coefficient (a) is calculated for the proposed composition as 1.28, when burning rate is in mm/s and pressure is expressed in kg/cm^2. Throat diameter is calculated for the using equation 1 for the burning area. As maximum pressure is attained at maximum burning area, it is apt to equate burning area for both the configuration to arrive at the throat diameter.

$$113.5^{(1-0.25)} \text{x} 44.5^2/(2.28156 \text{x} 1680 \text{x} 1526) = 113.5^{(1-0.44)} \text{x} D_t^2/(1.28 \text{x} 1760 \text{x} 1600)$$

$$113.5^{(1-0.44)} x D_t^2/(1.28 x 1760 x 1600) = 0.0117725$$
$$D_t = 54.759323 \text{ mm}$$

In the calculation, it is clear that value of throat diameter is dependent on value of equality of pressure. If this equality is run at 100 kg/cm^2, the calculated throat diameter is 54.1 mm. This indicates that lowering of equal pressure for two compositions, leads to lowering of throat diameter also. Contrary to this, lower values of burning rate coefficient, burning rate index, density and C* give higher value of throat diameter needed for same pressure values. Most of the time, in practical situation, peak pressure is kept same because it is decided by hardware design consideration and any enhancement in peak pressure may reduce margin of safety available with the rocket motor hardware. Since rocket hardware is generally fabricated at very low margin of safety, reduction in margin of safety is not permitted. Lowering of peak pressure is always attempted. However, on other side, pressure should not be so low that propellant combustion becomes non-smooth.

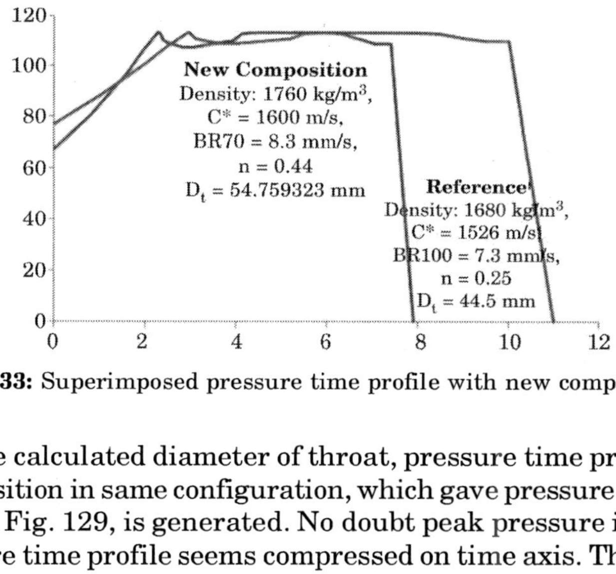

Fig. 133: Superimposed pressure time profile with new composition

With the calculated diameter of throat, pressure time profile for the new composition in same configuration, which gave pressure time profile depicted in Fig. 129, is generated. No doubt peak pressure is same, but the pressure time profile seems compressed on time axis. The reference pressure time profile is also superimposed and both curves are shown as Fig. 133.

Burning area versus web burnt profile is same for both the curves, and volume of propellant is also same for both the curves, but propellant mass in both the situations is different (different density). Since new composition has higher density, the mass of propellant with new composition in same configuration is higher (114.6 kg against 109.4 kg in reference composition). So, with new composition, propellant weight is high. New propellant has more burning rate, so burning time of propellant is less, which is depicted in the superimposed curve. The new composition is more energetic, as depicted by its higher characteristic velocity (C* = 1600 m/s as against C* = 1526 m/s). This higher energy generates more pressure and enlargement of throat is necessary. One of the important points to note in the superimposed curves is that, area under pressure time profile is less for the new configuration. In fact, numerically, area is reduced to around 71.39% of the area under pressure time profile of the reference curve. This is natural outcome because of higher mass flow rate (higher propellant mass is to be discharged in less time at high burning rate) and higher energy of the propellant. In other words, throat enlargement takes care of higher mass flow rate, higher energy and lower area under pressure time profile. In general, it seems contradictory to have higher C* but lower area under pressure time profile, because generally, C* is considered proportional to the area under pressure time profile. Since maximum pressure is maintained same, and pressure time profile has same nature, the average pressure for the two situations is almost equal. This leads to reduced burning time, proportionally.

The value of ∫Pdt is calculated from the static firing result as W x C* / A_t. Since propellant volume is same in the two cases, the weight can be depicted as varying with density of propellant, only.

$$\int Pdt \propto \rho \times C^* / A_t$$

$$\int Pdt \propto \rho \times C^* / D_t^2$$

$$[\int Pdt]_{New}/[\int Pdt]_{Ref}=(1760/1680)x(1600/1526)x(44.5/54.759323)^2$$

$$= 1.047619 \times 1.04849 \times 0.660395$$

$$= 1.09842 \times 0.660395 = 0.725392$$

$$[\int Pdt]_{New} = 0.725392 \times [\int Pdt]_{Ref}$$

$$[BT]_{New} = 0.725392 \times [BT]_{Ref}$$

The above calculation indicates that, if burning time of the reference firing (composition) is 10.59 seconds, the same with new composition will be 7.68 seconds, only. This is well depicted by the curves in Fig. 133.

If instead of peak pressure, it is attempted to get same burning time with new propellant composition, then required throat diameter will change. From the above indicated calculation, it is clear, that peak pressure for maintaining same time leads to lowering of pressure levels. From the òPdt calculation shown above, it is observed that if burning time is maintained same, then pressure follows similar relation. Peak pressure is taken as reference for calculation. From obtained peak pressure with new composition, the value of throat diameter will also change and calculation is depicted below to give throat diameter of 67.171 mm. The peak pressure is also calculated with modified throat. The superimposed curve for this modified throat diameter for the new composition is shown as Fig. 134.

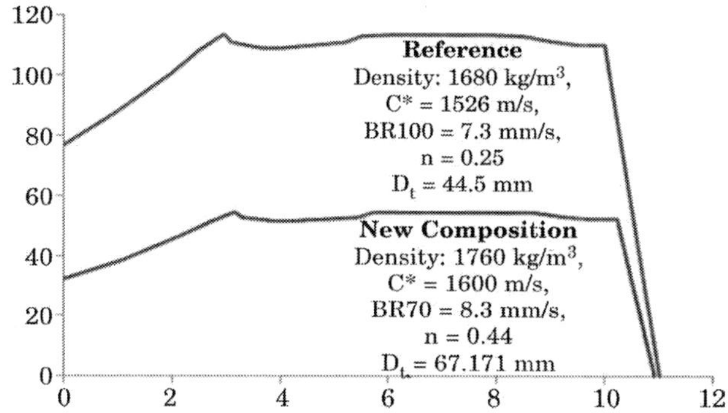

Fig. 134: Prediction for equal total burning time

P_{New}=1.09842x44.5^2 P_{Ref}/D_t^2=2175.1462xP_{Ref}/D_t^2=246879.09/ D_t^2

[246879.09/D_t^2]$^{(1-0.44)}$xD_t^2/(1.28x1760x1600) = 0.0117725

246879.09$^{0.56}$ x $D_t^{0.88}$ = 42433.741

D_t = 67.171 mm.

P_{New} = 246879.09/ D_t^2 = 54.7168 kg/cm^2

The peak pressure realized in this case is 54.7 kg/cm^2, as predicted by calculation. Since throat diameter is enlarged further (from 54.759323 mm to 67.171 mm), the value of area under pressure time profile will reduce further. In this case, the value of ∫Pdt is only 47.45% of the area under pressure time profile of the reference curve. The reduction can be simulated from the ratio of nozzle throat area. With throat diameter of 54.759323 mm the reduction was 71.39%, so, with throat diameter of 67.171, reduction in *∫Pdt is 71.39 x 54.759323^2 / 67.171^2 = 47.45%.*

However, most of the time, rocket motor is kept same and any variation in throat diameter demand change in the production line of the rocket components. So, for any new energetic composition, attempt is made to tailor burning rate of the propellant, accordingly and maintain same peak pressure (or lower peak pressure) and same throat diameter. Practical situation always demand estimation of burning rate for new composition for same throat diameter and peak pressure. For the given situation, the burning rate requirement of new composition to give same peak pressure and throat diameter is estimated. The superimposed curve for burning rate of new propellant composition is shown as Fig. 135.

$113.5^{(1-0.25)}$x44.5^2/(2.28156x1680x1526)=$113.5^{(1-0.44)}$x44.5^2/(a x1760x1600)

So, burning rate coefficient, a = 0.8453

Burning rate at 70 kg/cm^2 = 0.8453 x $70^{0.44}$ mm/s = 5.481 mm/s.

Obviously, the burning rate requirement to maintain same peak pressure and throat diameter for energetic propellant will be low. Lower burning rate enhances burning time and maintaining same peak pressure give higher value of area under pressure time profile. For the given situation, the value of area under pressure time profile is 8.11% higher as compared to reference value of òPdt. The lower pressure in the beginning for the new composition is because of lower burning rate and higher burning rate index. This is already explained earlier.

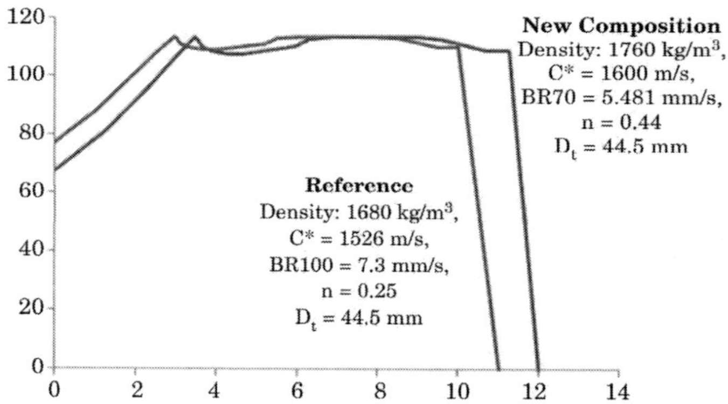

Fig. 135: Superimposed pressure time profiles with varied burning rate

The analysis given in this section may be performance for any pressure time profile and is a very useful way to arrive at the prediction of pressure time profiles for unknown configurations, for which pressure time profiles from static firing are available. Many a time, after finalization of configuration of propellant, propellant compositions are modified leading to variations in propellant properties (density, burning rate, burning rate index, characteristic velocity etc)since burning area versus web burnt profile is unaltered, the prediction can be made based on available pressure time profile with earlier known propellant compositions. This helps in research on propellant formulation. In fact without firing propellant, predictions can be made and compositions can be sorted. Calculation of throat diameter for new composition for desired output criteria can be obtained by the calculation strategies discussed in this section.

REVIEW QUESTIONS

1. How nozzle-erosion affects deviation from prediction in rocket firing?
2. What is significance of 10% of maximum pressure in pressure-time profile obtained from rocket?
3. How is 10% of maximum pressure chosen as reference for calculation of ballistic parameters in static firing of rockets?
4. What are ballistic parameters derived from the pressure-time profile in static firing of rockets?
5. Why is there a variation in pressure recorded at head end and nozzle end of rockets?
6. What is strategy for calculation of burning rate of propellant in end-inhibited neutral burning tubular propellants in static firing of rockets?
7. What is strategy for calculation of burning rate of progressive or regressive burning propellant grains in static firings of rockets?
8. How is web consumed calculated from volume consumed in (i) progressive (ii) regressive burning tubular propellant grain in static firing of rockets?
9. Compare burning rate calculation strategy for progressive and regressive burning tubular propellant grains in static firing of rockets?
10. How is burning area calculated from the pressure-time profile in static firing of rockets?
11. How is correctness of burning area-web consumed profile, derived from pressure time profile, ascertained from the static firing of rockets?
12. What is utility of calculation of burning surface area versus web consumed profile in static firing of rockets?
13. What is effect of changing propellant burning rate on the pressure-time profile?
14. How is effect of changing throat diameter calculated for a given rocket motor?
15. Given pressure-time profile, can thrust-time profile for the rocket propellant, predicted?

13

Ballistics from Thrust-time Curves

CHAPTER SUMMARY

This chapter covers the analysis of thrust, as the main ballistic parameter of rockets. Thrust is the main ballistic parameters because rocket acceleration, velocity, trajectory etc are derived from thrust data. Thrust–time profiles, obtained from static firing are analyzed and practical evaluation of gamma is attempted in this section for the exhaust gases. This chapter covers the following:

- Importance of thrust-time profile from static firing of rockets
- Analysis of thrust-time profile from static firing of rockets
- Features of thrust-time profile from static firing of rockets
- Derivation of ratio of specific heats for exhaust gases of rockets
- Calculation of pressure-time profile from thrust-time profiles in rockets

INTRODUCTION

Rocket motors are designed to generate thrust and system specification is expressed in terms of this parameter only. Pressure is a rocket propellant design criteria and is limited for a given rocket from safety considerations. No doubt pressure is a replica of thrust and proportionality is also observed but nature of pressure and thrust may not be similar for particular cases like nozzleless boosters. Since thrust is the system requirement, from which velocity, acceleration, trajectory of the rocket in flight in simulated, static evaluation demands measurement of thrust versus time also. Such measurements need placement of load cell between a rigid stationary thrust wall and the rocket. Such measurement measures thrust as compressive force. Unlike pressure time profile, where location of sensor affects numerical values of pressures, in thrust measurement, single value of thrust is obtained.

ANALYSIS OF THRUST TIME PROFILE

Thrust time data is obtained during static evaluation in similar fashion as the pressure time profile. Most of the time thrust time profile is exact replica of pressure time profile with vertical axis plotted with some scaling. In certain cases, deviations are observed in both the curve. This becomes subject matter of discussion and further exploration. However, for normal rocket firing, parameters from thrust time profile are similar but have different significance.

Thrust from a rocket is never constant through-out combustion of propellant and has variations. Maximum thrust is one of the important output parameter and it is representative of peak acceleration. In case of rockets, mass is continuously decreasing and thrust may not represent acceleration directly. However, if peak thrust occurs at the end of firing (when mass of rocket is less), acceleration realized is maximum. If similar thrust levels are realized in the beginning of firing, corresponding acceleration will be less. The requirement for peak acceleration varies from system to system. In general, acceleration beyond certain level is avoided in rockets. So progressive burning profile, where thrust is high at the end of propellant consumption, is not preferred. This may be the reason that progressive burning profile of solid rocket propellant is not used in most of the operational system except for particular cases of cartridge actuated devices, gas generators or gun propulsion systems.

Similar to pressure time profile, lower values of thrust has more noise. Hence, for the calculation of ignition delay 10% of maximum thrust is considered. In the rising side, wherever this thrust levels are achieved, ignition delay is assumed at that location. Similarly, during end of propellant combustion, achievement of 10% of maximum thrust is considered as end of burning. Any thrust beyond this point is considered insignificant and neglected in calculation. This residual portion of thrust time profile represents depressurization phase and is not at all contributing to the thrust values. The difference between, two times representing 10% of maximum thrust on rising and falling curve is called burning time of the rocket firing. This is similar to burning time represented on a pressure time profile.

Area under thrust time profile is called total impulse and is representative of energy content of the system. It is directly related to delivered energy and is used for the measurement of specific impulse also. During calculation of total impulse, thrust during burning time only (10% of maximum thrust) is considered. The total impulse is divided by burning time to give average thrust during rocket motor operation. Total impulse divided by mass of the propellant is called specific impulse and is a performance parameter of the complete rocket. No doubt, thrust time profile is system designer need, while pressure time profile is propellant designer need.

A typical rocket firing example is considered, where pressure time and thrust time profiles are recorded. The sampling rate of the data acquisition is 2000 samples per seconds and data is recorded at every 0.5 seconds interval. Although burning time is very near to 3 seconds but recording is carried out till 5 seconds. The propellant weight is 9.39 kg and throat diameter is 30.4 mm (throat area = 725.83 mm^2). The exit diameter of nozzle is 56.92 mm (area = 2544.6 mm^2) Pressure time profile is shown on secondary axis (axis on right hand side) and thrust time profile is shown on primary axis (axis on left hand side) in Fig. 136.

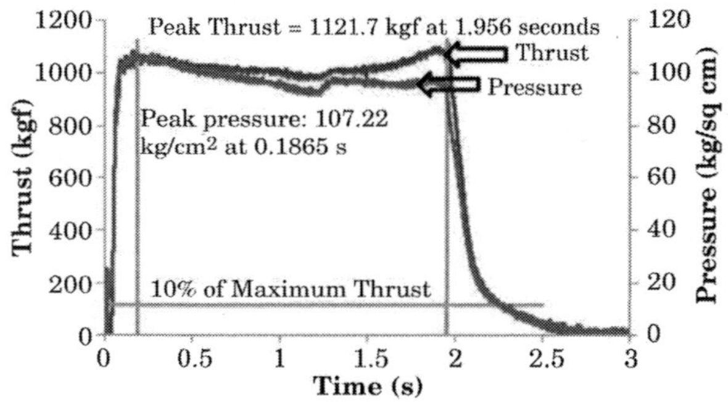

Fig. 136: Typical rocket evaluation curves

This curve is chosen to show that it is not mandatory for thrust time profile to follow pressure time profile exactly. During initial phases of combustion, both the curves have similar trend but pressure time profile is less progressive (neutral) at the end of burning than thrust time profile. On analysis of pressure time data (as discussed in previous chapter), following results are observed:

Maximum pressure	:	107.22 kg/cm^2
Ignition delay	:	0.0200s
Time to maximum pressure	:	0.1865s
Burning time	:	2.2275s
Area under curve	:	197.16 kg-s/cm^2
Average pressure	:	88.51 kg/cm^2
Characteristic velocity	:	1495.1 m/s

The way analysis is carried out for the pressure time profile (in previous chapter), thrust time profile is also analyzed. The thrust at 0.1865 s, corresponding to peak pressure is 1069.4 kgf, which is lower than thrust at the end of burning. Peak thrust is obtained as 1121.7

kgf at 1.956 seconds 10% of maximum thrust is 112.17 kgf and on the rising curve, it occurs at 0.0455 seconds. This clearly indicates that ignition delay on thrust time profile is at a slightly later stage than ignition delay in pressure time profile. On the falling curve 10% of maximum thrust occurs at 2.266 seconds giving a burning time of 2.2205 seconds. Area under thrust time profile is evaluated as 2064.67 kgf-s. This is total impulse of the propellant system (including performance of nozzle and hardware). Average thrust is (2064.67/2.2205 kgf =) 929.8 kgf. Total impulse divided by weight of propellant gives specific impulse and for the displayed firing curve, specific impulse is (2064.67/9.39 s =) 219.87 seconds.

During static evaluation, measurement of side thrust is also possible, if rocket motor is fired in handing condition. For this six-component test bed is needed and measurement of side thrusts needs placement of load cells at the sides also. The system becomes more complicated, but for assessing thrust loss due to swiveling of nozzle, thrust vector control, gimbaled nozzle, canted nozzle scarf nozzle etc such testing facilities are needed.

Although single value (not multiple values as obtained for pressure) of axial (longitudinal) thrust is obtained during rocket operation, but during static firing, the trolley, mounting fixtures, pre-loading, overhang of rocket motor etc play a major role and result in noisy thrust time data. Trolley movement should be frictionless and before static firing, rocket should be touching the thrust block. If any gap exists, rocket will move back and forth resulting in oscillations in the obtained results. One mechanism to avoid such physical oscillations of rocket is to pre-load it. Generally pre-loading to the tune of 10% of the predicted thrust is applied. Restricting all movements of rocket during static firing is main objective and any lateral or other type of movement may result in variation in measured thrust data. Since exhaust from rocket has natural fluctuation of flow, the overhang of nozzle end side of the rocket during static firing should be restricted to reduce any physical oscillatory movement of nozzle. The lower the overhang, lower will be the oscillations in pressure data.

Thrust has a very complex dependence on throat diameter. Since thrust is a function of thrust coefficient, pressure and throat area (equation 17 of chapter 3), the variations in throat diameter is worth exploration. If throat diameter is reduced, pressure is going to increase. In fact, the product of pressure and throat area, which is controlling factor for the determination of thrust, is proportional to $A_t^{(-2n)/(1-n)}$. So this product (P_i x A_t) is lower for higher value of throat diameter. This product may be called as thrust equivalent (controlling product). For comparison, if all parameters are same, then increase in throat diameter from 40 mm to 50 mm results in a reduction in the controlling product by around 17.5% for burning rate coefficient as (n =) 0.30. So, if thrust coefficient is constant, thrust at higher throat diameter (50 mm) will be 82.5% of thrust at lower diameter (40 mm). Since the value of reduction in thrust by alteration of throat diameter is dependent on burning rate index of the propellant, this effect is also explored. For burning rate coefficient as (n =) 0.40, the reduction in thrust equivalent product is by 25.7%, which is higher than previous case. This means that at throat diameter of 50 mm, the thrust will be 74.3% of thrust obtained at throat diameter of 40 mm. This indicates that higher burning rate index, makes thrust variation more sensitive to the variation in throat diameter. Effect of other factors like density, characteristic velocity, burning area etc on throat area dependent thrust equivalent is negligible. This means that, a propellant of higher density or characteristic velocity will exhibit same percentage change in the value of thrust equivalent, if throat area is changed by known amount.

In addition to these two factors, effects of thrust coefficients are also worth exploration. Higher pressure in rocket motor for same nozzle area ratio results in higher value of thrust coefficient for optimum expansion (figure 8/10 of chapter 3). However, the rate of increase slows down at higher value of chamber pressure realized with lowering of throat diameter. So, reduction of throat diameter increases both pressure and thrust coefficient. As a result, thrust generally increases with decrease in throat diameter.

One obvious advantage of acquiring thrust data is its direct relevance for the system-design. It gives acceleration, velocity and trajectory of rockets directly. However, one practical advantage is that thrust data recording is a non-invasive technique and rocket motor need not be prepared separately with any thrust adaption ports. In case of pressure recording, rocket motor must possess a vent to mount pressure gauge and vent transfer pressure inside rocket motor chambers to pressure sensors. Sometimes, these locations of additional port, becomes a point of failure in static evaluations. Contrary to this thrust data is recorded through a thrust frame, which is mounted at the head end portion as a cap and no vent or connection to internal combustion gases is needed. The whole motor is subjected to pressure and local effects are also not present during thrust recording. It is not mounting alone, which is relatively simpler, the data on thrust channel are always recorded. The pressure port may be choked by particulate matters, but thrust channels do not have this type of possibility. It is only arresting the jerks or impact type of motion of the rocket motor during static evaluation, which is restricted to get clean thrust time data. For this rocket is pressed against thrust wall by pre-loading. So overall thrust time data has the following features:

- Thrust gives an overall performance of the rocket motor.
- Thrust gives specific impulse, which has direct relevance for the propulsion systems.
- Thrust time profile is similar to pressure time profiles for most of the firings.
- Thrust is directly connected to system requirements.
- Thrust recording is a non-invasive technique.
- Thrust does not interfere with combustion gases.
- Thrust is not local phenomena inside rocket, like pressure.
- Thrust has unique value for a rocket firing at a given time and is independent of spatial location of sensors.
- Thrust sensors are not choked by particulate matters.
- Thrust is primary data for system designers, while pressure is primary data for propellant and propulsion designers.
- Thrust data is relevant for system performance but pressure data is relevant for safety of rocket chambers.
- Thrust coefficient and nozzle efficiency introduces uncertainty in thrust data.
- Thrust generally decreases with increase in throat diameter.
- Thrust variation with throat diameter is more for higher burning rate index of used propellants.
- Thrust generation starts only after flow of combustion gases through nozzle is established.
- Mechanical vibrations of structure and combustion instability are reflected in measured thrust.

Evaluation of Gamma

Although value of gamma is not very important from practical point of view and for propellant or propellant systems, characteristic velocity, burning rate and specific impulse are sufficient to characterize the system, an attempt can be made to calculate value of gamma from practical firing curves. This is possible for those firings for which both pressure time and thrust time data are recorded. Since for the firing curve given above as Fig. 136, both the profiles are available, evaluation of gamma can be attempted. The input data taken are as follows:

Average pressure	=	88.51 kg/cm^2
Average thrust	=	929.8 kgf
Throat diameter	=	30.40 mm
Exit diameter	=	52.96 mm
Area ratio, A_e/A_t	=	$(56.92/30.4)^2$ = 3.50576

Area ratio is a function of gamma and pressure ratio as per the expression derived in chapter 3. As a reverse case, if area ratio is known, then for a given gamma, pressure ratio can be obtained. Best fit curve is obtained for this and resulting expression is given below. The expression is similar to equation 17 of chapter 3, but with some variation for ease of calculation.

$$P_e/P_c = 0.29873g^{-0.55421} \times (A_e/A_t)^{-1.00623 \times g - 0.124}$$

As firing is carried out at ambient condition, outside atmospheric pressure can be taken as 1.0332 kg/cm^2. For known pressure ratio, exit pressure can be calculated. Since both thrust time and pressure time profiles are available, there are certain parameters, which can be derived from the calculation using data of both the curves. One such parameter is thrust coefficient ($C_f = F/P \times A_t$). For the representative firing, thrust coefficient is calculated.

$$C_f = F/P \times A_t = 929.8/(88.51 \times 7.2583) = 1.44723$$

From the expression derived in chapter 3, it is clear that thrust coefficient is a function of gamma, pressure ratio, chamber pressure, atmospheric pressure and nozzle area ratio. Out of the given parameters, gamma is assumed earlier and other parameters are already derived. So, thrust coefficient can be calculated. The calculated thrust coefficient can be compared with realized thrust coefficient. Any deviation is compensated by changing the value of gamma till matching thrust coefficient is realized from both practical data and theoretical calculation. However, the theoretical calculation depicts ideal situation and nozzle efficiency has to be assumed accordingly. For the current firing nozzle efficiency is taken as 94%.

$$\text{Ideal value of thrust coefficient, } C_f = 1.44723/0.94 = 1.5396$$

For estimation of gamma, initial assumption of gamma is taken as 1.19. For this value of gamma and area ratio of 3.50576, pressure ratio from best fit curve is calculated. From that exit pressure can be calculated. The value of thrust coefficient is calculated using equation 18 of chapter 3.

Pressure ratio, $P_e/P_c = 0.051704$

Exit pressure, P_e = 0.051704 x 88.51 kg/cm^2 = 4.576 kg/cm^2
Thrust ratio, C_f = 1.5459

The value of thrust coefficient is higher than realized thrust coeff-icient and value of gamma is to be increased to get matching value of thrust coefficient. The calculation is repeated for gamma equal to 1.21.

Pressure ratio, P_e/P_c = 0.04995
Exit pressure, P_e = 0.04995 x 88.51 kg/cm^2 = 4.4212 kg/cm^2
Thrust ratio, C_f = 1.5411
The calculation is repeated fro gamma equal to 1.22.
Pressure ratio, P_e/P_c = 0.0491
Exit pressure, P_e = 0.0491 x 88.51 kg/cm^2 = 4.346 kg/cm^2
Thrust ratio, C_f = 1.5388

So, actual value of gamma, lie between 1.21and 1.22. By successive iteration and approximation, the value of gamma is obtained as 1.21629 and for this value the resulting calculations are as follows.

Pressure ratio, P_e/P_c = 0.04941
Exit pressure, P_e = 0.04941 x 88.51 kg/cm^2 = 4.3737 kg/cm^2
Thrust ratio, C_f = 1.5396

The crux of the calculation is that from geometry, nozzle area ratio is known for a rocket firing. Average pressure during firing and atmospheric pressure during firing is also known. Thrust coefficient can be derived from static firing data, if both pressure time and thrust time profile is recorded. Nozzle efficiency can be taken between 90% and 98%. In the expressions for area ratio (equation 17 of chapter 3) and thrust coefficient (equation 18 of chapter 3), only two parameters are unknown – gamma and exit pressure. Although, explicit relation for calculation cannot be derived but two expressions and two unknown should give unique solutions for both the variables.

Prediction of Thrust Time Profile from Pressure Data

As mentioned earlier, thrust measurement is definite and data is non-invasive, as well as free from any influence from particulate or other obstructions in data acquisition. Generation of thrust time profile from pressure time data is relatively simpler. Since pressure and throat diameter are already known, the problem is reduced to estimation of thrust coefficient. Theoretical estimation of thrust coefficient needs nozzle area ratio, throat diameter and gamma. Out of these, first two are known and value of gamma is generally taken as 1.19 for composite propellants and 1.21 for double base propellants. With these assumed values of gamma, pressure ratio for given nozzle area ratio can be estimated (equation 16 of chapter 3). As pressure at nozzle inlet is known, the value of exit plane (of nozzle) pressure can be calculated. With known, pressure-time variation, pressure ratio, gamma, nozzle area ratio, and exit plane pressure, the thrust coefficient with time can be estimated, and finally, thrust versus time profile can be generated. This sequence of operation is explained in section 2 of the book with different configurations. However, the calculation is independent of propellant configuration and is a universal mechanism of calculation.

A typical pressure time profile is shown as Fig. 137 for a rocket fired with throat diameter and exit diameter of 19.8 mm and 56.5 mm, respectively. The area ratio is calculated to be 8.1426. Using equation depicted by regression analysis in the previous section (in lieu of equation 17 of chapter 3), the value of pressure ratio is obtained as 0.016128. For the propellant used in the rocket firing, the value of gamma is taken as 1.21. The calculation is repeated for other values of gamma, also and there is no effect of these values on the calculated value of thrust coefficient.

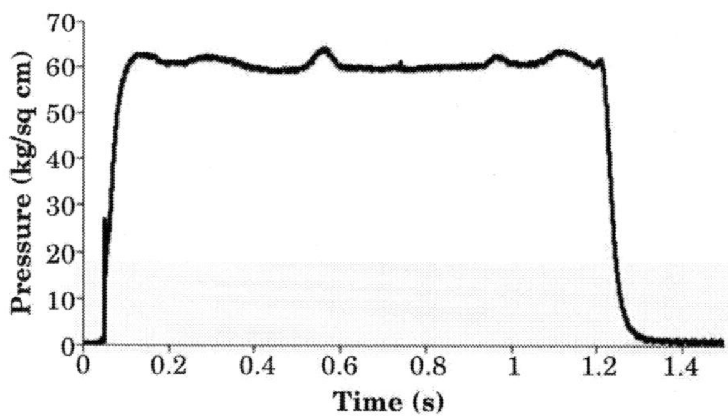

Fig. 137: A typical pressure time profile

Thrust coefficient contains two parts. One is derived from momentum part of thrust and other is pressure part. In fact, some part of the expression (equation 18 of chapter 3) is constant and independent of pressure inside rocket chamber.

$$C_{f1} = \sqrt{[2/(g + 1)]^{(g+1)/(g-1)} [2g^2/(-1)]x[1-(P_e/P_i)^{(g-1)/g}]} = 1.57914$$

$$C_{f2} = C_{f1} + (P_e/P_i)(A_e/A_t) = 1.71046$$

$$C_f = C_{f2} - (P_a / P_i)(A_e/A_t)$$

As instantaneous pressure with time is known, the value of thrust coefficient for each time interval is known. The atmospheric pressure

(P_a) in this case is taken as 1.0332 kg/cm^2. Using equation 17 of chapter 3, the value of thrust corresponding to pressure in rocket chamber can be obtained. The predicted thrust time profile from the pressure time profile is shown as Fig. 138.

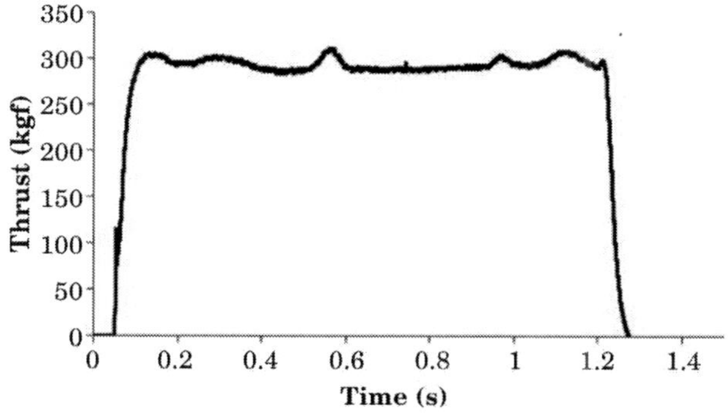

Fig. 138: Generated thrust time profile

The curve seems to be exact replica of pressure time profile, except for some scaling on y-axis for numerical values. The scaling factor is basically product of thrust coefficient and throat area. In a rocket firing, throat area is considered fixed (some enlargement due to erosion or reduction due to deposition is possible) and only variable is thrust coefficient. Variation of thrust coefficient with time is plotted as Fig. 139. It is clear from the figure that the value of thrust coefficient is not affected by small variations in the value of pressure and its value is relatively stable for this firing. In the ignition delay period, the value of thrust coefficient is very high, but they are spurious results and should be neglected for calculation. The stable value of thrust coefficient for this firing is around 1.56, which is lower but near to the value of C_{f1}, calculated earlier.

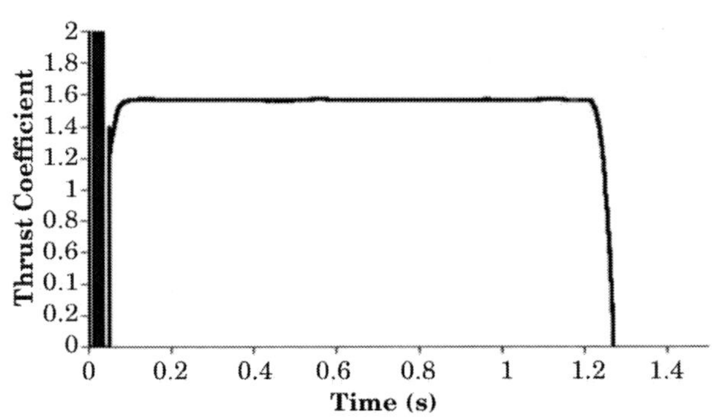

Fig. 139: Variation of thrust coefficient with time

For the subject firing, thrust time profile is also available and prediction from pressure time profile is superimposed on actual thrust time profile as Fig. 140. The prediction of thrust is more than actual thrust. However, ignition delay and nature of the curve matches exactly. This indicates that value of thrust coefficient is higher for prediction and value of thrust coefficient is to be reduced. This indirectly states that rocket nozzle is not 100% efficient and there are certain thrust losses in flow through the nozzle.

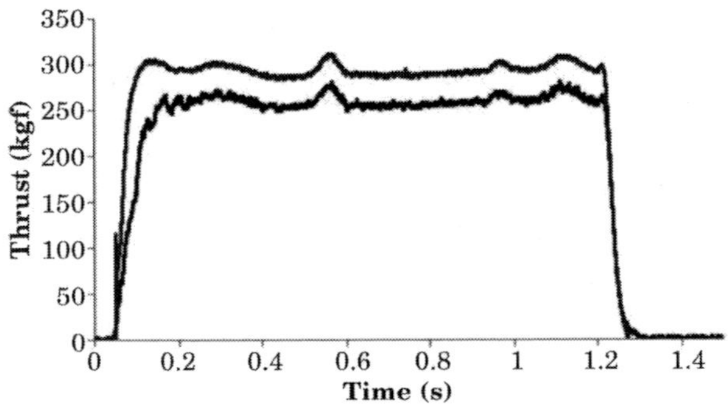

Fig. 140: Superimposed predicted and actual thrust time profiles

Since there is no mechanism to theoretically determine nozzle efficiency, the value of nozzle efficiency is a practical parameter, used for matching pressure time and thrust time profile. In the current situation, various nozzle efficiencies are considered and for nozzle efficiency of 88%, the thrust time profile matches exactly with the recorded thrust time profile for most part of the curve (Fig. 141). However, the matching is poor in the ignition transient zone (in the beginning of firing). This basically represents chamber filling, port pressurization, flame spreading, nozzle closure opening phenomena and is transient as well as unpredictable. So, the prediction of thrust time profile from pressure time profile can be made with limited accuracy

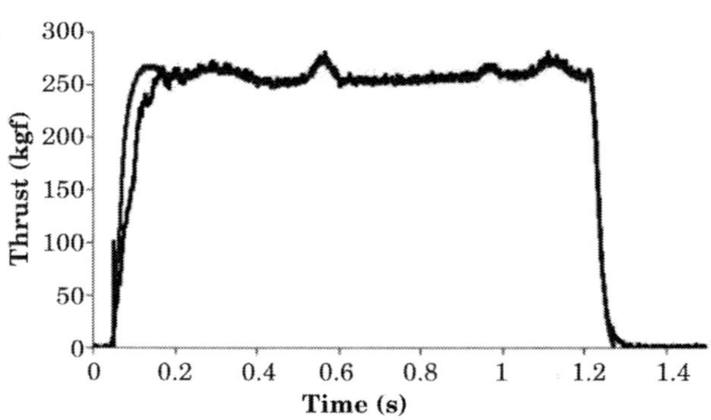

Fig. 141: Prediction of thrust time profile with nozzle efficiency (88%)

and nozzle efficiency estimates introduce difficulties in such calculations. Similar variation is observed in previous section also for the calculation of gamma from thrust time profiles.

The reverse calculation is slightly difficult to arrive at. If thrust time profile is available for a rocket firing, the development of pressure time profile again demand similar practice as discussed earlier, *i.e.*, estimation of thrust coefficient. Since area ratio is known and gamma can be assumed conveniently, the pressure ratio can be arrived at. But estimation of pressure inside rocket motor chamber (or nozzle inlet pressure) is still not easy, which needs estimation of chamber pressure. For nozzle with optimum expansion, the expression for thrust coefficient depends on gamma and pressure ratio, only. For such situations, calculation of thrust coefficient and pressure time profile is easy. However, for most of the rockets, the nozzle area ratio is sub-optimal. Most of the time area ratio is less than ideal and exit plane pressure is higher than atmospheric pressure, giving rise to pressure thrust and altered thrust coefficient. For such situations, the value of thrust coefficient can be obtained from experience or from previous firing data of similar propellant, in similar configurations. Once thrust coefficient is known, conversion of thrust time to pressure time and vice-versa is possible.

REVIEW QUESTIONS

1. What is importance of thrust-time profile of rockets?
2. What does thrust in a rocket indicate?
3. What are advantages of thrust-time profile in rockets?
4. What are parameters derived for rockets from thrust-time profile?
5. Why is thrust measurement preferred over pressure measurement in a rocket?
6. What is strategy for the measurement of ratio of specific heats of exhaust gases of propellants in rocket?
7. Whether thrust-time profile, alone can be used to get ratio of specific heats of exhaust gases of propellants?
8. How is pressure-time profile derived from thrust-time profile?
9. What is effect of changing throat diameter on thrust-time profile in rocket?
10. Can thrust in a rocket be negative?

14

Advanced Evaluations of Firing Curves

CHAPTER SUMMARY

Advancement in any field is a relative term, which is dependent on current competence level. So, this chapter is an open chapter, ready to accommodate more and more topics of relevance. Currently this chapter covers the following.

- Calculation of burning rate of multi-perforated propellant configuration
- Analysis of nozzleless booster rocket
- Analysis of Dual-thrust rocket firing

INTRODUCTION

When conventional techniques fail to characterize propellants by static evaluation, advanced techniques are developed on case-to-case basis for analysis of firing curves and arrive at additional propellant parameters. In addition to this, despite all precautions, when propellant is evaluated in static evaluation in rockets, the realized or measured curves many-a-time do not match the predictions and sometimes surprises are also observed. It is clear that prediction for propellant performance has several other factors, in-built, which are generally neglected or idealized in prediction and sometimes they become prominent during rocket firing. The assumption of smooth combustion, uninterrupted flow through port, integrity of nozzles/thermal insulation, erosive burning, pressure oscillations are some of the effects observed during firing and some of these curves, where variation from prediction is noted, are discussed in this chapter. So the chapter discusses two aspects – first is advanced methods for propellant characterization and second is handling anomalous behavior of propellants in static firing.

Evaluation of Burning Rate

Although evaluation of burning rate from static evaluation of propellants in ballistic evaluation motors (BEM) for tubular propellant configuration (both progressive and regressive modes) are discussed, earlier, sometimes burning rate evaluation of multi-tubular or multi-perforated configurations are also attempted using obtained pressure time profile.

For multi-perforated configuration, the burning area profile is linear till first web transition. The nature of pressure-time profile resembles burning area variation profile. So, burning area can be obtained from the pressure time profile till first web transition. Assume that propellant outer radius is 'R', propellant length is 'L' and hole radius is 'r'. Total number of holes are 'n'. Propellant weight (W) and throat area (A_t) of the rocket motor is known before firing. As initial dimension of the propellant is known, the value of density [$\rho = W/\pi(R^2\text{-}nr^2)L$] of propellant is also known, beforehand. Using area under pressure time profile value of C* (= $A_t \int_0^T Pdt/W$) can be obtained. Here integration is valid for the entire burning duration (T) of the pressure time profile. The initial progressive portion of pressure time profile is marked on at least 10 locations, such that for those marked times (t), pressure (P) and area under pressure-time profile till that time ($\int_0^t Pdt$) can be calculated. The calculation of burning rate (r) at each time step is given by the following formula. By calculating burning rate at different pressures, the value of constant in the burning rate law ($r = aP^n$) can also be obtained.

Burning rate, r = H / (AB), where
$H = PA_t/\rho C^*$, $A = 2\pi nL$, $B = \sqrt{r^2+2S/A}$, $S = (W/\rho)x(\int^t Pdt)/(\int^T Pdt)$.

Here 'A' is a constant, where as H, B and S are dependent on the time considered for calculation. 'A' and 'B' are purely geometrical parameters, while 'H' includes ballistic parameters. Pressure time profile of 10-hole multi-perforated configuration, which is discussed in chapter on multi-perforated propellant configuration, is taken for analysis. The dimensions are same as discussed in an earlier chapter and are reproduced below for reference. The propellant weight is (W =) 12.395 kg and the throat diameter is 41.8 mm.

R = 132.75 mm, L = 149 mm, r = 9.25 mm, n = 10, p_1 = 48.5 mm, p_2 = 108.5 mm.

Density of propellant, ñ = 12.395 x 10^6 /ð (132.75^2 – 10 x 9.25^2) x 149 g/cc = 1.579 g/cc

Throat area, A_t = ð$x41.8^2/4$ mm^2 = 1372.279 mm^2 = 13.72279 cm^2

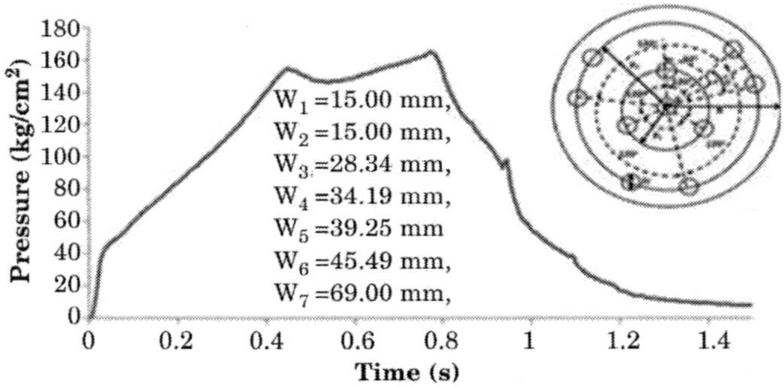

Fig. 142: Pressure-time profile for multi-perforfated configuration

The pressure time profile is depicted as Fig. 142. The sampling rate of data acquisition is 5 ms. The curve is similar to Fig. 85 of chapter 8 and has 'M'-shape clearly visible. The first peak occurs at 0.45 seconds and the value of pressure is 154.64 kg/cm^2. The second peak occurs at 0.775 seconds and the value of pressure is 165.21 kg/cm^2. Second peak depicts peak pressure of the firing. 10% of the peak pressure is 16.521 kg/cm^2. In the rising curve, the 10% of peak pressure is attained at 0.02 seconds, which is ignition delay of the firing. Similarly, on the falling side, this pressure is crossed at 1.205 seconds. Burning time of the firing is (1.205 – 0.02 seconds =) 1.185 seconds. Area under pressure time profile during duration of firing (from 0.02 seconds to 1.205 seconds) is computed to be 122.2113 kg-s/cm^2.

$C^* = 13.72279 \text{ x } 122.2113 \text{ x } 9.81 / 12.395 \text{ m/s} = 1327.32 \text{ m/s}$

$H = P\text{x}1000 \text{ x } 9.81\text{x}13.72279/(1.579\text{x}1327.32) = 64232.3P \text{ mm}^3/\text{s}$

$A = 2\pi \text{ x } 10 \text{ x } 149 \text{ mm} = 9361.946 \text{ mm}$

$S = (12395/1.579) \text{ x } 1000 \text{ x } (\int_0^t Pdt)/122.2113\text{mm}^3 = 64232.236 (\int_0^t Pdt) \text{ mm}^3$

Since first web transition occurs at 15 mm, this corresponds to first peak at 0.45 seconds. The calculation is to be conducted till this point only. Although calculations are repeated for many points between 0.02 seconds and 0.45 seconds, for illustration purpose, calculation is shown for time equal to 0.2 seconds. At this point, pressure is 84.45 kg/cm^2. Area till this point is 11.19388 kg-s/cm^2.

$H = 64232.3 \text{ x } 84.45 \text{ mm}^3/\text{s} = 5424417.7 \text{ mm}^3/\text{s}$

$A = 9361.946 \text{ mm}$

$S = 64232.236 \text{ x } 11.19388 \text{ mm}^3 = 719007.94 \text{ mm}^3$

$B = \sqrt{9.25^2 + (2 \text{ x } 719007.94/9361.946)} \text{ mm} = 15.465 \text{ mm}$

Burning rate, $r = 5424417.7/(9361.946\text{x}15.465) \text{ mm/s} = 37.466 \text{ mm/s}$

The calculations are repeated for the entire curve and Pressure versus burning rate profile is shown in Fig. 143. The curve resembles pressure-time profile with a right angle rotation of the profile. However, the developed mathematical formulation is valid for the initial progressive section only. So, burning rate depicted in after first peak is irrelevant. Data is extracted between pressure-range from 60 kg/cm^2 to 150 kg/cm^2 and burning rate is calculated using formulation above. Burning rate law is fitted as $r = 16.22P^{0.187}$. Value of burning rate at 84.45 kg/cm^2 from this best fit burning rate law is evaluated as 37.18 mm/s.

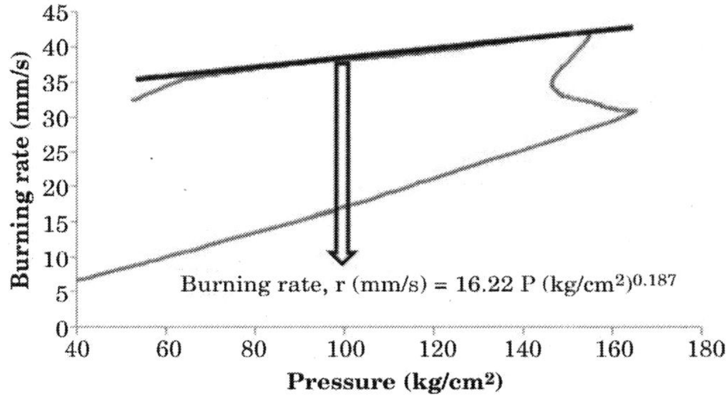

Fig. 143: Burning rate calculation for multi-perforated configuration

With the generated burning rate law, pressure time profile is predicted and superimposed over the pressure time profile from static evaluation. The superimposition is shown as Figure 144.

It is clear that superimposition is perfect for the initial progressive section. However, peak pressure is not simulated correctly by the generated burning rate law. In fact, in the calculation of burning rate law, a very wide range of pressure is covered. Practically, propellant show higher burning rate and burning rate index at higher pressures,

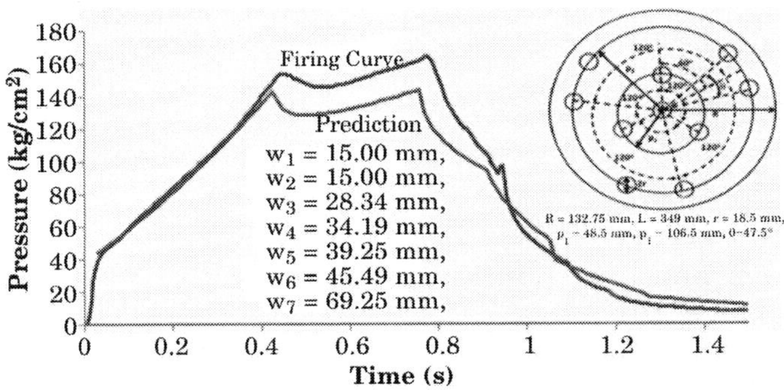

Fig. 144: Superimposed pressure time profiles for multi-perforated configuration

which is unfortunately not captured in this formulation. For the pressure range from 110 kg/cm^2 to 150 kg/cm^2, the burning rate law is derived as $r = 11.2P^{0.263}$. From the uniform burning rate law, burning rate at 110 kg/cm^2 is 39.065 mm/s while with selected burning law valid for selected region, burning rate at 110 kg/cm^2 is 38.55 mm/s. But burning rate at 150 kg/cm^2, by the two burning laws are 41.398 mm/s and 41.834 mm/s, respectively. In the mentioned formulation, for the entire burning uniform burning rate law and index is considered. Probably, this may be a reason for not attaining same numerical values of pressure in the predicted pressure time profile. Although transitions are captured correctly by prediction but numerical values are not matching. Another reason for concern is increase in volume available for the combustion gases. In the rocket motor, initial volume available for the combustion gases is equal to volume of 10 holes = $10 \times \pi \times 9.25^2 \times 149$ mm^3 = 400515.76 mm^3. However, at the end of burning the volume available becomes $\pi \times 132.75^2 \times 149$ mm^3 = 8249074 mm^3, which is 20 times higher than the initial available volume. This also leads to error in calculation, as steady state analysis neglects change in available volume with respect to time.

Nozzleless Propellant Configuration

Most of the propellant configurations discussed earlier in this book or available otherwise has identical shape of the pressure-time and thrust-time profiles. However, in case of nozzleless booster configurations, the nature of pressure-time and thrust-time profiles is non-identical. The configurations of nozzleless booster is independent of heavy separate nozzle block (containing convergent and divergent sections), attached to rocket motor casing. The propellant end is shaped at the exhaust end in the shape of nozzle divergent. No spare rigid throat section is available and propellant itself forms nozzle. The salient features of this configuration are as follows:

- Inert nozzle section is dispensed with and hardware weight is reduced.
- For integrated ramjet application, it can act as booster, whose throat can enlarge for sustainer firing without any separation piece for nozzle-enlargement.
- The propellant configurations are simple.
- Since propellant port and nozzle throat enlarges during combustion, the propellant needs high burning rate.
- The activities, requiring rigid nozzle, like thrust vector control, secondary injection etc could not be executed by this configuration.

- The variation in nozzle throat during rocket operation requires separate ballistic calculation strategy.
- The pressure in propellant port reduces from head end towards nozzle end side.
- Since rigid nozzle is not available and pressure is changing along propellant port, the propellant retention inside rocket motor during rocket operation require specific feature at the rocket motor internal surface.
- The calculation of internal ballistic parameters is difficult, because nozzle throat area is continuously varying.
- Nozzleless configuration cannot be designed for liquid propellants.
- The axial differential pressure pushes propellant in rearward direction during propellant operation and propellant must be configured in case-bonded mode (if grids are not present).
- The propellant configuration cannot be thought in the form of cartridge-loaded configuration due to differential axial pressure.
- The propellant combustion undergo very high erosive burning, due to velocity gain by the combustion gases during flow in the propellant port.

A typical rocket motor with nozzleless propellant configuration is depicted as Fig. 145. The Rocket motor is insulated at the inner walls. The igniter is positioned at the head-end and configuration is case-bonded in nature. The nozzle end section of rocket motor has inward projection for neutralizing the axial differential pressure, which has tendency to push propellant rearward. During rocket operation, the throat diameter continues to increase, but the insulated metallic back-up provides some fixed throat diameter at the end of firing.

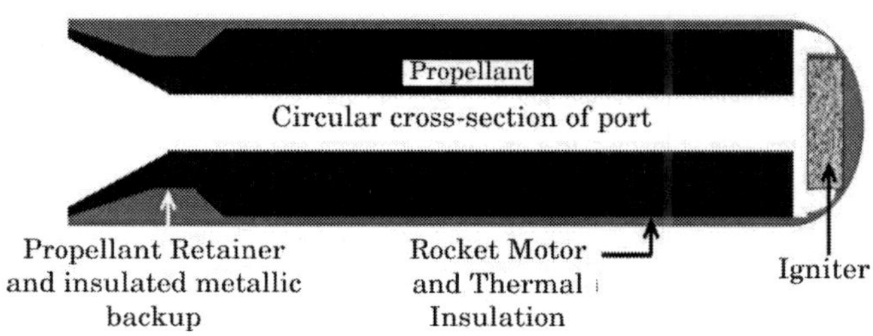

Fig. 145: Nozzleless propellant configuration

Nozzleless propellant configuration with port diameter of 50 mm and total propellant weight of 39.4 kg is considered as test case for static evaluation. The insulated metallic back-up for the retention of propellant has internal diameter of 118 mm and the outer diameter of the propellant is 160 mm.

The propellant configuration seems to be similar to progressive burning tubular propellant. The port area is continuously increasing and the burning area versus web consumed profile is progressive in nature. However, the throat area is not constant and instantaneous throat area is increasing during rocket operation and for the pressure calculation, ratio of burning area and throat area is considered. Hence the variation of pressure may not be following the variation of either burning area or throat area, but that of ratio of burning area and throat area. So, the pressure time profile will be different from burning area variation profile. The recorded pressure-time (at head end) and thrust-time profiles are depicted as Fig. 146.

First look at the recorded curve reveals that natures of pressure-time and thrust-time profile are not identical, contrary to the observations in almost all the previous configurations. The pressure is continuously reducing with time, but thrust is having a relatively neutral profile. The peak and average values of pressure is observed as 116.8 kg/cm^2 and 46.29 kg/cm^2, respectively. The value of peak and average thrust is obtained as 4167 kgf and 3351 kgf, respectively. The area under pressure and thrust profiles are obtained as 101.3 kg.s/cm^2 and 7901.6 kgf.s, respectively. The ignition delay and burn time for the curves are in the range of around 30 ms and 2.2 s, respectively. As far as practical calculation of specific impulse is concerned, it is given by dividing area

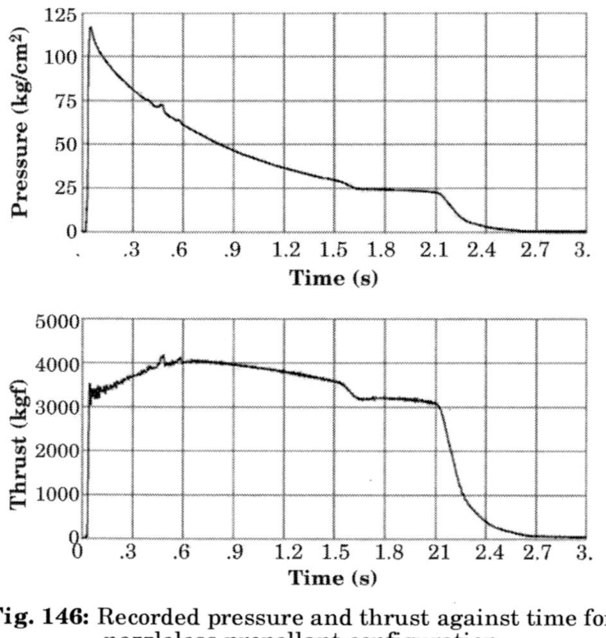

Fig. 146: Recorded pressure and thrust against time for nozzleless propellant configuration

under thrust-time profile by the weight of the propellant. In this case, it is (7901.6/39.4 =) 200.5 s. However, calculation of characteristic velocity (C*) has difficulty, due to varying throat diameter.

C* is given by area under pressure-time profile multiplied by throat diameter and divided by the weight of the propellant. The varying throat diameter poses problem with this calculation. If initial throat diameter of 50 mm is considered as throat diameter, the value of C* comes out to be 495.24 m/s. This value is very low and is practically not possible for propellants in rockets. Calculation of thrust coefficient (C_f = Isp/C*) from this value of C* gives a value of 3.9716. This value of thrust coefficient is not possible. If throat diameter of 118 mm is considered in calculation as throat diameter, the value of C* comes out to be 2758.3 m/s. This gives thrust coefficient of 0.713. With these numerical results, two conclusions are possible – (i) Calculation of C* needs some different strategy (ii) the calculation of thrust coefficient needs alternate method from the conventional way for the configuration.

Although nature of pressure-time and thrust-time profiles are different, but it is clear that after around 1.6 seconds, both the profiles has neutral values. This may be corresponding to situation when propellant in the nozzle end insulated metallic backup portion is consumed and throat diameter is constant at 118 mm during this period. For the calculation, the value of various parameters at 1.65 is depicted in Fig. 147. The value of thrust coefficient at this time is 1.1819, obtained by instantaneous thrust divided by product of instantaneous pressure and throat area. (F/PxA_t).

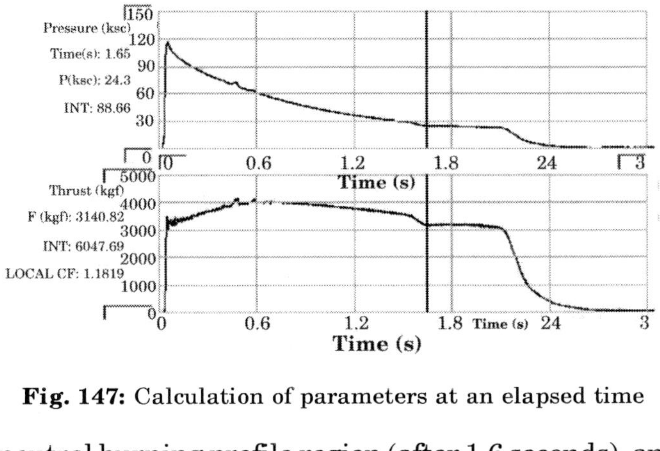

Fig. 147: Calculation of parameters at an elapsed time

In the neutral burning profile region (after 1.6 seconds), an arbitrary point is selected and value of thrust coefficient is calculated at that point. At 1.9735 seconds, thrust coefficient is calculated as 1.25898.

Considering this thrust coefficient valid for the entire rocket operation, the throat diameter variation can be calculated. This is depicted as Fig. 148.

It is clear from the figure that actual initial throat diameter of 50 mm is obtained in back-calculation. This somehow ensures correctness of calculation to a certain extent. For the calculation of C*, the instantaneous value of nozzle throat diameter is considered and is calculated to be 1547 for the configuration. Thus for the nozzleless propellant configuration, calculation of specific impulse from thrust time profile is similar to earlier configurations, but calculation of C* require, determination of instantaneous throat diameter by calculation from a reference constant throat diameter instant in rocket operation.

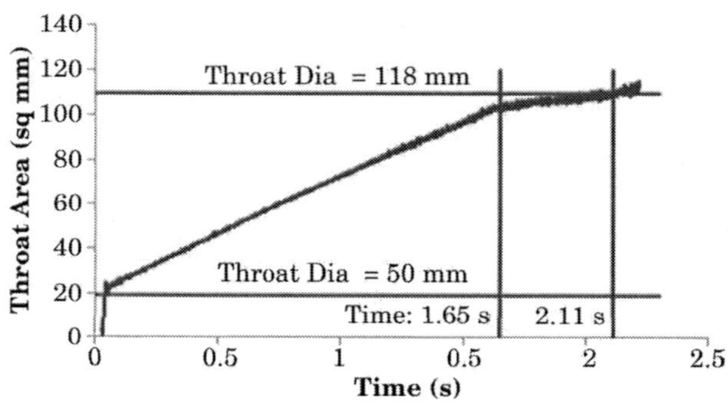

Fig. 148: Variation throat diameter in nozzleless propellant configuration

If variation of throat (port) diameter is available, then from burning area profile, pressure time profile can be calculated in normal way. However, constant thrust coefficient is an assumption, which needs validation. Another problem with this configuration is availability of certain propellant weight in divergent section. The propellant in divergent section may be contributing for the generation of thrust but their contribution in pressure is debatable and the values of C* may need certain fine tuning, by excluding mass of propellant lying outside the throat section. Although, this seems to be a difficult configuration to handle, through internal ballistic calculations, it has more practical utility for integrated rocket ramjet applications for air-launched rockets.

Dual Thrust Rocket Firing

Firing of single-chamber single-nozzle rocket configuration with two separate levels of thrust has been realized in several ways to accomplish specific tasks. In such firings two separate levels of near neutral thrusts are available. In dual thrust configurations, *initial high thrust for small duration (called booster)*, followed by *small thrust levels for prolonged duration (called sustainer)* is generally realized. For ASLV, additional thrust in the initial phases of lift-off in imparted by strap-on boosters (chapter 4). This may be considered as a typical example of multi-thrust configurations. It is clear that reverse dual thrust, where initial lower thrust, followed by high thrust, is rarely found in practical rockets.

The ratio of higher thrust to lower thrust is called *thrust ratio* and it is always more than 1. In fact, pressure variation is also realized and term like *booster-sustainer pressure ratio* may also be floated for such configurations. Booster-sustainer is added before pressure ratio to differentiate it from the pressure ratio (P_i/P_e) defined in chapter 3, as ratio of chamber pressure and exit plane pressure. As higher pressure ratio may lead to extinction of propellant due to rapid depressurization, a limit of 10 on thrust ratio is kept. However, firings of propellants with even higher thrust ratios are also successfully executed. One type of propellant shape, able to realize dual-thrust is depicted in chapter 6 under blind-hole configuration.

As per equation 17 of chapter 6, *Thrust*, $F = C_f P_i A_t$. This equation is valid for both booster and sustainer phase of dual-thrust rocket burning. If subscripts 'b' and 's' are used to represent booster and sustainer respectively, then thrust ratio for a dual-thrust configuration is given by product of thrust coefficient ratio, booster-sustainer pressure ratio and throat diameter ratio. So, thrust ratio $F_b / F_s = (C_{fb} / C_{fs}) \times (P_b / P_s) \times (A_{tb} / A_{ts})$. In a very simple case, if throat area remains invariant, then pressure realized in booster phase is higher than that in the sustainer phase. At high pressures, temperature is also high. At high temperature, quantity of high molecular weight combustion products is low and overall mean molecular weight of combustion product reduces. So, value of ratio of specific heat (g) increases. As throat area is constant, higher ratio of specific heats gives higher pressure ratio (Figure 16 of chapter 3). So, in booster phase higher pressure ratio (P_i/P_e) is realized, across nozzle. High value of pressure-ratio (P_i/P_e) for same ratio of specific heats (g) gives high thrust coefficient (figure 19 of chapter 3). It is clear that both pressure and pressure ratios are high in booster phase. At higher value of ratio of specific heats (g), for same pressure ratio (P_i/P_e), thrust coefficient (C_f) is lower (Fig. 19 of chapter 3). But if pressure ratio also increases, it is difficult to understand the bahaviour and comment on variation or values of thrust coefficient in booster and sustainer phases is difficult to make. So, booster-sustainer pressure ratio (P_b/P_s) may be higher or lower than thrust ratio (F_b/F_s).

However, for firing rocket in non-vacuum conditions (atmospheric pressure, $P_a > 0$), the equation for thrust is written as *Thrust,* $F = C_f P_i A_t - P_a A_e$. For this expression for thrust, thrust ratio $F_b / F_s = [(P_b - P_a e C_{fb}) / (P_s - P_a e C_{fs})$, *where* $e = A_e / A_t$. This clearly indicates that even if thrust coefficient is same, the thrust ratio exceeds booster-sustainer pressure ratio. Suppose booster-sustainer pressure ratio (P_b / P_s) is 5 and the deducted term from pressure $(P_a e C_{fb})$ is 0.3 times sustainer average pressure (P_s), then thrust ratio is equal to 6.71, which is more than booster-sustainer pressure ratio (5). If atmospheric pressure is zero, then booster-sustainer pressure ratio is equal to thrust ratio.

The dual thrust configuration can be conceived in two broad ways à (i) single-chamber configuration (ii) two-chamber configuration. For both the cases, variation in number and throat diameter of nozzle may play a vital role. Obviously, there are several methods of getting dual thrust in rocket firings. If rocket nozzle is considered, single nozzle and multi-nozzle configurations can produce dual thrust combustion profiles. In single nozzle single chamber configuration following features can be thought off:

- The rocket configuration is simple and assembly of such rockets is also easy. However, propellant processing, ageing and compatibility issues must be addressed properly.
- Single nozzle single chamber configuration relies on variation of burning area and burning rate of propellant during combustion for achievement of booster-sustainer pressure ratio and thrust ratio.
- Since rocket is operating at two different levels of pressure with constant area ratio of nozzle, the configuration cannot be optimized for both the pressure levels. If nozzle is designed for booster phase (higher pressure), nozzle area ratio will be high and it results in poor performance during sustainer phase. If nozzle is designed for sustainer phase, nozzle area ratio will be lower than optimum and booster operation will be sub-optimal.
- As requirement of high thrust ratio demands higher values for booster-phase pressure, there is associated weight penalty on hardware for very high values. Higher pressure requires thicker rocket motor casing and pressure during booster phase cannot be increased beyond a certain level. This restricts achievement of higher thrust ratio in single chamber configurations. In fact, during sustainer phase, thicker rocket casing is unavoidable weight penalty on the system due to higher booster phase pressure.
- Another alternative for higher thrust ratio is to reduce sustainer phase pressure. But low pressure combustion limit of propellants prevents further reduction of sustainer phase pressure. As

described in chapter 2, propellant burning is smooth, when operating pressure is above their LPCL. Obviously, Composite propellant allows operation of sustainer phase at lower pressures (12-15 atm), thus permitting larger (booster-sustainer pressure ratio) thrust ratio for same booster phase pressure.

- Overall, booster phase pressure enhancement is limited by hardware weight increment and sustainer phase pressure reduction is restricted due to combustion limits of propellants, thrust ratio higher than 8 is practically difficult to achieve. But rocket operating at more than 20 thrust ratio are also practically demonstrated.
- One variant of dual thrust rocket configuration may be obtained by use of propellants with two compositions, having different burning rates burning. In this configuration, there is no change in nozzle-action and same nozzle is operated during both the phases of thrust generation. The propellant with higher burning rate is ignited first, which gives high thrust and after consumption of this propellant, propellant with lower burning rate is ignited. In such configurations, transition of burning from propellant of one burning rate to the other, becomes critical. Another problem of this type of configuration is compatibility of mechanical properties, coefficient of thermal expansion, composition and processing steps in propellants of different burning rates. One of the major problems is encountered due to formation of propellant-propellant interface at the junction of propellants of different burning rates. However, such configuration is practically realized in propellants of anti-tank missiles. Some of the dual-thrust propellant configurations with propellants of two different burning rates are depicted as Fig. 149.

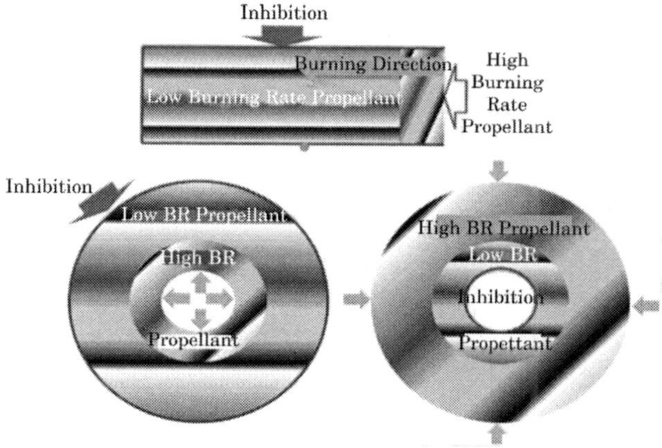

Fig. 149: Dual thrust propellant configurations with two propellants

The propellant may be burning in longitudinal direction or it may be burning in radial outward/inward direction.

- o For burning in longitudinal direction, the propellant behaves as solid sustainer and burning surface area of the propellant is constant. Throat area is invariant and propellant density and characteristic velocity may be assumed to have negligible variation in two phases of thrust. Hence, pressure ratio is dependent on burning rate ratio and burning rate index (n). Suppose, burning rate ratio of two propellants is two and burning rate index for both the propellant is 0.3, then booster-sustainer pressure ratio 2.69. If same propellant has higher value of burning rate index (say 0.6), then booster-sustainer pressure ratio becomes 5.66 (high). Obviously, for higher values of burning rate index, small burning rate ratios may result in higher booster-sustainer pressure ratio, leading to higher thrust ratio.
- o For cylindrical end inhibited propellant, inhibition can be applied at the inner diameter (radial inward burning configuration) or outer diameter (radial outward burning configuration). If higher burning rate propellant is placed at lower diameters and burning pattern is radial outward, then effect of initial higher burning rate is offset by initial lower burning surface area and full potential of burning rate ratio cannot be realized. However, in radial inward burning, higher burning rate propellant is at outer diameter (higher surface area) and better thrust ratio can be realized. However, radial burning propellant grains are either progressive or regressive and neutral thrust levels in booster and sustainer phases is difficult to achieve in such configurations.

- Another variation is possible, where burning surface area is tailored by geometry and inhibition to give two separately identifiable levels. Higher burning area corresponds to higher thrust phase and lower burning area corresponds to sustainer phase. Under such circumstances, same propellant acts in both the phases and stable burning of propellant during low and high pressures must be ensured, before use. This is depicted in blind-hole configurations of chapter 6. Complete theoretical analysis is presented in that chapter.
- Sometimes in a single nozzle single-chamber configuration, two different propellants grains are used in such a way that in the initial phases of burning, both th4e propellant grains are burning, giving as higher value of thrust (Booster). One of the propellant is configured for faster consumption and second propellant continues to burn, alone, as sustainer in later phases of rocket operation. This type of configuration can also be conceived.

Change of configuration or composition of propellant for achieving dual-thrust ballistics is an easier option, where configuration of rocket is simple. However, sudden change in nozzle throat diameter or operating multiple rockets in tandem may be useful in some of the practical situations. Special features in hardware are needed to accommodate variation in nozzle-configuration during firing. Multiple nozzle configurations in dual thrust mode may give following options for analysis.

- In first type of configuration, propellant grain is designed to give neutral burning profile. However, in course of burning, at certain instant, rocket nozzle is enlarged by additional means. This enlargement of nozzle throat diameter results in lowering of equilibrium chamber pressure (and thrust). The enlargement of nozzle has to be programmed and ejection is to be planned properly and instantaneously. In such configuration, enlargement of nozzle results in lowering of thrust, if same nature of burning continues. In ramjet configurations of typical missiles (KVADRATH of Russia, AKASH of India), the sustainer phase require intake air and to accommodate higher quantity of exhaust gas, nozzle is enlarged, which is accompanied with lowering of chamber pressure. The extinction of propellant must be checked for such configurations, if single propellant grain is used in both the phases. In this case propellant during booster and sustainer remains same, but area ratio of nozzle throat during booster and sustainer phase can be correlated with booster-sustainer pressure ratio for a given burning rate index. If burning rate index for the propellant is 0.3 and ratio of nozzle throat diameters during sustainer and booster phases (throat diameter during sustainer is larger than that during booster phase) are 2, then booster-sustainer pressure ratio will be 7.246. This enhancement is more than same ratio of propellant burning rate because nozzle area is square of nozzle throat diameter. This configuration may be considered as two-nozzle single chamber configuration.
- In another type of configuration, two different propellants are located in different rocket chambers and have different nozzles. One motor operates as booster, while other acts as sustainer. However, packaging of such system, needs that rocket motor in the front, should have central blast tube with nozzle at the end, while the other rocket must have multiple nozzles, located around central nozzle. The blast tube of front rocket must be located through propellant port of rear rocket. The blast tube must be insulated on both outer and inner side and it is exposed to hot combustion gases of propellants on both the sides. This type of

configuration may be known as two-nozzle two chamber configuration. There is no limitation of thrust ratio and practically 50:1 is also possible.

For dual thrust configuration, correlation between (burning surface) area ratio and booster-sustainer pressure ratio is derived based on theoretical ballistics. However, relation between booster-sustainer pressure ratio and thrust ratio is slightly cumbersome. This is because of variation in thrust coefficient, mainly. One booster-sustainer firing curve is depicted as Fig. 150. Obtained pressure-time and thrust-time profiles for the configuration is shown in the figure. This is obtained by firing single nozzle single chamber configuration, where propellants of different burning rates but constant burning surface area are used. The throat diameter of the rocket is 3.8 mm, which gives nozzle throat area as 11.34 mm^2.

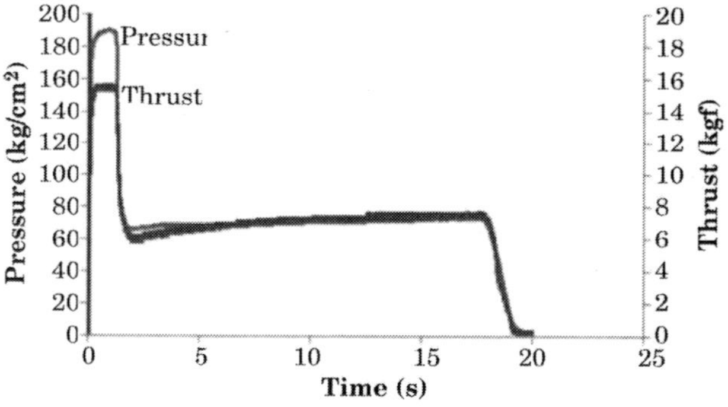

Fig. 150: Firing curves for dual thrust rocket

From the figure, average pressures, during booster and sustainer phase are found to be approximately 187.85 and 70.61 kg/cm^2, respectively. The value of booster-sustainer pressure ratio is for this firing is thus 2.659. Similarly, the values of average thrust, during booster and sustainer phases, are 15.3 and 7.2 kgf, respectively. The value of thrust ratio for this firing is 2.125.

To get an insight into the value of thrust coefficient, variation of thrust coefficient with time is plotted as Fig. 151. It is clear that thrust coefficient is almost invariant during booster and sustainer phases, but their values during these two phases are different. The values of thrust coefficient for booster and sustainer phases are 0.71827 and 0.899103, respectively.

For the cited example, it is found that thrust coefficient during booster phase is lower than that in the sustainer phase. This is contrary to the calculated values from the equations. It is also observed that booster-sustainer pressure ratio is higher than thrust ratio. So, a universal fact about booster-sustainer pressure ratio and thrust ratio cannot be conclusively derived for the dual thrust rocket motor in practical firings. These results changes and prediction about thrust ratio or booster-sustainer pressure ratio for booster or sustainer phases cannot be made. This needs actual firing of rockets in those configurations.

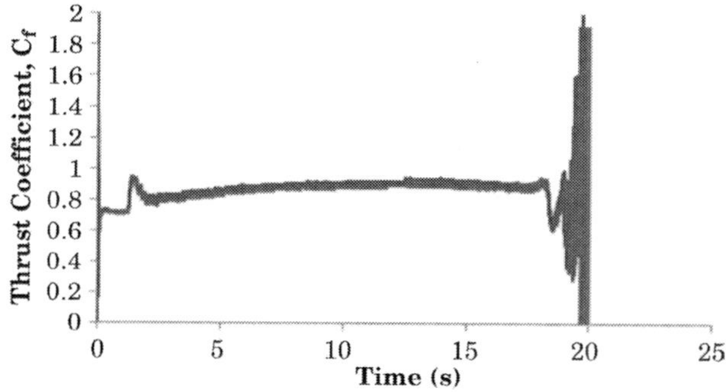

Fig. 151: Variation of thrust coefficient in booster and sustainer phases

REVIEW QUESTIONS

1. How is burning profile of multi-perforated propellant different from progressive tubular propellant configuration?
2. Derive expression for the evaluation of burning rate from pressure-time profile of multi-perforated rocket propellant combustion?
3. Why is M-shaped curve obtained for the 2-pitch multi-perforated configurations?
4. What are the advantages of nozzleless booster?
5. Why is pressure-time profile of the nozzleless booster configuration regressive, although burning area variation is progressive?
6. Why is thrust-time profile of nozzleless booster configuration neutral, despite pressure-time profile being regressive?
7. What is strategy for the calculation of ballistics for nozzleless booster rockets?
8. How is dual thrust attained in rockets?
9. What are possible methods of attaining dual thrust in single-chamber configurations?
10. How does thrust coefficient during booster and sustainer phases vary?
11. Comment on thrust ratio in dual-thrust rocket configurations.

15

Combustion Instability in Solid Rocket Propellants

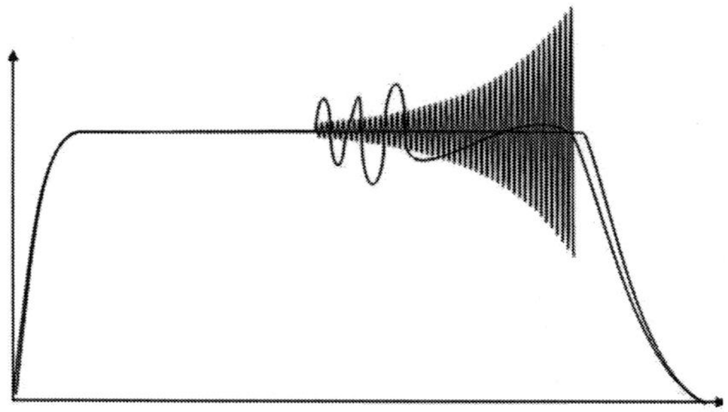

CHAPTER SUMMARY

Despite presence of several claims to predict combustion instability in rocket propulsion, it remains a grey area in rocketry. Claims for presence of combustion instability, and their suppression mechanisms are highly probabilistic and no solution is anticipated in near future. However, a brief about various aspects of combustion instability is presented in this chapter to understand and investigate them in this advanced area of rocketry. Combustion instability is present in rockets based on solid and liquid propellants. However, in this chapter, keeping central theme of book intact, combustion instability in solid propellant based rockets is only discussed. This chapter covers the following aspects:

- Introduction to combustion instability
- Historical perspectives in combustion instability
- Classification of combustion instability
- Suppression of combustion instability
- Current approaches to study combustion instability

INTRODUCTION

In any rocket operation working with solid propellants, following events are simultaneously occurring.

- Mass addition to propellant port by propellant combustion
- State change by surface burning of propellant
- Expansion of combustion gases inside chamber
- Chemical reactions inside propellant port
- Burning of inert materials like insulator, inhibitor and liners
- Heat transfer to propellant from combustion gases
- Radiation heating of propellant from high temperature combustion gases
- Heat losses to walls and their conduction
- Local energy imbalances due to absorption and emission
- Two-phase viscous flow due to presence of condense phase
- Eddies formation at longitudinal irregularities and nozzle entrance
- Turbulence in flow due to various non-streamlined flow
- Discharge of combustion gases from nozzle continuously at varying rates
- Mass imbalance due to flow rate variation
- Acoustic waves emanating from all rigid surfaces

The entire process mentioned above occurs simultaneously and it is highly optimistic to expect smooth pressure-time profile in course of rocket operation. However, rockets do oblige with smooth pressure time profiles and predictable performance from bulk parameter prediction strategy discussed in chapter 5. In certain other situations, erosive burning is observed, where burning rate of propellant is locally enhanced due to flow of combustion gases over the propellant burning surface. Erosive burning results in longer tail-off and introduction of regressivity in the pressure time profile. Yet another non-steady mechanism is combustion instability, where pressure inside rocket chamber fluctuates about a mean value with amplitude of oscillations as high as 20% of average instantaneous steady-state pressure. These oscillations subsides in certain cases and in other cases, it may aggravate. Sometimes marked increase in mean pressure is also observed due to combustion instability. The current chapter is dedicated to the phenomena of combustion instability with caution that neither it can be artificially created, nor it can be studied by scaled down models, nor it is understood completely. Various studies are in progress to understand these practical non-steady variations in pressure-time profile of the rockets. Usually oscillations are unforeseen and too complicated to predict beforehand. Combustion instability (CI) is encountered late in motor development programmes leading to limited options for any design changes. Tricks of the trade employed rather than scientific approach to solve the problem of combustion instability.

Features of Combustion Instability

Combustion instability is rapid fluctuating pressure wave inside rocket motor and is a function of many inherent parameters. One concurrent term is combustor instability, which refers to pressure oscillations related to combustion chamber and its geometry. Another is *combustion instability, which occurs due to positive combustion response to the oscillations in the combustion chamber*. So combustion instability is indirectly related to burning rate of propellants. In fact, there are many other important features of combustion instability, which makes its study multi-disciplinary phenomena. Knowledge of chemistry, physics, thermo-chemistry, thermodynamics, chemical kinetics, fluid-dynamics, aerodynamics, hydrodynamics, structural engineering are needed for the study of the following facets of combustion instability.

- Propellant composition and combustion wave structure.
- Acoustic in a non-regular geometry.
- Continuity equations with mass addition.
- Viscous two-phase flow in chamber.
- Energy transfer with wall heat addition.
- Port flow gas dynamics

The exact nature of combustion instability can be mathematically observed by considering complex variation of pressure inside rocket chamber.

$P^* = P\,e^{-iwt}$, where P is steady state pressure and w = frequency defined as f+ia.

$P^* = [P\,e^{at}]\,e^{-ift}$, where $[P\,e^{at}]$ is amplitude of pressure oscillations.

If $\alpha < 0$ pressure amplitude decays exponentially and system becomes stable

If $\alpha > 0$ pressure amplitude grows exponentially and system becomes unstable

So, to predict nature of oscillations, imaginary part of the frequency is to be considered. Unfortunately, this imaginary part cannot be measured by oscilloscope. Hence nature of combustion instability is unavailable, theoretically. Acoustic field is always present during rocket operation due to presence of compressible media. This field always moves back and forth within rocket motor. If the acoustic field does not influence flame/combustion, flame will continue to have its natural frequency. However, if flame is affected by acoustic wave, new coupled frequency will emerge, which is different from homogenous natural frequency. This coupling aspect is governed by imaginary part of complex frequency (α). The behavior of pressure for different values of amplification coefficient is shown as Fig. 152.

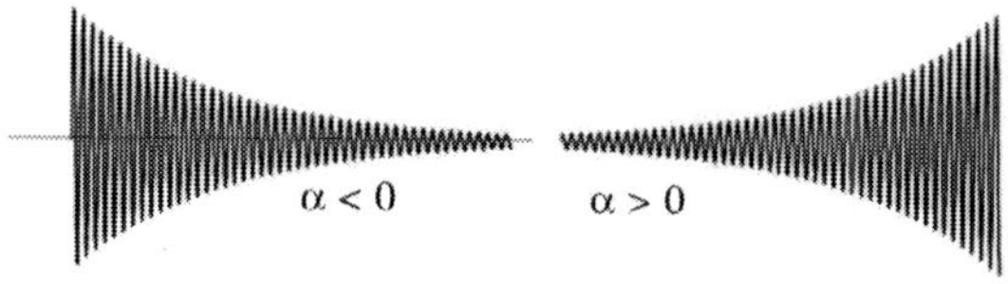

Fig. 152: Nature of combustion instability

In many literature, *the growth constant*, a is divided into two parts – *Driving factors and damping factors*. Driving factors are responsible for enhancing growth factor and making unstable for operation. Three factors are generally listed as driving factors – *Pressure coupling, Velocity coupling and Distributed combustion*. Pressure and velocity coupling arises due to response of propellant combustion to pressure and velocity fluctuations. The distributed combustion refers to delayed combustion in the propellant port and it mainly involves metal-combustion, which may result in generation of locations for initiation of combustion instability. Another type of factors are damping factors, which are responsible for reducing combustion instability inside rocket chamber. Main contributors are enlisted as – *Particulate damping, Flow turning, Structural vibration and Nozzles*. The particles of condensed phase in propellant port interfere with the movement of waves inside rocket chamber and ultimately damping effects are seen. Any flow turning results in change in physical dimension of the flow path and damping effect is initiated. The vibration of rocket casing or internal baffles may results in damping, but they may be driving factors also. The radiation loss from nozzle leads to loss of energy from longitudinal mode of vibration. This dampens combustion instability.

Pressure coupled combustion instability is present in longitudinal mode and has frequency in the range of 100-200 Hz. It is valid for rockets with lower change in pressure in course of burning. *Velocity coupling* is 90^0 out of phase with pressure coupling. Combustion of burning surface is sensitive to velocity parallel to the surface and velocity coupling in unsteady flow is prevalent (Erosive burning occurs in steady flow). As combustion process is dependent on magnitude of velocity and not direction of velocity, the velocity coupling may be treated as non-linear. The node of pressure wave coincides with anti-nodes of velocity coupling. Velocity coupling is dominant in transverse mode and is a term varying with location inside rocket motor. No means for measurement of velocity coupled response of propellant combustion is present. *Distributed combustion* is present at the junction of segments and is associated with vortex shedding and flow reversal.

Among damping effects, *particulate damping* is a predominant effect. Addition of aluminium (Al-powder) in propellant composition is considered an effective means of suppressing combustion instability. The generation of Al_2O_3 (alumina) is responsible for such damping and size of alumina matters. Around 15% of Al can generate 40% alumina. Particles and droplets are 500-5000 times denser than combustion gases, generated in rockets. Bigger particles do not block the flow and has less damping potential. Similarly, smaller particles move along with the flow, so they show lower damping. In fact, each size of particulate matter in flow field suppresses one frequency of combustion instability. Optimum particle size for suppressing given frequency of combustion instability, f is $\sqrt{\frac{9\mu}{\pi f \rho}}$, *where* μ = *viscosity of combustion gases,* ρ = *density of the particle in flow*. Optimum particle size decreases, if particles are denser. In fact, use of denser particles like silicon carbide, zirconium carbide is also reported for suppressing given frequency of combustion instability. Each particle size suppresses one frequency effectively. At fixed frequency, damping increases with mass of particle present. Temperature has no effect on particulate damping. Natural frequency of large motors is low (< 1000 Hz) and effective damping is possible by particle size of 8-30 micron. Since, majority of alumina particle lie in < 2 micron range, it is not possible to suppress low frequency oscillation in such motors. Transverse mode oscillations have high frequency and smaller particle size (3-5 micron) is needed for suppression of combustion instability. Large amount of fine alumina can dampen this mode of oscillations. Transverse modes are suppressed by adding Al, but longitudinal modes are unaffected. Coagulation of alumina particles modifies size distribution considerably, giving unpredictable results. As Al increases at the cost of oxidizer, wave strength and fluctuation reduces. As Al increases for higher burn rate propellants, burn rate reduces. Faster burning propellants are stable at highest operating pressure. Finer Oxidizer (AP) favors high burning rate and lower severity of instability. As binder rises, burn rate reduces, severity of instability reduces.

Flow turning has damping effect on combustion instability. The gasified components of solid propellants are released in port (free volume) in radial direction. As flow in the propellant port is longitudinal, flow turning is a natural phenomena, which consumes energy. This leads to suppression of combustion instability. Cavity geometry, bulk damping, boundary and shear layer dissipation affects this factor. Another damping factor is *nozzle discharge*. Acoustic waves in rocket motor are reflected from open end (nozzle), in similar fashion, as they reflect from the rigid walls of the rocket motors. During this, transverse waves are released to environment, resulting in damping effects. Reduction in throat diameter makes motor unstable, but with consumption of propellant, (A_p/A_t) and K_n (A_b/A_t) rises, nozzle damping reduces.

The effect of combustion instability on mission, sub-components and environment is detrimental. If it is only resulting in pressure oscillations in rocket motor, there is no necessity of reducing, eliminating or controlling it, but its adverse devastating effects require proper understanding of mechanism to suppress the occurrences of combustion instability in rockets. For example, combustion instability in solid rocket boosters, as low as 15 Hz is capable of giving a jittery sensation to the astronauts. Re-ignition of solid propellant grain is possible due to rapid heating under combustion instability. Sometimes burning rate enhances, sometimes pressure levels changes, sometimes thrust direction and magnitude varies resulting in very uncontrollable situations. The effect of combustion instability on various components of propulsion is summarized as features of combustion instability.

- ***Effect on propulsion*** – Combustion instability inside rocket motor leads to variations in operating pressures and burning rate. Although pressure only oscillates and marked variation is observed only in case of DC shift, burning rate of propellant invariably changes significantly. In fact, pressure oscillations may lead to very high variation in thrust-oscillations. For example, 0.1 bar oscillations over 50 bar average pressure results in 5-10 times thrust oscillations. Sometimes the values may go out of acceptable limits of the mission. The variations mentioned may also result in compromise on reliability and performance. Ultimately, cost is at stake. The structural integrity of rocket casing is greatly affected by pressure and heat transfer and repetitive high pressure fluctuating loading on casing results in premature failure. It is not casing alone, but propellant is also stressed beyond their capabilities. The fluctuating pressurization results in enhancement of burning rate by rapid heating of local surfaces of the propellants and prediction is offset. Difficulty of ignition and sustaining combustion of propellant is also encountered as an outcome of combustion instability.

- ***Effect on vehicle*** – The fluctuating pressure waves generates fluctuation in thrust. During this fluctuation of thrust, acceleration levels may reach a level where premature actuation of warhead fuze may result in warhead function before intended time. The guidance program is unnecessarily loaded with more correcting torques to counter the variations due to combustion instability. This may result in unscheduled maneuvers and sometimes vehicle break-up. The control system is also adversely affected and may result in jamming of electronic switches. The components of vehicle and structures may be subjected to excessive vibrations.
- ***Effect on environment*** – Any rocket interacts with environment through discharge of exhaust. The combustion instability results in fluctuations in exhaust and any control in afterburner may not function correctly under fluctuating exhaust. The signature characteristic of exhaust will change. More discharge of condensed phase is envisaged and longitudinal mode (discussed later) of instability may send out sound waves, which interferes with test-site and launch-site functioning.

Types of Combustion Instability

The classification of combustion instability is based on several critical parameters. One such classification is based on role of acoustic or sound waves in the flow through rocket motor. Two types of combustion instabilities are generally prevalent in the rocket motors – Acoustic and Non-acoustic.

- ***Acoustic instability*** – Each object has a natural frequency of vibration and if applied frequency has same frequency as natural frequency of the object, resonance occurs and object oscillates with very high amplitude. Excitation of natural frequency of the object is one significant mode of instability. In addition to this, spatial variation of pressure due to localized effects may also generate sound waves, which can trigger acoustic instability. The frequency of acoustic instability in rocket may vary from 10 Hz to 10 kHz. As burning rate and pressure in rocket motor changes with time, time dependent or time-independent oscillatory burning is possible in rockets. The pressure inside rocket motor and velocity of flow inside rocket motor is responsible for adding combustion instability. This velocity coupled or pressure coupled oscillatory combustion leads to rise in local burning rate, heat transfer rate and operating pressure. Depending on movement of oscillatory wave in rocket chamber, three modes of combustion instability are prevalent.

- o *Longitudinal* – During longitudinal mode of instability, high pressure zone traverses length of rocket motor at certain velocity in both back and forth directions. For the fundamental mode of oscillations, this high pressure zones goes from head end to nozzle end and returns thereafter. Any pressure gauge located at any position on the rocket, will observe periodically rising and falling value of pressures. Along with axial movement, this wave has constant pressure amplitude. The most basic feature of this mode is constant (not changing with time) frequency of oscillations. The frequency of oscillations is given by expression na/2L, where n = Integer indicating mode number, a = velocity of sound in the rocket motor, L = length of rocket motor. So, smaller motor has higher value of basic frequency. The compression wave movement through the free space inside rocket in longitudinal mode of oscillations is shown as Fig. 153.

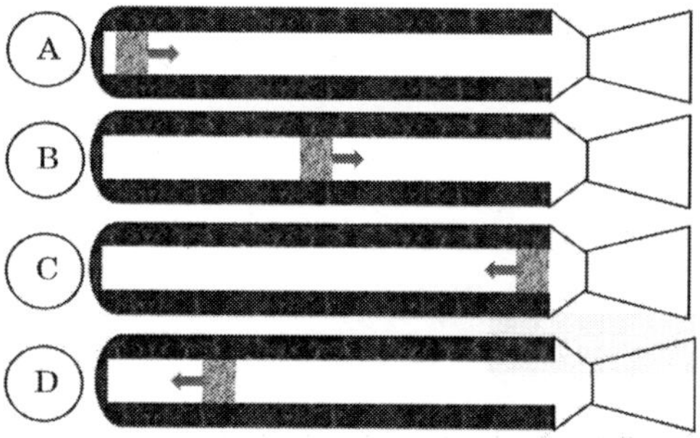

Fig. 153: Longitudinal mode combustion instability

It is clear from the figure that a compression zone created during combustion in rocket motor at the head end (Fig. A), generates high amplitude of pressure at the pressure gauge located at head end. This move towards nozzle end (Fig. B) and any pressure gauge experiences pressure surge at that instant, when it reaches location of mounted pressure gauges. It goes to nozzle end (Fig. C) and reflected back towards head end (Fig. D). This is basic mode of oscillations. Sometimes, compression wave is generated simultaneously at both the ends (A and C together at same time). Both move towards center, collide there and then reflect back. This mode of vibration has higher frequency (lower time period) and lower amplitude, generally.

Radial – This is first type of transverse mode of combustion instability, where surge of compression wave moves in radial outward direction from center and after reaching rocket casing or internal insulation, it returns back. The functioning is shown pictorially in Fig. 154. This is

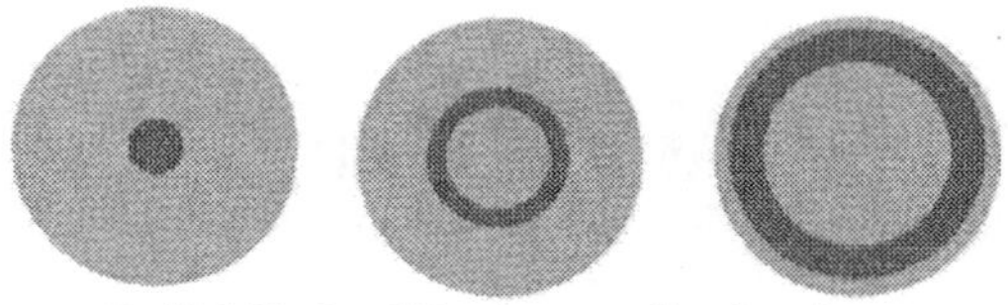

Radial Mode of Transverse Combustion Instability

Circumferential Mode of Transverse Combustion Instability

Fig. 154: Transverse mode of combustion instability

difficult to detect by normal pressure measurement tools. This type of oscillations has very high frequency due to relatively shorter length for reversal. As propellant geometry changes in course of burning, the frequency of oscillation may not be constant. However, thermal insulation may dampen the oscillations.

- o *Circumferential* – In this mode of transverse oscillations, compression waves move in peripheral direction in circular pattern (Fig. 154). This has frequency lying between longitudinal and radial mode of combustion instability. This mode has more or less constant frequency of oscillations and is present for cartridge-loaded configurations in external ports. This mode of oscillations is also difficult to measure.

- ***Non-acoustic instability*** – Any instability other than movement of waves or disturbances inside rocket chamber is responsible for the generation of non-acoustic combustion instability. It may be initiated as local event but may capture entire rocket motor and their amplitude may also magnify in course of burning. These non-acoustic combustion instabilities are triggered due to various phenomena.

  - o *Slag ejection* – Rocket is continuously ejecting hot combustion gases from the nozzle. However, in addition to normal ejections, if condensed solid mass is ejected out of nozzle, disturbance towards head end travels from the nozzle. This may result in combustion instability in rocket operation. The ejected mass may be a piece of igniter canister, it may be slag deposited on rocket nozzle or it may be a piece of nozzle (convergent or throat), insulator, inhibitor or propellant etc. All ejected particles may temporarily block the nozzle, resulting in pressure kink and sometimes local reduction in thrust. The pressure kink sets oscillating waves in the combustion flow field inside rocket motor, which may start combustion instability.

- *T* combustion instability* – Flame temperature in rocket always below adiabatic flame temperature because of incomplete combustion, cooling due to expansion, heat loss to walls and dilution. Most of the time, actual temperature follows power law, similar to one followed by burning rate of propellant with pressure exponent as 'm'. The equation is written as $T^* = b \times P^{2m}$. If burning rate index is 'n', then criteria for stable burning is $m+n < 1$ and not $n<1$, as depicted earlier in chapter 5. The burning rate dependent criteria may not be sufficient for stable burning of propellants. In high pressure regions, value of $m \simeq 0$ and stability with $n<1$ is sufficient. At lower pressure region ($P < 1$ MPa), the value of m as high as 0.35 is possible, which limits value of n for stable combustion to 0.65. In low pressure region, high burning rate index (n) of double base propellants leads to chuffing. The propellant extinguishes and reignites leading to unstable combustion. This T* instability is responsible for low pressure combustion limit of propellants. This mode of combustion instability is present for propellants with high burning rate index (n) and high pressure exponent (m) of flame temperatures (m). In this case sum of 'n' and 'm' exceeds 1 and combustion becomes unstable.
- *L* combustion instability* – L* is defined for a rocket motor as ratio of free volume inside rocket chamber and throat area of the rocket nozzle. The value of L* in an end burning solid sustainer rocket propellant grain can vary from 4 to 20. Low L* leads to low frequency oscillations in the rocket motor. This

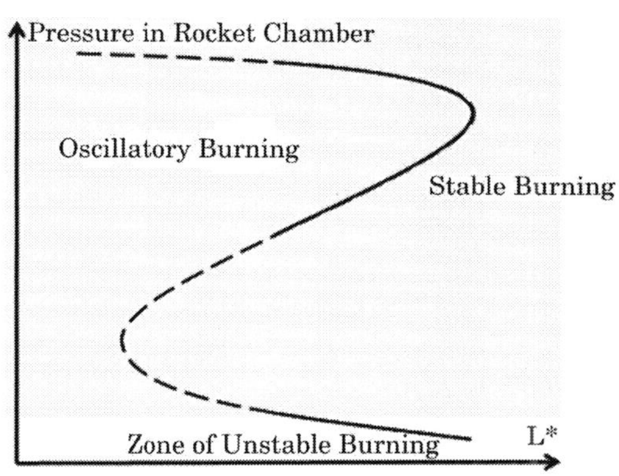

Fig. 155: Combustion behaviour in different operating pressures and L*

instability depends on burning rate index (n) and free volume in the rocket chamber. L* instability has higher probability, if 'n' is near unity. For low L*, the highest amplitude of oscillations may be as high as 20* of average pressure, but as L* increases in course of burning, the amplitude of oscillations decreases. Contrary to T*, L* is a high pressure phenomena. At low pressures, even if 'n' is high, no combustion instability is observed. Instead of this in this region, mass discharge rate exceeds mass generation rate and unstable burning occurs. The propellant burning behavior in rocket is depicted in Fig. 155, along with different zones of combustion. The oscillatory burning zone is relevant for combustion instability and is possible for low L* and moderate pressures in rocket chamber.

If combustion instability is so devastating, the source of devastation must be estimated. Suppose combustion gases with ratio of specific heat (g) as 1.2 is present in rocket chamber with mean operating pressure (P) of 70 bar. If 10% of mean pressure is assumed as amplitude of pressure oscillations, then energy required for generating the frequency in a rocket chamber of 300 mm cross-sectional area (A) and 1000 mm length (L) is $PV\,(dP/P)^2/4g = 70 \times 300 \times 1000 \times 0.1^2/(4 \times 1.2 \times 100)$ *kg-mm* = *437.5 kg-mm* = *4.2918 J* = *1.025 cal*. Such small energy is needed for generating 10% pressure oscillations. Considering calorific value of propellant as 1000 cal/g, *the amount of propellant required to impart this energy is 1.025 mg, only*. So, energy source is available inside rocket chamber for the initiation of combustion instability.

One of the peculiarities of combustion instability is difficulty with scaling. Combustion instability occurring in larger rockets could not be replicated by scale-down rockets. Suppose axial mode combustion instability is triggered in 1 m long cavity. Then basic natural frequency of oscillations is (a/2 L) 600 Hz, under assumption that velocity of sound is 1200 m/s. Since combustion instability require, generation of similar frequency in scale down model. For this, if half model is considered, length reduces to ½ m and to maintain basic natural frequency, sonic velocity in the combustion gases must be reduced equally (to 600 m/s). This requires reduction in flame temperature of gases and ultimately some other cooler propellants should be used in the scaled down model. Thus scaling of CI is impossible. Larger motors are less prone to CI, but CI is observed in TITAN-III and SPACE-SHUTTLE booster due to low dynamic response of combustion at low characteristic frequencies. Such effects cannot be duplicated in smaller rockets, because alumina may act as CI suppressant. For 600-1800 Hz and characteristic dimensions of 1/3 m to 1 m, competition between two-phase flow and combustion response makes prediction of stability more difficult.

Historical Perspectives

The history of combustion instability moved parallel to the history of solid propellant development. Initial days, sophisticated tools for proper diagnosis of combustion instability were not available. However from various signatures, unpredictable performance of solid propellants has been recorded in open literature. Double base propellants have been in initial days, all earlier records of combustion instability involves mention of this type of propellant.

The first record of erratic behavior of propellant combustion was recorded in 1937 for cordite propellants. The propellant was internal burning type and unavailability of pressure measurement resulted in inconclusive analysis. The differentiation between combustion instability from other forms of problems like break-up of propellant charge, plugged nozzles, non-reproducible propellant properties etc could not be ascertained. However, combustion instability may be one of the culprits for the failure. In 1940, testing of extruded double base (EDB) propellant at California Institute of Technology, USA reported a peculiar observation about ejected pieces of propellants from rocket nozzle. The ejected mass was having heated surfaces and rippled surfaces showed gas flow patterns. The alteration of burning rate is also envisaged. In 1940, another incident of combustion instability is sighted with solvent extruded tubular propellant grain in cluster formation. The combustion instability is suppressed by insertion of rod. It is clear that rod cannot interact with longitudinal mode of combustion instability. So, transverse mode of combustion instability was observed and suppressed in this exercise.

It was earlier thought that NC-NG based propellants are susceptible to combustion instability, only. But later by 1950, composite propellants with aluminium as one of its ingredients were found to show combustion instability. By 1960, all rockets exhibiting CI were compiled by Price, but efforts to suppress CI were not undertaken. The particulate damping of alumina was understood. One of the significant development was during development of torpedo tube launched surface to surface weapon "SUBROC" rocket motor (ϕ20"x120") by US NAVY in 1959. The rocket used AP-Polyurethane-Al based propellant and was supposed to be free from CI. No problem was observed in ¼ or ½ or full scale firings. However, in one full scale firing oscillations of 180 Hz was observed, accompanied with DC shift of 30% in pressure. Similar trend of exhibiting CI after many successful firing continued. Such sporadic evidence of CI questioned reliability of rocket. Complexities of scaling CI were observed during measures taken for suppressing CI. Presence of Al-powder in no way guaranteed absence of CI. Problem was solved by taking two measures – (i) retention of igniter casing was planned in place of earlier system of ejection of igniter casing and (ii) Finer Al-powder in place of existing Al-powder was used. Which of the mentioned measures helped in solving the problem was never known or investigated.

Minuteman (1961) missile developed is a unique example of staying with combustion instability. The missile propulsion was accepted with combustion instability at 450 Hz. However, during flight trial, oscillations rose to 1250 Hz and remedial measures became mandatory. The problem was identified as change in Al-supplier. New Al production plant was cooling aluminium at lower temperature, thus giving amorphous cxide skin of Al-particles in place of crystalline skin. Amorphous oxides were soft and Al can be penetrate more effectively. Agglomeration size in combustion product changed and ineffective damping resulted in enhanced oscillations in flight. As, earlier Al-supply could not be restored, problem solved by re-designing the control system. Tartar rocket (1965) contained concentric propellant grains to show boost-sustain configuration in 13" diameter. Resonant burning was observed at 270 Hz during sustainer phase. The sustainer composition contained AP+NQ+PU and sustainer pressure rose 2 to 3 times nominal. The oscillations were suppressed by reducing throat. During 1967-68, f2.75"x27" rocket motor showed low amplitude 600 Hz oscillations in 1st second of firing, accompanied with 3% shift in pressure. Port was modified to have a flare end, but occasional thrust oscillations at 800 Hz along with large shift in mean pressure continued to occur. Another rocket motor, working in dual mode exhibited axial mode instability at 650 Hz during transition from booster to sustainer phase. This propellant was containing around 16% Al-powder.

During development of ATR rocket motor (1974), based on smokeless-CP, many geometries (3-blade aft finocyl, Dogbone) and propellant compositions (9 different compositions tried) were tried. Dogbone shaped propellant showed axial mode oscillations at -54°C but at 74°C, only tangential mode of oscillations observed. During development of ARROW rocket, as part of SEAS program (1976), 4-blade aft finocyl grain showed axial mode oscillations with large DC shift. First fins were modified, and then shape was changed to Dog-Bone. Axial oscillations disappeared, but tangential mode of oscillations started appearing. Finally compositions were modified to get rid of combustion instability. During CHAPARRAL rocket development, low smoke HTPB based propellant in 5" diameter port showed axial mode frequency of 300 Hz.

In 1980, MK-36 rocket showed axial mode of oscillations at 320 Hz. In 1994, SPACE-SHUTTLE showed pressure and thrust spikes. The rocket had submerged nozzle, where slag accumulation with subsequent ejection during TVC action occurred. The sudden occurrence of oscillations was attributed to change of Ammonium perchlorate supplier. IN 1997, retro motor of Mars-mission PATHFINDER was designed with 2% Al-powder, so as to minimize pollution of Martian atmosphere. However, combustion instability occurred and final rocket design was conceived with 18% Al-powder.

Among Indian rockets, PSLV has been accepted with oscillations of 28-34 Hz. The booster rocket of pilotless target aircraft (PTA) Lakshya, based on EDB propellants (in cluster of 7 tube configuration) showed CI at 510 Hz. It is accepted with CI, provided amplitude is within 10% of average pressure.

Advances in Combustion Instability

Suppression of combustion instability is attempted through various means in the past in absence of any known theoretical principles.

- Drill radial holes in tubular propellant grains.
- Insert solid rigid non-combustible rod in the propellant port.
- Add baffles or flow turning internal features.
- Add metal powder in non-metalized propellant compositions.
- Provide odd number of fins, star-points for propellant configurations.
- Change operating pressure levels in rocket operation.

But later on, exhaustive studies gave some insight into the complex phenomena of combustion instability. The advancements are due to developments in the area of advanced diagnostic tools, new multi-disciplinary theories and instrumented measurement support.

The thumb rules indicated above are refined after detailed diagnosis, investigation, study, theoretical developments and practical evaluations. Drilling radial holes is considered suitable to suppress longitudinal combustion instability, only when holes are located at the anti-node of the oscillations. This helps in creation of nodes, at the drilled-hole-locations and alteration in the frequency of oscillations. The amplitude of oscillations is also reduced and damping factor for the new frequency of oscillations may be predominant. Additionally, drilled holes act as locations of cross-flow locations, which help in reduction of CI. Thus CI is suppressed. Inserting solid rigid objects in the port can always change frequency and amplitude of oscillations and reduction in frequency and amplitude is envisaged.

Adding metal powder is one of the effective means for damping combustion instability oscillations. It is mainly aluminium-powder, which contributed to the phenomenon. It must be noted that each particle size have capabilities to suppress one frequency of oscillations, only. So, in the flow field, particle size of optimized dimensions must be generated. Although particles in flow filed are alumina, generated by combustion of aluminium-powder and particle size of alumina is governed to certain extent by the particle size of Aluminium-powder, but it is observed that particle size of alumina is given more importance. As there is no mechanism to measure alumina-particle size, it is not possible to give a give certify means to suppress combustion instability. There are other inert heavy particulate matters like silica, zirconium silicate etc, which are used for particulate damping and are ejected in same particle size in which they are added in the propellants, but they reduce performance of the rockets.

Providing odd number of fins are recorded as one way to suppress CI, but rocket with odd number of fins are also found to show CI. So, conclusive evidence on this aspect is still not available. Change in operating pressure levels of rocket by changing nozzle, throat, burning rate of propellant or propellant grain design may help in CI suppression, but prediction on suitable pressure levels for a given composition and configuration of propellant is unavailable.

Combustion instability is said to be studied using T-Burner. In T-Burner, equal mass, dimension and shape of propellant closes both ends of head of T. The vertical stem is connection to surge tank, which is used for pressurization and depressurization of the T-Burner. Pressure measurement occurs at both ends of the head of the T by fast response pressure transducers. Both propellants are ignited simultaneously after pressurization of burner to preset levels. The combustion wave moves from both ends towards center and generated oscillations is captured by pressure transducers. The length of head of the T has provision for change of length for changing different natural frequency. This system can be utilized to certain extent for studying combustion response of propellants.

Another study is through pulsing of rocket at preplanned time during operation. In this case, sudden pressure spike is generated by passing rigid mass of known dimension through nozzle. This spike may result in triggering oscillations or the spike remains an instantaneous feature in rocket firing without any consequent oscillations. This is a more realistic method for assessing combustion instability without changing configuration or composition.

Aluminised propellants also show combustion instability. Scaling of velocity of sound (lower frequency) needed for simulating CI at smaller scale. Combustion instability is to be suppressed, after understanding the mechanism of their generation through experimental techniques.

REVIEW QUESTIONS

1. What factors are responsible for the deviation from steady behavior in rocket operations?
2. Why rocket operation is not as per predictions in certain cases?
3. What is combustion instability?
4. What are detrimental effects of combustion instability?
5. Why is there a need to suppress combustion instability?
6. What are driving factors for combustion instability?
7. What are damping factors of combustion instability?
8. How is particulate damping different from other damping factors of combustion instability?
9. Classify combustion instability and explain their features.
10. What is acoustic mode of combustion instability?
11. What is non-acoustic mode of combustion instability?
12. What is T* instability?
13. What is L* instability?
14. Why it is difficult to scale CI?
15. What is source of energy for the generation of CI?
16. Differentiate between
    a. Longitudinal and transverse mode of combustion instability
    b. Acoustic and non-acoustic mode of combustion instability
17. How combustion instability in rocket is detected in earlier days in absence of proper diagnostic tools?
18. What are modern methods for studying combustion instability?

Appendix 1

Tables of Internal Ballistic Parameters

Values of $[1-(P_e/P_c)^{(\gamma-1)/\gamma}]$ for $P_e = 0.101325$ MPa

Chamber pressure (MPa)	*Values of γ for the Combustion Gases of Rocket Propellants*					
	1.13	*1.15*	*1.19*	*1.21*	*1.25*	*1.31*
3	0.32279	0.357197	0.417803	0.444566	0.492169	0.551456
4	0.344836	0.380871	0.44394	0.471617	0.520563	0.580976
5	0.361441	0.398631	0.463403	0.491689	0.541489	0.602528
6	0.374695	0.412764	0.478798	0.507521	0.557908	0.619312
7	0.385687	0.424453	0.491469	0.520522	0.571329	0.632949
8	0.395052	0.434391	0.502196	0.531506	0.582626	0.644366
9	0.403194	0.443014	0.51147	0.540986	0.592343	0.654142
10	0.410384	0.450616	0.51962	0.549303	0.600843	0.662658
11	0.416814	0.457403	0.526875	0.556697	0.60838	0.670181
12	0.422623	0.463527	0.533402	0.563341	0.615136	0.676903
13	0.427915	0.469099	0.539327	0.569365	0.621248	0.682965
14	0.432772	0.474206	0.544746	0.574868	0.626821	0.688477
15	0.437256	0.478916	0.549734	0.579928	0.631935	0.693522
20	0.455576	0.498107	0.569948	0.600387	0.652514	0.713692
25	0.469374	0.512504	0.585	0.615567	0.667681	0.728418
30	0.480388	0.523961	0.596906	0.627541	0.67958	0.739886

Value of A_e/A_t of Nozzle with P_e = 0.101325 MPa

Chamber Pressure	*Values of γ for the Combustion Gases of Rocket Propellants*					
(MPa)	*1.13*	*1.15*	*1.19*	*1.21*	*1.25*	*1.31*
3	5.371162	5.193556	4.871926	4.726015	4.460003	4.115776
4	6.703276	6.459107	6.018869	5.82001	5.458918	4.994607
5	7.976931	7.665584	7.106157	6.854318	6.398473	5.815254
6	9.206346	8.827412	8.148484	7.843706	7.293487	6.59244
7	10.40049	9.953684	9.155088	8.79745	8.153245	7.335355
8	11.56551	11.05061	10.13227	9.721862	8.984055	8.050221
9	12.70582	12.12267	11.08456	10.62149	9.790451	8.7415
10	13.82476	13.17325	12.01538	11.49976	10.57584	9.412532
11	14.92493	14.20496	12.92738	12.3593	11.34285	10.0659
12	16.00842	15.21991	13.82267	13.20225	12.09358	10.70364
13	17.0769	16.21979	14.70296	14.03031	12.82973	11.32743
14	18.13177	17.20603	15.56969	14.8449	13.55271	11.93864
15	19.17423	18.17981	16.42405	15.64723	14.2637	12.5384
20	24.23097	22.89286	20.54137	19.50582	17.66951	15.39536
25	29.0839	27.40199	24.45715	23.16503	20.88156	18.06891
30	33.7822	31.75666	28.22081	26.67403	23.94815	20.60552

Values of Thrust Coefficient for Optimum Expansion (P_e=P_a = 0.101325 MPa)

Chamber pressure	*Values of γ for the Combustion Gases of Rocket Propellants*					
(MPa)	*1.13*	*1.15*	*1.19*	*1.21*	*1.25*	*1.31*
3	1.50323	1.494608	1.479181	1.472281	1.459909	1.444417
4	1.553717	1.543342	1.524746	1.516413	1.501431	1.482573
5	1.590685	1.578915	1.55781	1.548345	1.531312	1.509822
6	1.619588	1.60666	1.583476	1.573076	1.554353	1.530706
7	1.643171	1.629251	1.604292	1.593097	1.572939	1.547467
8	1.663001	1.648214	1.621706	1.609818	1.588414	1.561361
9	1.68005	1.664492	1.636612	1.62411	1.601605	1.57316
10	1.694965	1.678713	1.649599	1.636547	1.613055	1.583368
11	1.708191	1.691309	1.661074	1.647525	1.623141	1.592331
12	1.720052	1.702592	1.671332	1.657327	1.632128	1.600296
13	1.730789	1.712795	1.680589	1.666165	1.640217	1.607446
14	1.740583	1.722093	1.689011	1.674197	1.647556	1.613919
15	1.749578	1.730625	1.696725	1.68155	1.654264	1.619821
20	1.785853	1.764958	1.727638	1.710953	1.680984	1.643207
25	1.812696	1.790284	1.750303	1.732448	1.700408	1.660074
30	1.83384	1.810183	1.768025	1.749217	1.715494	1.673091

Values of Vacuum Thrust Coefficient for A_e/A_t tailored for P_e=0.101325 MPa

Chamber pressure (MPa)	*Values of γ for the Combustion Gases of Rocket Propellants*					
	1.13	*1.15*	*1.19*	*1.21*	*1.25*	*1.31*
3	1.684641	1.67002	1.64373	1.631902	1.610546	1.583427
4	1.723519	1.706959	1.677212	1.663841	1.639712	1.609092
5	1.752337	1.734259	1.701817	1.687248	1.660977	1.627668
6	1.77506	1.755733	1.721083	1.705537	1.677522	1.642036
7	1.793718	1.773331	1.736812	1.72044	1.690957	1.653646
8	1.809485	1.788176	1.750038	1.732951	1.702202	1.663322
9	1.823097	1.800974	1.761405	1.74369	1.711829	1.671575
10	1.835044	1.812191	1.771344	1.753068	1.720215	1.67874
11	1.84567	1.822156	1.780153	1.761371	1.727624	1.685051
12	1.855224	1.831105	1.788047	1.768803	1.734243	1.690675
13	1.86389	1.839215	1.795188	1.77552	1.740215	1.695735
14	1.871812	1.846622	1.801697	1.781637	1.745644	1.700325
15	1.8791	1.853429	1.80767	1.787247	1.750615	1.704518
20	1.908613	1.880939	1.831706	1.809774	1.770502	1.721204
25	1.930573	1.901344	1.849428	1.826336	1.785041	1.733307
30	1.94794	1.917441	1.863341	1.839308	1.796378	1.742686

Values of Characteristic Velocity, C* (m/s) for Different Combustion Gas Properties

Combustion gas Properties		*Values of γ for the Combustion Gases of Rocket Propellants*					
T (K)	*Mol Wt*	*1.13*	*1.15*	*1.19*	*1.21*	*1.25*	*1.31*
3500	18	2003.677	1990.929	1966.468	1954.727	1932.155	1900.393
3000	18	1855.045	1843.242	1820.596	1809.726	1788.828	1759.422
2500	18	1693.416	1682.642	1661.969	1652.046	1632.969	1606.125
2000	18	1514.638	1505.001	1486.51	1477.635	1460.572	1436.562
3500	20	1900.855	1888.761	1865.556	1854.417	1833.003	1802.871
3000	20	1759.85	1748.653	1727.169	1716.856	1697.031	1669.134
2500	20	1606.516	1596.294	1576.682	1567.268	1549.17	1523.704
2000	20	1436.912	1427.769	1410.227	1401.807	1385.62	1362.842
3500	22	1812.394	1800.863	1778.737	1768.117	1747.7	1718.97
3000	22	1677.951	1667.275	1646.791	1636.958	1618.056	1591.457
2500	22	1531.753	1522.007	1503.307	1494.332	1477.076	1452.795
2000	22	1370.041	1361.324	1344.599	1336.571	1321.137	1299.419
3500	25	1700.177	1689.359	1668.604	1658.641	1639.488	1612.537
3000	25	1574.058	1564.042	1544.827	1535.603	1517.871	1492.919
2500	25	1436.912	1427.769	1410.227	1401.807	1385.62	1362.842

Table *(Contd...)*

Table (*Contd...*)

Combustion gas Properties T (K)	Mol Wt	*Values of γ for the Combustion Gases of Rocket Propellants* 1.13	1.15	1.19	1.21	1.25	1.31
2000	25	1285.213	1277.035	1261.346	1253.815	1239.336	1218.963
3500	28	1606.516	1596.294	1576.682	1567.268	1549.17	1523.704
3000	28	1487.345	1477.881	1459.724	1451.008	1434.253	1410.676
2500	28	1357.754	1349.115	1332.54	1324.583	1309.288	1287.765
2000	28	1214.412	1206.685	1191.86	1184.743	1171.063	1151.812
3500	30	1552.042	1542.167	1523.22	1514.125	1496.641	1472.038
3000	30	1436.912	1427.769	1410.227	1401.807	1385.62	1362.842
2500	30	1311.715	1303.369	1287.356	1279.669	1264.892	1244.099
2000	30	1173.233	1165.768	1151.446	1144.571	1131.354	1112.756
3500	32	1502.758	1493.196	1474.851	1466.045	1449.116	1425.294
3000	32	1391.284	1382.431	1365.447	1357.294	1341.621	1319.566
2500	32	1270.062	1261.981	1246.477	1239.034	1224.727	1204.594
2000	32	1135.978	1128.75	1114.883	1108.226	1095.429	1077.421
3500	35	1436.912	1427.769	1410.227	1401.807	1385.62	1362.842
3000	35	1330.322	1321.857	1305.617	1297.821	1282.835	1261.747
2500	35	1214.412	1206.685	1191.86	1184.743	1171.063	1151.812
2000	35	1086.203	1079.292	1066.032	1059.667	1047.43	1030.212

Values of Specific Impulse, Isp (s) for T/M = 60

Chamber pressure (MPa)	*Values of γ for the Combustion Gases of Rocket Propellants* 1.13	1.15	1.19	1.21	1.25	1.31
3	170.5541	168.4969	164.7089	162.9618	159.7264	155.4335
4	176.2822	173.9909	169.7826	167.8465	164.2692	159.5395
5	180.4766	178.0014	173.4644	171.381	167.5384	162.4718
6	183.7559	181.1292	176.3223	174.1184	170.0594	164.7192
7	186.4316	183.6761	178.6402	176.3344	172.0928	166.5228
8	188.6814	185.8138	180.5793	178.1852	173.7859	168.0179
9	190.6158	187.6491	182.239	179.7672	175.2291	169.2876
10	192.308	189.2522	183.6851	181.1438	176.4819	170.386
11	193.8086	190.6722	184.963	182.3589	177.5853	171.3505
12	195.1544	191.9443	186.1052	183.4438	178.5686	172.2077
13	196.3725	193.0945	187.136	184.4221	179.4535	172.9771
14	197.4838	194.1427	188.0738	185.3112	180.2566	173.6736
15	198.5043	195.1046	188.9328	186.125	180.9904	174.3088
20	202.62	198.9752	192.375	189.3795	183.9138	176.8253
25	205.6656	201.8304	194.8987	191.7587	186.039	178.6403
30	208.0646	204.0737	196.8721	193.6148	187.6895	180.0411

Values of Specific Impulse, Isp (s) for T/M = 80

Chamber pressure (MPa)	*Values of γ for the Combustion Gases of Rocket Propellants*					
	1.13	*1.15*	*1.19*	*1.21*	*1.25*	*1.31*
3	196.9389	194.5634	190.1894	188.1721	184.4361	179.4792
4	203.5531	200.9074	196.0481	193.8125	189.6818	184.2203
5	208.3964	205.5383	200.2994	197.8937	193.4567	187.6062
6	212.183	209.15	203.5994	201.0546	196.3677	190.2013
7	215.2726	212.0909	206.276	203.6135	198.7157	192.2839
8	217.8705	214.5593	208.515	205.7506	200.6706	194.0104
9	220.1042	216.6785	210.4315	207.5773	202.3371	195.4765
10	222.0581	218.5297	212.1013	209.1668	203.7837	196.7448
11	223.7909	220.1693	213.5769	210.5699	205.0578	197.8585
12	225.3449	221.6381	214.8958	211.8227	206.1933	198.8483
13	226.7515	222.9663	216.0861	212.9522	207.2151	199.7368
14	228.0346	224.1767	217.1689	213.9789	208.1423	200.541
15	229.213	225.2874	218.1608	214.9186	208.9897	201.2744
20	233.9654	229.7568	222.1355	218.6767	212.3654	204.1803
25	237.4821	233.0536	225.0497	221.4239	214.8193	206.2761
30	240.2523	235.644	227.3284	223.5671	216.7251	207.8936

Values of Specific Impulse, Isp (s) for T/M = 100

Chamber pressure (MPa)	*Values of γ for the Combustion Gases of Rocket Propellants*					
	1.13	*1.15*	*1.19*	*1.21*	*1.25*	*1.31*
3	220.1843	217.5285	212.6382	210.3827	206.2059	200.6638
4	227.5793	224.6213	219.1884	216.6889	212.0707	205.9646
5	232.9942	229.7988	223.9416	221.2519	216.2912	209.7502
6	237.2278	233.8368	227.6311	224.7859	219.5457	212.6515
7	240.6821	237.1248	230.6236	227.6468	222.1709	214.98
8	243.5866	239.8846	233.1269	230.0361	224.3566	216.9102
9	246.084	242.2539	235.2696	232.0785	226.2197	218.5493
10	248.2685	244.3236	237.1365	233.8556	227.8371	219.9674
11	250.2059	246.1568	238.7862	235.4243	229.2616	221.2126
12	251.9432	247.799	240.2608	236.825	230.5311	222.3192
13	253.5158	249.2839	241.5916	238.0878	231.6735	223.3125
14	254.9504	250.6372	242.8022	239.2357	232.7102	224.2117
15	256.2679	251.8789	243.9111	240.2863	233.6576	225.0317
20	261.5813	256.8759	248.355	244.4879	237.4317	228.2805
25	265.5131	260.5619	251.6132	247.5595	240.1753	230.6237
30	268.6102	263.458	254.1608	249.9556	242.3061	232.4321

Values of Specific Impulse, Isp (s) for T/M = 120

Chamber pressure (MPa)	*Values of γ for the Combustion Gases of Rocket Propellants*					
	1.13	*1.15*	*1.19*	*1.21*	*1.25*	*1.31*
3	241.1999	238.2906	232.9335	230.4628	225.8872	219.8162
4	249.3007	246.0603	240.1089	237.3708	232.3118	225.6229
5	255.2324	251.732	245.3157	242.3694	236.9351	229.7698
6	259.87	256.1554	249.3574	246.2406	240.5003	232.9481
7	263.6541	259.7572	252.6355	249.3745	243.376	235.4987
8	266.8358	262.7804	255.3777	251.992	245.7703	237.6132
9	269.5715	265.3758	257.7249	254.2292	247.8113	239.4088
10	271.9645	267.6431	259.77	256.176	249.5831	240.9622
11	274.0868	269.6513	261.5772	257.8944	251.1435	242.3262
12	275.99	271.4502	263.1925	259.4288	252.5341	243.5384
13	277.7127	273.0768	264.6503	260.8122	253.7856	244.6266
14	279.2842	274.5593	265.9765	262.0696	254.9213	245.6116
15	280.7275	275.9195	267.1913	263.2205	255.9591	246.5098
20	286.548	281.3935	272.0593	267.8231	260.0934	250.0688
25	290.855	285.4312	275.6284	271.1878	263.0988	252.6356
30	294.2477	288.6038	278.4192	273.8127	265.433	254.6166

Values of Specific Impulse, Isp (s) for T/M = 140

Chamber pressure (MPa)	*Values of γ for the Combustion Gases of Rocket Propellants*					
	1.13	*1.15*	*1.19*	*1.21*	*1.25*	*1.31*
3	260.5256	257.3832	251.597	248.9282	243.9861	237.4286
4	269.2755	265.7755	259.3473	256.3898	250.9254	243.7006
5	275.6825	271.9016	264.9712	261.7888	255.9192	248.1797
6	280.6917	276.6795	269.3367	265.9703	259.77	251.6127
7	284.7789	280.5699	272.8775	269.3553	262.8761	254.3677
8	288.2156	283.8353	275.8394	272.1824	265.4623	256.6516
9	291.1705	286.6387	278.3747	274.5989	267.6668	258.5911
10	293.7553	289.0876	280.5837	276.7017	269.5805	260.269
11	296.0476	291.2567	282.5356	278.5578	271.266	261.7423
12	298.1033	293.1997	284.2804	280.2151	272.768	263.0516
13	299.964	294.9567	285.855	281.7093	274.1198	264.2269
14	301.6614	296.5579	287.2874	283.0675	275.3464	265.2909
15	303.2203	298.0272	288.5996	284.3106	276.4674	266.2611
20	309.5072	303.9397	293.8576	289.282	280.933	270.1052
25	314.1593	308.301	297.7127	292.9163	284.1792	272.8776
30	317.8239	311.7278	300.7271	295.7515	286.7004	275.0173

Values of Specific Impulse, Isp (s) for T/M = 160

Chamber pressure (MPa)	*Values of γ for the Combustion Gases of Rocket Propellants* 1.13	1.15	1.19	1.21	1.25	1.31
3	278.5136	275.1543	268.9685	266.1155	260.8321	253.8219
4	287.8676	284.126	277.2539	274.0922	268.2505	260.5269
5	294.717	290.6751	283.2662	279.864	273.5891	265.3153
6	300.0721	295.7828	287.9331	284.3342	277.7058	268.9852
7	304.4415	299.9418	291.7183	287.9529	281.0264	271.9305
8	308.1154	303.4327	294.8847	290.9753	283.7911	274.3721
9	311.2743	306.4296	297.5951	293.5586	286.1479	276.4455
10	314.0376	309.0476	299.9566	295.8066	288.1937	278.2392
11	316.4882	311.3665	302.0433	297.7908	289.9955	279.8142
12	318.6858	313.4437	303.9086	299.5625	291.6013	281.214
13	320.675	315.3219	305.5918	301.16	293.0464	282.4704
14	322.4896	317.0337	307.1232	302.6119	294.3577	283.6079
15	324.1561	318.6044	308.5259	303.9408	295.5561	284.645
20	330.8771	324.9252	314.147	309.2555	300.33	288.7546
25	335.8504	329.5876	318.2683	313.1407	303.8004	291.7184
30	339.768	333.251	321.4908	316.1716	306.4956	294.0059

Values of Specific Impulse, Isp (s) for T/M = 180

Chamber pressure (MPa)	*Values of γ for the Combustion Gases of Rocket Propellants* 1.13	1.15	1.19	1.21	1.25	1.31
3	295.4083	291.8452	285.2841	282.2581	276.6542	269.2188
4	305.3297	301.3611	294.0722	290.7187	284.5226	276.3305
5	312.5946	308.3075	300.4491	296.8406	290.1851	281.4094
6	318.2745	313.725	305.3992	301.5819	294.5515	285.3019
7	322.909	318.1363	309.414	305.4202	298.0735	288.4259
8	326.8058	321.839	312.7725	308.6259	301.0059	291.0156
9	330.1563	325.0177	315.6473	311.3659	303.5056	293.2147
10	333.0872	327.7945	318.152	313.7503	305.6756	295.1173
11	335.6864	330.254	320.3653	315.8548	307.5867	296.7878
12	338.0173	332.4572	322.3437	317.7341	309.2899	298.2724
13	340.1272	334.4494	324.1291	319.4284	310.8227	299.6051
14	342.0519	336.2651	325.7533	320.9684	312.2135	300.8116
15	343.8195	337.9311	327.2411	322.3779	313.4845	301.9117
20	350.9482	344.6352	333.2032	328.015	318.5481	306.2705
25	356.2232	349.5804	337.5745	332.1359	322.229	309.4141
30	360.3784	353.466	340.9925	335.3507	325.0877	311.8403

Values of Specific Impulse, Isp (s) for T/M = 200

Chamber pressure	*Values of γ for the Combustion Gases of Rocket Propellants*					
(MPa)	*1.13*	*1.15*	*1.19*	*1.21*	*1.25*	*1.31*
3	311.3877	307.6318	300.7159	297.5261	291.6191	283.7815
4	321.8458	317.6625	309.9793	306.4444	299.9132	291.2779
5	329.5036	324.9846	316.7012	312.8975	305.8819	296.6315
6	335.4908	330.6952	321.919	317.8953	310.4845	300.7347
7	340.3759	335.3451	326.151	321.9412	314.1971	304.0276
8	344.4835	339.2481	329.6912	325.3202	317.2881	306.7573
9	348.0153	342.5987	332.7214	328.2085	319.923	309.0754
10	351.1047	345.5257	335.3617	330.7218	322.2103	311.0809
11	353.8445	348.1183	337.6947	332.9402	324.2249	312.8418
12	356.3015	350.4407	339.7801	334.9211	326.0202	314.4068
13	358.5255	352.5406	341.6621	336.7071	327.6359	315.8115
14	360.5544	354.4545	343.3741	338.3304	329.1019	317.0832
15	362.4176	356.2106	344.9424	339.8162	330.4417	318.2428
20	369.9318	363.2774	351.2271	345.7582	335.7792	322.8374
25	375.4922	368.4901	355.8347	350.1019	339.6591	326.1511
30	379.8722	372.5859	359.4377	353.4906	342.6725	328.7086

Appendix 2

Numerical Example for Design of a Rocket

PROBLEM STATEMENT

Design rocket propellant in 800 mm length with inner diameter of rocket casing as 250 mm to deliver total impulse of 80000 N.s in 2 seconds. The rocket motor is designed for operating pressure of 10 MPa.

Assumptions

- Rocket is operating at nominal atmospheric condition, giving atmospheric pressure of 0.101325 MPa.
- Optimum expansion is assumed at the exit plane of the rocket nozzle.
- Pressure at the exit plane of the rocket nozzle is 0.101325 MPa.
- Nozzle efficiency is 95%.
- Burning rate law exhibited by propellant is Vieille's power law.
- Burning rate of propellant at 7 MPa is 10 mm/s with burning rate index of 0.3.
- Propellant can be selected as double base or composite with the following properties.

Properties	γ	C^*	ρ
Unit		m/s	kg/m^3
Double Base Propellant	1.21	1350	1625
Composite Propellant	1.19	1540	1750

- Internal Ballistic equations derived in chapter 3 are utilized.
- Equations are not reproduced here and the same can be referred from chapter 3.
- Solid sustainer configuration require burning rate of (800/2=) 400 mm/s, which is not feasible.

- Cartridge-loaded tubular propellant configuration with end inhibition, giving neutral burning profile, is considered for design.
- Cartridge-loaded multi-tubular configuration (in cluster of tubular shaped grains) with end inhibited can also be used.
- The propellant grain is characterized by three geometrical parameters – outer diameter (D), inner diameter (d) and Length (L).
- Case-bonded end-inhibited star port configuration is also attempted in this design with composite propellants.
- Nozzle is conical with half divergent angle of the nozzle is 15°.

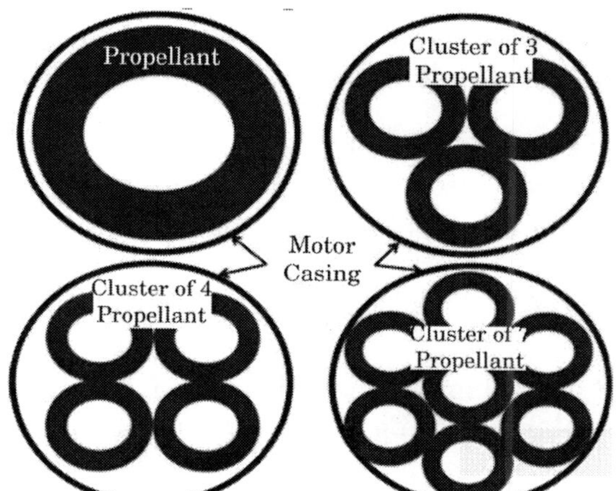

General Calculation

Total impulse, $\int F.dt$ = 80000 N.s
Burning time, T = 2 s
Average thrust, F = 80000/2 N = 40000 N
Burning rate at operating pressure, $r_{10}=7x(10/7)^{0.3}$ mm/s = 11.129 mm/s
Web needed, w = (D-d)/4 = 11.129 x 2 mm = 22.258 mm
Difference of diameters, D-d = 89.035 mm
Pressure ratio, P_e/P_c = 0.0101325
Length of propellant grain, L = 800 mm
Rocket motor inner diameter = 250 mm

Design with Double Base Propellant (Equation numbers refers to chapter 3)

Nozzle area ratio, A_e/A_t (Eqn 16) = 11.49976
Thrust coefficient, C_f (Eqn 18) = 1.636547
Specific Impulse, I_{sp} (Eqn 20) = 213.95 s

Mass Flow Rate, dm/dt (Eqn 20) = 19.06 kg/s

Weight of Propellant, W = 38.12 kg

Throat Area of the Nozzle, A_t (Ch 5) = 2444.17 mm^2

Throat Diameter of the Nozzle, D_t = 55.78 mm

Exit diameter of the Nozzle, D_e = 189.17 mm

Length of the Nozzle, L_n = 248.91 mm

Burning area needed, A_b (Ch 5) = 1053780 mm^2

Cartridge-loaded Configurations

If single propellant grain is conceived, then this burning area is obtained from lateral burning of tubular propellant grain and is given by π(D+d)L.

Sum of the propellant diameters, D+d = 419.286 mm

Outer diameter of propellant, D = 254.16 mm

Inner diameter of propellant, d = 165.12 mm

As outer diameter exceeds, rocket motor internal diameter, the geometry is not feasible.

If cluster of 3 propellant grains is conceived, then this burning area is obtained from lateral burning of tubular propellant grain and is given by 3π (D+d)L.

Sum of the propellant diameters, D+d = 139.762 mm

Outer diameter of propellant, D = 114.39 mm

Inner diameter of propellant, d = 25.36 mm

Cluster of 3 propellant grains of given outer diameter D can be accommodated in rocket motor ID of 2x(1/2+1/$\sqrt{3}$)D = 2.1547D = 246.5 mm. As this is less than available rocket motor inner diameter, *propellant grain design is feasible.*

If cluster of 4 propellant grains is conceived, then this burning area is obtained from lateral burning of tubular propellant grain and is given by 4π(D+d)L.

Sum of the propellant diameters, D+d = 104.82 mm

Outer diameter of propellant, D = 96.93 mm

Inner diameter of propellant, d = 7.89 mm

Cluster of 3 propellant grains of given outer diameter D can be accommodated in rocket motor ID of ($\sqrt{2}$+1)D = 2.141D = 234 mm. As this is less than available rocket motor inner diameter, *propellant grain design is feasible*. However, the small inner diameter of the propellant grain may have practical problem is handling.

If cluster of 7 propellant grains is conceived, then this burning area is obtained from lateral burning of tubular propellant grain and is given by $7\pi(D+d)L$.

Sum of the propellant diameters, $D+d = 59.89$ mm

As sum of diameter is obtained less than difference of diameters, this *propellant grain design is not feasible*. Such configurations will result in negative inner diameter of propellant.

Case-bonded End-inhibited Star Configuration

Since end-inhibited tubular propellant grain burns at both internal and external surfaces, the burning area, available, is very high. Contrary to this, case-bonded end-inhibited propellant configurations, burns in radial outward direction from inner surface, the burning area is relatively smaller. Even after making more number of star points, it is very difficult to match the burning area and only resort, applied, is enhancement of burning rate to such an extent that mass generation rate is matched. As mass flow rate required is 19.06 kg/s with density of propellant as 1625 kg/m^3, the product of burning rate and burning area should be 11729231 mm^3/s at the operating pressure of 10 MPa. As lower burning area is obtained for the star-port configuration, burning rate must be enhanced. The value of operating pressure (10 MPa), throat diameter (55.78 mm), propellant density (1625 kg/m^3), characteristic velocity (1350 m/s), burning rate index (0.3) and propellant length (800 mm) are kept same. Many star configurations are generated and after analyzing more than 100 alternate configurations, following geometrical parameters may be considered for the design of end-inhibited star-port double base propellant configuration.

Propellant outer diameter = 240 mm (Considering 5 mm external inhibition)

Number of star-points = 8 (to give high initial burning area)

Outer diameter of star-point = 145 mm

Inner (Lobe) diameter of star point = 90 mm

Star half-opening angle = 25°

Star fillet radius = 6 mm

Star tip radius = 3 mm

This configuration results in the following output parameters

Angular fraction = 0.6768

Initial burning area = 550864.8 mm^2

Weight of propellant = 41.978 kg

Volume of the propellant = 25832629.9 mm^3

Web available = (240-145)/2 mm = 47.5 mm

Average burning area = 543844.84 mm^2

Burning rate at 10 MPa = 11729231/543844.84 mm/s = 21.567 mm/s

Burning rate at 7 MPa = 21.567 (7/10)$^{0.3}$ mm/s = 19.378 mm/s

Pressure-time profile is generated and the values calculated are shown below.

Initial pressure = 11.07 MPa

Maximum pressure = 12.47 MPa

Web burning time = 2.15 s

Web based average pressure = 10.45 MPa

Web average thrust for the configuration is calculated as 1.636547 x 10.45 x 2444.17 N = 41799.99 N, which meets the requirements. Although constant thrust for 2 seconds is not observed, but the requirement of total impulse can be met by the designed configuration. The propellant design has much lower tail-off portion. The pressure-

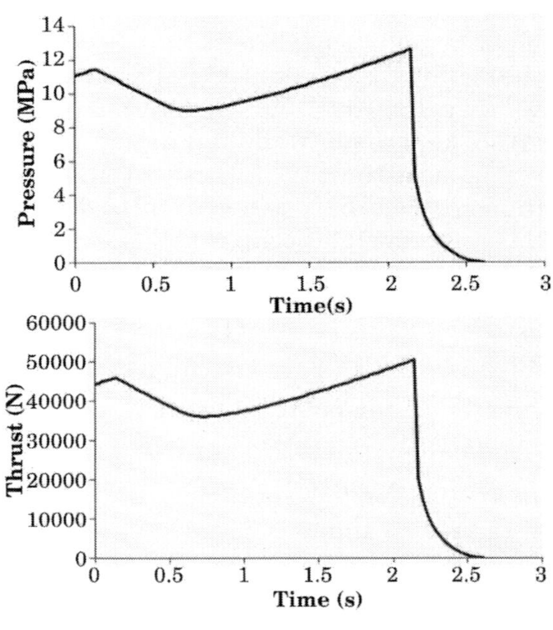

time and thrust time profiles are shown below for the designed configuration.

Design with Composite Propellant (Equation numbers refers to chapter 3)

Nozzle area ratio, A_e/A_t (Eqn 16) = 12.01538

Thrust coefficient, C_f (Eqn 18) = 1.649599

Specific Impulse, Isp (Eqn 20) = 246.01 s

Mass Flow Rate, dm/dt (Eqn 20) = 16.57 kg/s

Weight of Propellant, W = 33.15 kg

Throat Area of the Nozzle, A_t (Ch 5) = 2424.83 mm^2

Throat Diameter of the Nozzle, D_t = 55.56 mm

Exit diameter of the Nozzle, D_e = 192.6 mm

Length of the Nozzle, L_n = 255.7 mm

Burning area needed, A_b (Ch 5) = 850998.5 mm^2

Cartridge-loaded Configurations

If single propellant grain is conceived, then this burning area is obtained from lateral burning of tubular propellant grain and is given by $\pi(D+d)L$.

Sum of the propellant diameters, D+d = 338.6 mm

Outer diameter of propellant, D = 213.8 mm

Inner diameter of propellant, d = 124.8 mm

As outer diameter is less than rocket motor internal diameter, the geometry is feasible.

If cluster of 3 propellant grains is conceived, then this burning area is obtained from lateral burning of tubular propellant grain and is given by $3\pi(D+d)L$.

Sum of the propellant diameters, D+d = 112.86 mm

Outer diameter of propellant, D = 100.95 mm

Inner diameter of propellant, d = 11.92 mm

Cluster of 3 propellant grains of given outer diameter D can be accommodated in rocket motor ID of $2x(1/2+1/\sqrt{3})D = 2.1547D = 217.52$ mm. As this is less than available rocket motor inner diameter, propellant grain design is feasible. However, the small inner diameter of the propellant grain may have practical problem is handling.

If cluster of 4 propellant grains is conceived, then this burning area is obtained from lateral burning of tubular propellant grain and is given by $4\pi(D+d)L$.

Sum of the propellant diameters, $D+d = 84.65$ mm

As sum of diameter is obtained less than difference of diameters, this propellant grain design is not feasible. Such configurations will result in negative inner diameter of propellant.

Case-bonded End-inhibited Star Configuration

Since end-inhibited tubular propellant grain burns at both internal and external surfaces, the burning area, available, is very high. Contrary to this, case-bonded end-inhibited propellant configurations, burns in radial outward direction from inner surface, the burning area is relatively smaller. Even after making more number of star points, it is very difficult to match the burning area and only resort, applied, is enhancement of burning rate to such an extent that mass generation rate is matched. As mass flow rate required is 16.57 kg/s with density of propellant as 1750 kg/m^3, the product of burning rate and burning area should be 9468571.4 mm^3/s at the operating pressure of 10 MPa. As lower burning area is obtained for the star-port configuration, burning rate must be enhanced. The value of operating pressure (10 MPa), throat diameter (55.56 mm), propellant density (1750 kg/m^3), characteristic velocity (1540 m/s), burning rate index (0.3) and propellant length (800 mm) are kept same. Many star configurations are generated and after analyzing more than 100 alternate configurations, following geometrical parameters may be considered for the design of end-inhibited star-port double base propellant configuration. Although configuration is possible with 8 star-points, 6 star-points configurations are considered for composite propellant alternatives.

Propellant outer diameter = 240 mm (Considering 5 mm external inhibition)

Number of star-points = 6 (to give high initial burning area)

Outer diameter of star-point = 185 mm

Inner (Lobe) diameter of star point = 90 mm

Star half-opening angle = 32°

Star fillet radius = 6 mm

Star tip radius = 3 mm

This configuration results in the following output parameters

Angular fraction = 0.7015

Initial burning area = 671861.6 mm^2

Weight of propellant = 38.03 kg

Volume of the propellant = 21735814 mm^3

Web available = (240-185)/2 mm = 27.5 mm

Average burning area = 790393.23 mm^2

Burning rate at 10MPa = 9468571.4/790393.23 mm/s = 11.98 mm/s

Burning rate at 7 MPa = 11.98 $(7/10)^{0.3}$ mm/s = 10.76 mm/s

Pressure-time profile is generated with burning rate of 12 mm/s at 7 MPa and the values calculated are shown below.

Initial pressure = 10.04 MPa

Web burning time = 2.02 s

Web based average pressure = 10.2 MPa

Web average thrust for the configuration is calculated as 1.649599 x 10.2 x 2424.83 N = 40827.5 N, which meets the requirements. The pressure-time and thrust time profiles are shown below for the designed configuration. This configuration has longer tail-off section. However,

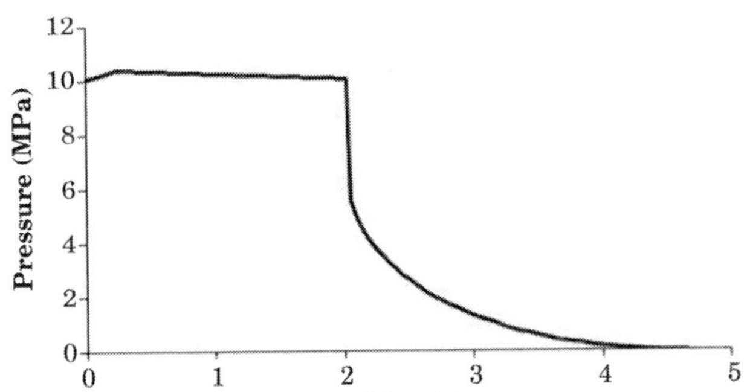

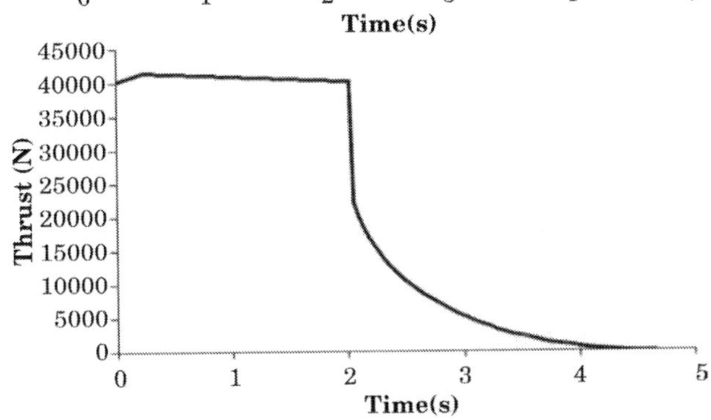

compared to design presented for double base propellants, this configuration has relatively neutral burning profile.

CONCLUSION

- Several cartridge-loaded designs meeting the requirements of the problem in multi-tubular configurations are possible. *Double base propellants in cluster of 3 configurations or mono-tubular composite propellant grain can meet the requirements with better handling provisions.* Double base propellants in cluster of 4 and composite propellant in cluster of 3 are also feasible designs, but the lower value of inner diameters of propellants may lead to handling issues in processing.

Propellant	***Grains***	***OD, D***	***ID, d***	***W***	I_{sp}	D_t	A_b
Type	no	mm	mm	kg	s	mm	mm^2
DBP	3	114.39	25.36	38.11	213.95	55.78	1053780
CP	1	213.8	124.8	33.14	246.01	55.56	850998.5

- Out of mono-tubular and cluster of 3-configurations, cluster of 3-configurations is preferred, because of ratio of burning area to port area. Both these configurations create at-least two ports in the cross-section of rocket motor – one inner (within central portion of tube) and another outer port (between propellant and rocket motor casing). For mono-tubular configuration, two ports are clearly identifiable. As far as, port area is concerned area of outer port is found as 13186.5 mm^2, while area of inner port is 12232.6 mm^2. Both the values are comparable. However, burning area of the outer port is 537338.01 mm^2 and that of the inner port is 313.656.61 mm^2. Burning area to port area ratio is 40.75 for outer port and 25.64 for the inner port. So, there are chances of generation of higher pressure in outer port area. However, for cluster configuration (double base propellants), the external ports are connected and have very complex situation of pressure mismatch. In practical ballistics, tubular propellants have radial holes (of diameter 5-10 mm) drilled at around 200 mm distance along length for pressure equalization.
- Although design is carried out for end inhibited tubular propellant shape for perfect neutrality, but *propellant without any end-inhibition is generally employed.* This makes *production of propellant faster (no inhibition step needed in processing) with a slight compromise on neutrality (regressive firing curve).*

- All configurations are not possible with all the inputs and tubular propellant grain is attempted in this example. The pressure time or thrust time profiles are not generated, because their variations are negligible and constant value throughout the burning time is observed. However, in practical situations, slight regressive nature and some tail-off is observed with tubular propellants. There are chances of rupture of propellant at the end of burning, resulting in a pressure kink at the end of burning profile. However, these are practical observations, which cannot be predicted.
- Star port propellant shapes in case-bonded configuration for both types of propellants are also attempted. For both the propellants, mass flow rates, throat diameters and required burning rates, are different. With same grain diameter (240 mm), length (800 mm), Lobe diameter (90 mm), Fillet radius (6 mm) and tip radius (3 mm), the grain designs are conceived by iterative process with more than 100 alternatives in each case. Dimensions and parameters for one of the probable configurations for each case is presented below.

Propel-lant	***Star-point***	***Star OD***	***Star half angle***	***Wt of Prop.***	***Bur-ning rate at 7 MPa***	***B.T.***	***Av. Pr.***	***Av. Thrust***
Type	no	mm	°	kg	mm/s	s	MPa	N
DBP	8	145	25	42.05	19.378	2.15	10.95	41799.9
CP	6	185	32	38.12	12.000	2.02	10.2	40827.5

- For any propellant design, mass flow rate requirements, pressure levels, propellant mass etc can be cross-checked.

Appendix 3

Feasibility of Neutral Configurations

Solid sustainer and end inhibited tubular configurations can produce neutral burning propellant configuration because of virtually no change in burning area in course of burning. However, there are certain practical limitations for these configurations. In this appendix practical limitations of these configurations are numerically evaluated. Propellant outer diameter is considered 300 mm for the calculations. For calculation, composite propellant configurations with following properties are considered.

Propellant outer diameter = 300 mm

Density of propellant = 1750 kg/m^3

Characteristic velocity = 1540 m/s

Specific impulse = 240 s

Burning rate index = 0.3

Thrust coefficient (calculated) = 240 x 9.81/1540 = 1.5288

Solid Sustainer Configuration

As discussed in chapter 6, solid sustainer propellant configuration is capable of giving low thrust for longer durations. So, for given input parameters, thrust more than given maximum value is not possible. If burning rate and throat diameter of propellant are considered variables, then maximum thrust is generated for minimum throat diameter and maximum burning rate. The range of throat diameters, considered for the analysis is given below. The variation of burning rate at 7 MPa, permitted for propellant in solid sustainer configuration is given below.

Throat diameter = 25 mm to 120 mm

Burning rate at 7 MPa = 5 to 40 mm/s

However, both pressure and thrust is generated for the entire range of throat diameter (from 25 mm to 120 mm). Two values of burning rates (40 mm/s and 20 mm/s) at 7 MPa are considered.

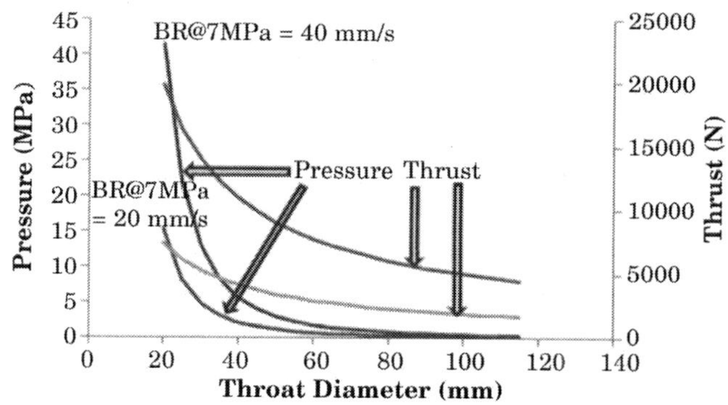

The variation of pressure and thrust for two burning rates for the entire range of feasible throat diameters is plotted above. At higher burning rates, same throat diameter gives higher pressure and thrust values. For reference burning rate of 40 mm/s at 7 MPa, thrust of 16389 N is realized at 20 mm throat diameter. At same throat diameter, lower burning rate (20 mm/s at 7 MPa) gives lower thrust (6088 N). With increase in throat diameter, thrust at both the burning rate levels falls. However, the trend indicates that at very high throat diameters, thrust level for both burning rates may be equal. As far as pressure is concerned, for burning rate of 40 mm/s at 7 MPa, 20 mm throat diameter can give a very high pressure of 22 MPa. The value of pressure drops to 3 MPa for throat diameter of 50 mm. This indirectly indicates high sensitivity of pressure to change of throat diameter in the range of 20 mm to 50 mm. Compared to thrust, pressure variation is more sensitive in this region of throat diameter. The increment of throat diameter beyond 50 mm, leads to marginal reduction in pressure values. Almost similar situation occurs with burning rate of 20 mm/s at 7 MPa. The pressures at throat diameters of 20 mm and 35 mm, are calculated to be 8 MPa and 3 MPa, respectively. Higher burning rate propellants are more susceptible to change, due to change in throat diameter.

Further calculations are repeated for propellant diameter of 300 mm. Let us limit the pressure inside rocket between 3 MPa and 15 MPa. This is nominal operating pressure limit of any rocket. Practical constraints are put on burning rate at 7 MPa (5 to 40 mm/s) and throat diameter (25 mm to 120 mm). The variations of pressure for these conditions for different throat diameters are plotted below.

From the above figure, it is clear that for the given range of throat diameter, burning rate (at 7 MPa) less than 10 mm/s cannot give operating pressure between 3 MPa and 15 MPa. For burning rate of 20 mm/s, the range of throat diameter to give operating pressure between 3 to 15 MPa is 25 mm to 35 mm, only. For burning rate of 30 mm, this

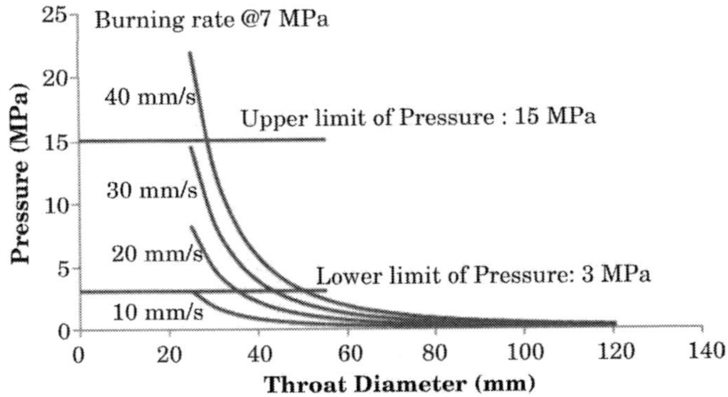

range changes from 25 mm to 43 mm. A peculiar situation occurs for higher burning rate. Here, at lower throat diameter (25 mm), pressure exceeds upper pressure limits. So, minimum throat diameter, in this case is restricted to 28.6 mm. At this burning rate, lower pressure limit is achieved for a throat diameter of 50 mm. With these constraints, the range of thrust achievable from 300 mm diameter propellant in solid sustainer mode is also very limited for different burning rates. The values of range of thrusts achievable, along with corresponding throat diameter, are tabulated below.

Burning rate at 7 MPa	*Lower Limit (kN)*	*Upper Limit (kN)*	*Range (kN)*
20	4.5 at 35 mm	6.1 at 25 mm	1.6
30	6.8 at 43 mm	10.8 at 25 mm	4.0
40	9.0 at 50 mm	14.6 at 28.6 mm	5.6

This clearly indicates that, all thrust values are not possible for rocket propellant of given diameter, burning in solid sustainer mode. Operational pressure, throat diameter and burning rate at standard pressure limit maximum value of thrust achievable from the rocket. Because of limitations, with 300 mm diameter propellant, thrust more than 14.6 kN is not possible, if throat diameter cannot be reduced beyond 25 mm, burning rate at 7 MPa cannot be raised beyond 40 mm/s and pressure cannot exceed 15 MPa. For achieving 15 MPa, in higher throat diameters, burning rate must exceed 40 mm/s at 7 MPa. For 30 mm throat diameter, for a pressure 15 MPa, burning rate required becomes 44.2 MPa and peak thrust achieved is 16.2 kN.

End Inhibited Tubular Configuration

Although analysis similar to that presented in previous section (for solid sustainer configuration) is possible for end inhibited tubular propellant configurations, but different form of analysis is presented on this neutral burning configuration for exploring other domains. End inhibited tubular propellant grain is discussed in chapter 7 and it is indicated that due to higher burning surface area, as compared to solid sustainer configurations, this propellant configuration can accommodate propellants with relatively lower burning rates also. This propellant configuration is capable of giving higher thrust for smaller duration. Limitations on throat diameter are not needed, but inner diameter of propellant less than 30 mm must be avoided. Throat diameter for this configuration is taken as 75 mm (one-fourth of propellant outer diameter) for numerical demonstration of feasibility. Length of propellant configuration is taken as 1000 mm. The burning rate of propellant is considered to be 10 mm/s at 7 MPa. Using calculation strategy described in chapter 5, the value of pressure and thrust can be calculated for different burning time. If perfectly neutral burning profile is assumed then for different burning time, pressure and thrust realizable can be calculated. Inner diameter of propellant is to be determined from ballistic parameters before proceeding further for the calculation of thrust. The input parameters for the calculation are reproduced below.

Propellant outer diameter = 300 mm

Propellant length = 1000 mm

Density of propellant = 1750 kg/m^3

Characteristic velocity = 1540 m/s

Specific impulse = 240 s

Burning rate of propellant at 7 MPa = 10 mm/s

Burning rate index = 0.3

Throat diameter = 75 mm

Thrust coefficient (calculated) = 240 x 9.81/1540 = 1.5288

The burning area (A_b) for the configuration is given by π(D+d)x L = π(300+d)x 1000 mm^2. The web available for complete consumption of propellant is (D-d)/4 = (300-d)/4 = 75 – (d/4). Suppose burning time is specified as ‘t’, then web divided by ‘t’ gives burning rate of propellant at the operating pressure of the rocket.

$$\text{Burning time, } t = (D-d)/4r = (D-d)/4aP^n$$

Operating pressure of the propellant is given by equation derived for power burning rate law in chapter 5.

$$P = [a\rho C^* A_b/A_t]^{1/(1-n)}$$

So, inner diameter of the propellant can be calculated by iterative means through the following equation.

$$(D\text{-}d) = 4atP^n = (D+d)^\eta\chi, \text{ where } \eta = n/(1\text{-}n),$$
$$\chi = [4a\rho C^*L\{4ta\}^{1/\eta}/D_t^{\,2}]^\eta$$

The variation of propellant inner diameter and pressure for various burning time is plotted below.

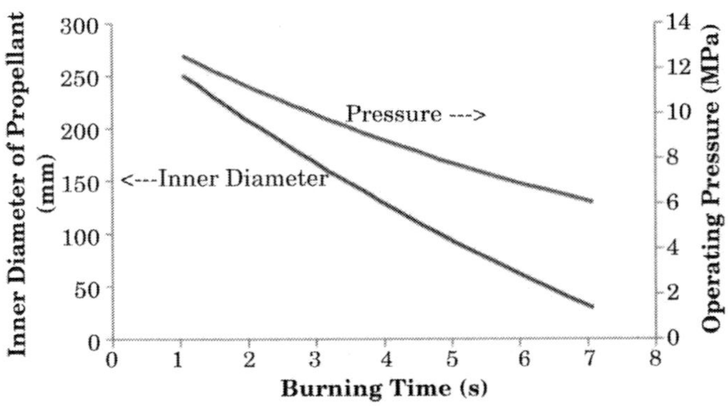

Increase in burning time requires a higher web, if burning rate (or indirectly operating pressure) is same. Here burning rate, throat diameter and characteristic velocity are constant. Higher burning time means higher web and lower inner diameter of the propellant (for constant outer diameter of propellant). The lower inner diameter of propellant leads to lower total burning area of propellant, resulting in lower operating pressure of the propellant. The same trend is depicted in the figure above. As burning time changes from 1 s to 7 s, the value of pressure changes from 12.5 MPa to 6 MPa. Lower burning time gives (i) higher web for complete consumption of propellant (ii) higher inner diameter of propellant (iii) higher burning area of propellant (iv) higher operating pressure of propellant and (v) higher burning rate of the propellant. This higher burning rate of the propellant reduces burning time.

If burning rate of the propellant at 7 MPa is increased to 20 mm/s, the inner diameter corresponding to 1 s burning time changes from 250 mm to 180 mm. Higher burning rate require, higher web resulting in lowering of inner diameter of propellant. Lower inner diameter gives lower burning area, but higher burning rate offsets this lowering and operating pressure at 1 s burning time changes from 12.5 MPa to 27.5 MPa. For burning rate of 20 mm/s at 7 MPa, with constraint on lower limit of inner diameter of propellant as 30 mm, burning time more than

2.6 s is not possible. With similar constraints, for burning rate of 10 mm/s at 7 MPa, maximum possible burning time is 7 s.

Similarly throat diameter can be increased to observe effects on pressure and inner diameters of the propellants. For throat diameters of 100 mm, the burning time possible cannot exceed 9 s (7 s for 75 mm throat diameter), if inner diameter is limited to 30 mm. For an upper limit of inner diameter of propellant as 250 mm, the burning time is 1.34 s in place of 1 s for 75 mm throat diameter. Increase in throat diameter reduces pressure and same is observed for this situation also.

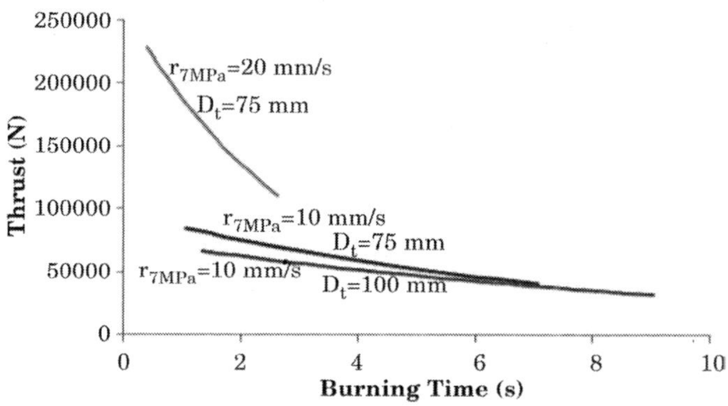

Thrust profile for different burning time is similar to pressure profile. The thrust variation for different burning time is plotted above. for nominal reference case, throat diameter is 75 mm and burning rate at 7 MPa is 10 mm/s. as inner diameter of propellant cannot be changed beyond 30 mm and 250 mm, the limitations on burning time is obvious. The thrust level varied between 40 kN to 85 kN. If throat diameter is raised to 100 mm, the thrust level achieved may be between lowered and the range is 32 kN to 66 kN. A higher change occurs by changing burning rate to 20 mm/s at 7 MPa. This gives a thrust ranging from 110 MPa to 228 MPa. The variation of burning time for different cases is already mentioned earlier, while discussing variation of pressure.

The most significant variation is observed for total impulse and although range of total impulse is same, but different burning time, results in achievement of different total impulse at different burning times. The variation of total impulse is shown in figure below. The burning time changes, but value of total impulse remains constant. It seems that the variation can be obtained by proper scaling along horizontal axis. The easier way to look at the variation of thrust and total impulse is by plotting their variation with respect to inner diameter of the propellant. It is clear that variation of total impulse is invariant due to change in burning rate or throat diameter, but thrust changes for different conditions. The same is depicted in figure below.

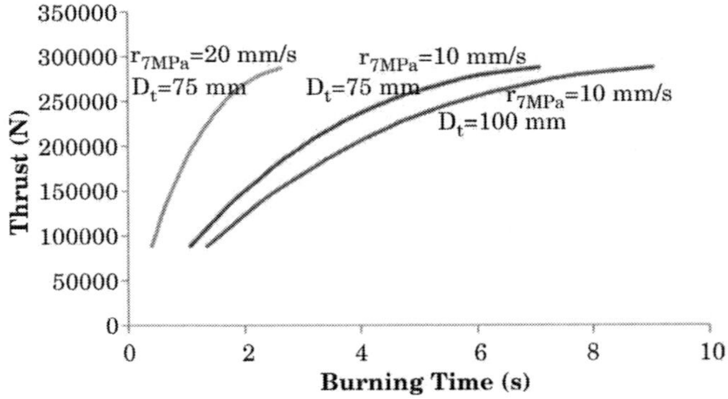

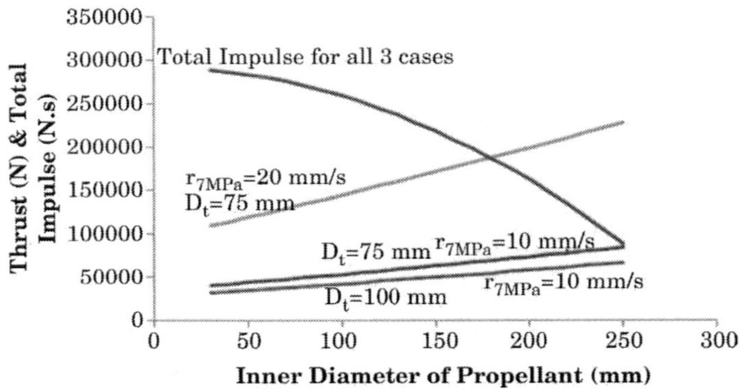

In the figure above, thrust for the three cases are plotted against inner diameter of propellant. Higher burning rate gives higher thrust levels and higher throat diameters give lower thrusts. This is depicted in the figure. Total impulse for the 3 cases depicted in the figure becomes single. So, the total impulse for a given configuration is invariant and it does not change due to the change in throat diameter or burning rate of propellants. As long as time is not considered, burning rate and throat variation does not affect total impulse. Time and thrust levels can be adjusted.

CONCLUSION

Although design for a given situation may be attempted, but for a given input condition, propellant configuration also restricts ballistics. Propellant configuration puts limitations on pressure, thrust, burning time and other ballistic parameters. The example in this appendix is meant to support this claim. So, selection of configuration needs due care in design of propellants.

Suggested Books for Further Reading

1. Rocket Propulsion Elements, 8th edition, George P. Sutton, Oscar Biblarz, John Wiley & Sons Inc, USA, 2010, ISBN 0-471-32642-9.
2. Introduction to Rocket Science and Engineering, Travis S. Taylor, CRC Press, Taylor & Francis Group, US, 2009, ISBN 978-1-4200-7528-1.
3. Science and Technology of Solid Rocket Propellants, Dr. Haridwar Singh, Himanshu Shekhar, Printwell, India, 2005.
4. Rocket and Spacecraft Propulsion: Principles, Practice and New Developments, Martin J. Turner, Springer-Praxis Series in Astronomy and Space Sciences, UK, 2000, ISBN 1-85233-105-4.
5. Principles of Guided Missile Design: Systems Preliminary Design, Grayson Merrill, D. Van Nostrand Company, 1960, USA.
6. Rocket Propellants, Francis A. Warren, Reinhold Publishing Corporation, Second print, 1960, USA.
7. Internal Ballistics of Solid Fuel Rockets, R.N. Wimpress, McGraw-Hill Book Company, 1950, USA.
8. Mathematical Theory of Rocket Flight, J. Barkley Rosser, Robert R. Newton, George L. Gross, McGraw-Hill Book Company, 1947, USA.

Publications by Author on Solid Rocket Propellants

1. Sreenadh Jayashree and Shekhar Himanshu (2016). "Studies on Burning Time of 40 kg Agni Ballistic Evaluation Motors (BEM) in Static Firing", DRDO International Women's Day Celebration Proceedings Compendium, SWADHARM-2016, NSTL, Vishakhapattanam, pp. 163–168.
2. Deshmukh, P.M., Erande, D.T., Raut, K.V., Sreenadh Jayashree, Kachi, G.T. and Shkehar Himanshu. Study on Extraction of Ballistic.
3. Parameters of 2-Chamber Integrated Rocket with Limited Instrumentation in Static Firing, Proceeding of 10th International High Energy Materials Conference and Exhibits HEMCE-2016, Vol. II, 840–843.
4. Kalal, R.K., Ropia, B., Babnsode, M.K., Jadhav, M.H., Shekhar, H., Gururaj Rao, C. and Alegaonkar, P.S. (2016). Study of heat fux inside rocket motor for the combustion of double base propellant, Proceeding of 10th International High Energy Materials Conference and Exhibits HEMCE, Vol. II, 844–847.
5. Ropia Balesh, Kalal, R.K., Sonawane, H.P., Gururaj Rao, C. and Shekhar, H. (2016). Design of proof rocket motor for heat flux measurement of various propellants, Proceeding of 10th International High Energy Materials Conference and Exhibits HEMCE Vol. II, 853–857.

6. Mishra, K.K., Babu, A.S. and Shekhar Himanshu (2016). Application of image processing for ballistic prediction of star port prismatic rocket propellants configurations, Proceeding of 10th International High Energy Materials Conference and Exhibits HEMCE Vol. II, 1079–1084.
7. Shekhar Himanshu (2015). Effects of the burning rate index on the pressure time profile of progressive burning tubular rocket propellant configurations. *Central European Journal of Energetic Materials*, 12(2): 145–158.
8. Shekhar Himanshu (2015). "Burning Rate Calculation from Static Firing of Rockets", Delivered at National Chiao Tung University, Taiwan. Shekhar,
9. Shekhar Himanshu (2015). "Star Perforated Propellant Configuration", Delivered at National Chiao Tung University, Taiwan.
10. Shekhar Himanshu (2015). "Static Evaluation and Validation of Predicted Ballistics", Delivered at National Chiao Tung University, Taiwan.
11. Shekhar Himanshu (2015). "Performance Prediction with Tubular Configuration of Solid Rocket Propellants", Delivered at National Chiao Tung University, Taiwan.
12. Shekhar Himanshu (2015). "Elements of Solid Propellant Rocket Propulsion", Delivered at National Chiao Tung University, Taiwan.
13. Erande, D.T., Kachi, G.T., Raut, K.V., Ms Jayashree, S., Deshmukh, P.M. and Shekhar, H. (2015). "Safety of Pressure Gauges in Static Firing of Rockets and Igniters". *Safety Bulletin, HEMRL*, 17: 27–31.
14. "Challenges for Engineers in Indigenous Rocket Development", Lecture at Army Institute of Technology, Dighi Hills, Pune, 03.03.2015.
15. "Challenges for Mechanical Engineers in Indigenous Development of Rockets", Lecture at AISSMS-COE, Pune, 24.02.2015.
16. "Propellants: Demure Depredation", Invited Talk to POINTS-19, DIAT(DU), Pune, 22.11.2014.
17. "Shelf life and Ageing Studies of Solid Rocket Propellants", Sardar Patel University, Gujarat, Academic Exchange of Teacher within India Program, 12 Nov 2014.
18. "Instrumentation for Static Evaluation of Rockets", Sardar Patel University, Gujarat, Academic Exchange of Teacher within India Program, 11 Nov 2014.
19. "Advances in Rocket Propellant Technology", Sardar Patel University, Gujarat, Academic Exchange of Teacher Within India Program, 10 Nov 2014.
20. "Life Assessment of Solid Rocket Propellant Grain", CEP course at ARDE, Pune, 24 Sept 14.
21. Shekhar Himanshu (2014). "Biodegradation of Solid Propellants for Rockets: A Review". *Bioglobia*, 1(1): 7–16, ISSN: 2339-5626.
22. Verma, P., Kalal, R.K., Ghavate, R.B., Bansode, M.K., Bhujbal, J.G., Shekhar Himanshu and Singh, R.V. (2013). "Use of Different Isocyanates in Boron-Viton-Polyurethane base Fuel Rich Propellant and its Effect on Thermal Expansion", Abstract Proceeding, THERMANS-2013 DAE-BRNS 19th Symposium and Workshop on Thermal analysis, BARC, Mumbai, India, H-5, p. 136.
23. "Performance Prediction of Composite Propellants in Static Evaluation", Invited Talk, Sardar Patel University, Ballabh Vidyanagar, Gujarat, 16.11.2013

24. "Simulated Dynamic Firing of Rockets, Warheads and Armament Systems", CEP at TBRL, Chandigarh, 15-19 July 2013.
25. Shekhar Himanshu (2012). "Mathematical Formulation and Validation of Muraour's Linear Burning Rate Law for Solid Rocket Propellants". *Central European Journal of Energetic Materials*, 9: 353–364.
26. "Life Assessment of Solid Rocket Propellants" (2012). Invited talk at CEP on Rocket Technology, ARDE, Pune.
27. "Mechanical Characterization and Structural Integrity of Solid Rocket Propellants", Invited talk at CEP on Rocket Technology, ARDE, Pune, 08.08.2012.
28. Shekhar Himanshu and Kankane, D.K. (2012). Viscoelastic characterization of different solid rocket propellants using the maxwell spring-dashpot model. *Central European Journal of Energetic Materials*, 9(3): 189–199.
29. Shekhar Himanshu (2012). "Studies on Stress-Strain Curves of Aged Composite Solid Rocket Propellants (CSRP)". *Defence Science Journal,* 62(2): 90–94.
30. Shekhar Himanshu and Sahasrabudhe, Anil D. (2011). "Longitudinal Strain Dependent Variation of Poisson's Ratio for HTPB Based Solid Rocket Propellants in Uni-axial Tensile Testing". *Propellants, Explosives and Pyrotechnics*, 36(6): P558–563.
31. Shekhar Himanshu, Shinde, P.D. and Ratheesh Jisha (2011). "Ageing Studies of Composite Solid Rocket Propellants (CSRP) Through Uni-axial Tensile Testing", Proceeding of 8th International High Energy Materials Conference and Exhibits, TBRL, Chandigarh, India, p. 345
32. Shekhar Himanshu, Shinde, P.D. and Ratheesh Jisha (2011). 'Studies on Effect of Temperature on Mechanical Properties of Different Classes of Solid Rocket Propellants", Proceeding of 8th International High Energy Materials Conference and Exhibits, TBRL, Chandigarh, India, pp. 343–344.
33. Shekhar Himanshu and Kankane, D.K. (2011). "Studies on Application of Maxwell Fluid Model to Tensile Testing Curves of Different Classes of Solid Rocket Propellants", Proceeding of 8th International High Energy Materials Conference and Exhibits, TBRL, Chandigarh, India, p. 333
34. Shekhar Himanshu (2011). "Prediction and Comparison of Shelf Life of of Solid Rocket Propellants Using Arrhenius and Berthelot Equations". *Propellants Explosives and Pyrotechnics*, 36(4): 356–359.
35. "Rocket and Gun Ballistics", Invited talk for POINTS-16 programme, DIAT(DU), Girinagar, Pune, 16 June 2011.
36. "Rocket Ballistics", Invited talk, HEMRL, 19 April 2011
37. Shekhar Himanshu (2011). "Burn Back Equations for High Volumetric Loading Single Grain Dual Thrust Rocket Propellant Configuration". *Defence Science*, 61(2): 165–170.
38. "Mathematical Simulation of Solid Rocket Propellants", Invited Talk, National Conference on Computational Engineering: Modelling, Simulation and Optimization (CEMSO-2010), 8 Dec 2010.
39. Shekhar Himanshu (2010). "Studies on Final Modulus of Solid Rocket Propellants in Uni-Axial Tensile Testing Curves", Science and Technology of Energetic Materials. *Japan Explosives Society*, 71(4): P106–110.

40. Shekhar Himanshu and Sahasrabudhe, Anil D. (2010). "Assessment of Poisson's Ratio for HTPB Based Solid Rocket Propellants". *Defence Science Journal,* 60(5): 497–501.

41. "Mechanical Engineering in Solid Propellant Rocket Propulsion", Lecture at College of Engineering Pune, Bhartiya Vidyapeeth University (COEP_BVU), 19 Aug 2010.

42. Shekhar Himanshu and Sahasrabudhe, Anil D. (2010). "Maxwell Fluid Model for Generation of Stress-Strain Curves of Viscoelastic Solid Rocket Propellants". *Propellants, Explosives and Pyrotechnics*, 35(4): P321–325.

43. Shekhar Himanshu and Sahasrabudhe, Anil D. (2010). "Viscoelastic Modeling of Solid Rocket Propellants at Different Strain Rates Using Maxwell Fluid Model". *Defence Science Journal*, 60(4): 423–427.

44. Shekhar Himanshu and Sahasrabudhe, Anil D. (2010). "Concept of Final Modulus for Complete Mechanical Characterization of Solid Rocket Propellants", Proceeding of International Conference on Recent Trends in Materials and Characterization, RETMAC 2010 (MA O-2), NIT, Surathkal, pp. 33–34.

45. Shekhar Himanshu and Sahasrabudhe, Anil D. (2010). "Complete Mechanical Characterization of Solid Rocket Propellants through Invariant Final Modulus (H)", International Conference on Emerging Trends in Chemistry, ETIC (IPA-O-19), UOP, Pune, p. 93.

46. Shekhar Himanshu and Sahasrabudhe, Anil D. (2009). "Stress Concentration Factor for Pre-cracked Composite Propellant", Proceeding of 7th International High Energy Materials Conference and Exhibits (7-HEMCE), Pune, India, pp. 38–40.

47. Shekhar Himanshu and Subhananda Rao, A. (2009). "Comparison of Arrhenius and Reaction Rate Models for Predicting Shelf-life of Solid Rocket Propellants", Proceeding of 7th International High Energy Materials Conference and Exhibits (7-HEMCE), Pune, India, pp. 103–105.

48. Shekhar Himanshu and Singh Haridwar (2009). "Factors affecting combustion behaviour of metal powders in fuel rich propellants for ram rockets: A fresh look", Eighth International Symposium on Special Topics in Chemical Propulsion, 8-ISICP, Cape Town, South Africa, P77–79.

49. Shekhar Himanshu (2009). "Estimation of Pressure Index and Temperature Sensitivity Coefficient of Solid Rocket Propellants by Static Evaluation". *Defence Science Journal,* 59(6): 666–669.

50. Shekhar Himanshu (2009). "Design of Funnel Port Tubular Propellant Grain for Neutral Burning Profile in Rockets". *Defence Science Journal*, 59(5): P494–498.

51. "Role of Mechanical Engineers in Solid Propellant Rocket Propulsion", Engineers Day Lecture at PS Modern College of engineering, 15 Sept 2009.

52. "Design of Rockets and Moon Mission", Delivered in DAV Public School, Pune on 20.07.2009

53. "Internal Ballistics of Solid rocket Propellants", Delivered at TBRL Chandigarh, 08-09 June 2009.

54. Shekhar Himanshu (2009). "Design og Gas Generator Propellant Grains for Missile Applications", Manthan – A quarterly Magazine by BiharBrains, ISSN No 09074-6331, 8: 8–11.

55. Shekhar Himanshu (2009). "Hazards of static Firing of rocket Motors with Cracked Solid Rocket Propellant". *Safety Bulletin, HEMRL,* 12: 8–13.
56. Shekhar Himanshu (2099). "Design of Gas Generator Propellant Grains for Missile Applications", Abstract and souvenir booklet of Second Bihar Science Conference, Patna, p. 5.
57. "Design of Gas Generator Propellant Grains for Missile Applications", Invited talk during Second Bihar Science Conference, Patna, Jan 30 – 01 Feb 2009.
58. Rao, Subhananda and Shekhar Himanshu (2008). "Solid Rocket Propulsion Technology in DRDO", International Workshop on Advances in Processing of Solid Propellant rocket Motors (IWAPS 2008), HEMRL, Pune, India.
59. "Solid Rocket Propulsion Technologies for Missiles: Indian Perspective", Delivered at College of Engineering Pune during MindSpark, 02 Oct 2008.
60. Shekhar Himanshu (2008). "Mechanical Characterization of Solid Rocket Propellants", Manthan – A quarterly Magazine by BiharBrains, p. 11.
61. "Mechanical Characterization of Solid Rocket Propellants", Invited talk during first Bihar Science Conference, Patna, 7-9 May 2008, P6 of Abstract and Souvenir Booklet.
62. Shekhar Himanshu (2008). "Failure of ballistic evaluation motor during static firing: A Case Study", Proceedings of National Conference on Failure Analysis, NCFA-2006, DMRL, Hyderabad, Excel India Publishers, ISBN: 978-81-907839-3-4, pp. 110–113.
63. "Ballistics of Solid Propellant Rockets", Delivered in NDT-MPT division on 08.10.2007.
64. Shekhar Himanshu (2007). "Close-form solution for burning surface evolution and performance prediction of finocyl shaped propellant grain", 2007 International Autumn Seminar on Propellant, Explosives and Pyrotechnics (2007 IASPEP), Nan Yang Hotel, Xi'an, China, pp. 907–911.
65. "Mechanical Properties of Solid rocket Propellants", Delivered at Advanced Centre of Research in High Energy Materials (ACRHEM), University of Hyderabad, Hyderabad, during their interactive seminar 06-08 September 2007.
66. "Solid Rocket Propulsion Technology", Delivered at Sinhgad Institute of Technology, Lonavala, during Technical Festival on Aviation and Space Technology, 20-21 Aug 2007.
67. "Development of Composite Rocket Motor Casing for Tactical Application", Technology Day, 27 Apr 2007, 3rd Prize.
68. Shekhar Himanshu (2007). "Studies on Peak Strain in Regressing Finocyl Port Propellant Grains under Pressure Loading". *Defence Science Journal*, 57(2): 259–266.
69. Shekhar Himanshu (2007). "Structural Margin of Safety for 122mm ERR Propellant", Safety Bulletin, HEMRL.
70. "Failure analysis of Ballistic Evaluation Motor during Static Firing: A Case Study", National Conference on Failure Analysis (NCFA-2006), Katriya Hotels and Towers, Somajiguda, Hyderabad, 22-23 Dec 2006.
71. "Performance Prediction of Solid Rocket Motors", CEP on Ballistics in Defence, 18-22 Sept 2006.

72. Shekhar Himanshu (2006). "Studies on prediction of maximum strain in finocyl port case bonded solid propellants under port pressurization". *Defence Science Journal,* 56(3): P417–422.
73. Shekhar Himanshu (2006). "Combustion Deflagration and Detonation". *Bi-Annual Magazine of HEMRL – "EXOTHERM"*, 3.
74. "Guided Missiles: An Overview", Delivered in COEP, Pune during Fervor 2006, 11.03.2006.
75. Shekhar Himanshu (2006). "Design Through Safety in Infrastructure for Processing Large Size Case Bonded Rocket Motors", CEP on Safety.
76. Shekhar Himanshu (2005). "Variation of peak strain during burning of pressurized finocyl port propellant grains", HEMCE 2005.
77. Shekhar Himanshu (2005). "Analysis of Ballistic Non-conformity during Static Evaluation of Motors", CEP on Quality Assurance.
78. Shekhar Himanshu (2005). "Parametric Studies on Star Port Propellant Grain for Ballistic Evaluations". *Defence Science Journal*, 55(4): 459–469.
79. "Combustion, Deflagration and Detonation", Delivered in Patwardhan Hall on 09.09.2005.
80. Singh Haridwar and Shekhar Himanshu (2005). "Combustion Mechanism of Solid Rocket Propellants", 5th International Seminar on Flame Structure, Russia.
81. "Systems Involved in Rocket Propulsion and Classification of Propellants", Delivered in a course on Rocket Propellants, Organised by NADP, Ambajhary at AF, Kirkee, 04-08 July 2005.
82. "Rocket Propulsion Elements", Delivered in a course on Rocket Propellants, Organised by NADP, Ambajhary at AF, Kirkee, 04-08 July 2005.
83. "Igniter and Ignition Systems", Delivered in a course on Rocket Propellants, Organised by NADP, Ambajhary at AF, Kirkee, 04-08 July 2005.
84. Shekhar Himanshu (2004). "Effect of Intermittent Supply of Compressed Hot Gas through Metal Pipes". *Process and Plant Engineering,* XXII(5): 73–76.
85. Shekhar Himanshu (2004). "A Case Study on Structural Analysis of Finocyl Port Case Bonded Solid Propellant under Pressure Loading", DRDO Workshop on Advanced Manufacturing Technology", DRDL, Hyderabad, P600-605.
86. "Modelling for Processing Solid rocket Propellants", Delivered in Patwardhan Hall on 15.04.2004.
87. Shekhar Himanshu (2004). "Firing of flawed propellant grain". *Safety Bulletin, HEMRL*, 7.
88. "Modeling and Sensitivity Analysis of Heat Transfer during Curing of polymeric Cylinder", Presented in CHEMCON 2003, 19-22 Dec 2003.
89. Shekhar Himanshu, Bhat, V.K. and Singh Haridwar (2003). "Heat Transfer and Solidification Process in Mould Cast Propellant Grains", Proceeding of 4th International HEMCE-2003, pp. 461–469.
90. Shekhar Himanshu (2003). "Studies on flow control during propellant casting through ball valve", Poster presentation in 4th International HEMCE-2003, p. 118 of abstract proceeding.
91. Shekhar Himanshu (2003). "Modelling for solution of 1D transient heat conduction problems", *Process and Plant Engineering*, XX1 No 2 (Part II): 91–92.

92. Shekhar Himanshu (2003). "Structural integrity analysis of case bonded propellant grains", CEP on NDT/MPT, pp. 113–120.
93. Shekhar Himanshu (2003). "Structural Integraity analysis of case bonded composite propellant", DRDO Science Spectrum, pp. 45–50.
94. "Structural Integrity Analysis of case bonded composite propellant grain", Science Day 28 Feb 2003, Sc day Oration Award of DRDO HQ.
95. Shekhar Himanshu (2002). "Analysis of vacuum casting of highly viscous polymer melt". *Process and Plant Engineering*, XX(2): 59–61.
96. Shekhar Himanshu (2002). "Safety Considerations in the design of Propellant Trimming Machine", *Safety Bulletin, HEMRL*, 5.
97. Shekhar Himanshu (2002). "Range Enhancement of Propulsion Unit by Mid-Flight Boost", Proceeding of Workshop on Propulsion Systems, HEMRL, Pune, pp. 44–47.
98. "Modelling of Temperature Profile in Propellant Slurry during Curing", Science Day 25 Feb 2002, 1st prize.
99. Shekhar Himanshu (2001). "Analysis of Flow through Ball Valve". *Process and Plant Engineering*, XVIII (4): 55–57.
100. Shekhar Himanshu (2000). "Design of spray nozzle for liner coating facility", HEMCE, pp. 283–285.
101. "Indigenisation of Liner Coating Nozzle", Technology Day, 8 may 2000, 2nd Prize.
102. Shekhar Himanshu (2000). "Modelling of Burning Surface Regression of Taper Convex Star Propellant Grain". *Defence Science Journal*, 50(2): pp. 207–211.
103. Shekhar Himanshu, Radhakrishnan, K.K., Narasimhan, V.L. and Nair, R.G.K. (2000). "Indigenization of Propulsion Units", Seminar on "Missile Maintenance Management" at INS Tunir.
104. "Fabrication Methods for Large Size Mandrels with Star Shaped Configuration", Science day, 26 Feb 2000.
105. Shekhar Himanshu (1999). "Case Preparation for Case Bonded Propulsion Units", Delivered in CEP (12-16 July) entitled "Propellants for Rockets and Missiles".
106. "Halon Based Fire Extinguishing System for Case Bonded Motor Evaluation", Technology Day, 10 May 1999, 3rd prize.
107. "Modelling of Burning Surface Regression of Propellant with Star Shaped Port", Science day, 26 FEB 1999, 1st prize.
108. Shekhar Himanshu (1998). "Fault Tree Analysis of Liner Coating Facility", Proceeding of National Seminar on Reliability Analysis and Engineering, CASSA, Bangalore, pp. 89–95.
109. Shekhar Himanshu (1998). "HAZAN in casting of composite propellant" *Safety Bulletin, HEMRL*, 1.

Subject Index

A

ABP 3, 80, 83
Acceleration 3, 75, 171
Accept 30
Accuracy 19, 245, 262
Acoustic 325, 330,
Action time 266, 272
Additives 47, 102
Adiabatic index 49
Advanced Propellant 30
Air 3, 14, 50, 321
Air drag 3
Air-breathing 18
Aircrafts 3, 20
Akash 17, 321
All burn point 82, 93
Altitude 26, 96, 99
Aluminium 18, 35, 106, 246
Aluminized 68, 246
Ammonium perchlorate 18, 35, 337
Analog 263
Angular fraction 203, 217, 221
Annular 109, 120, 135
Annular port 109
AP 36
Arc 132, 233, 240
Area 1, 48, 108
ARIANE 87
Atmosphere 4, 18, 34
Attitude correction 7, 124
Average pressure 111, 258, 275
Axial 52, 240, 313

B

Ballistic 1, 29, 48
Bare 83, 139, 141
BEM 103, 140, 255, 309
Beryllium 22, 30, 36
Binder 18, 32, 102
Bipropellant 21
Blow holes 245
Boiling 20, 31
Boiling point 20, 21
Boost 3, 81, 129
Booster 3, 86, 106
Boron 23, 36
Branch 231, 232
Brittle 34
Burning 18, 22, 46
Burning area 64, 101, 108
Burning area profile 177, 186, 193
Burning rate 22, 38, 101
Burning rate coefficient 104, 247, 281
Burning rate index 33, 103, 117
Burning surface 22, 52, 102
Burning time 89, 106, 116

C

Calculation 14, 48, 57
Calorific value 13, 22, 324
Canted nozzle 25, 28, 299
Carbon 30, 50, 71
Carbon dioxide 50,
Cartridge-loaded 34, 110, 121
Case-bonded 19, 41, 109
Casing 14, 41, 107
Cast Double Base 18, 29, 41
Casting 18, 33, 121
Casting fixture 45
Casting liquid 42, 43
Casting powder 42, 43
CDB 18, 29, 41,
Central hole 128, 163, 188
Chamber 2, 20, 64,
Chamber pressure 10, 50, 112
Characteristic velocity 39, 48, 67
Charcoal 22

Chemical 1, 15, 52
Chemical Bond 12, 14, 17
Chemical energy 2, 15, 83
Chemical Propulsion 12, 16, 18,
Choked 68, 300
Cigarette 19, 109, 122
Circular 120, 132, 157
Circumferential 218, 226, 269
Classification 2, 16, 25,
Cluster 161, 166, 335
CMDB 18
Coast 3, 96, 98
Colloidal 30, 31, 46
Combustion 2, 11, 21
Combustion chamber 2, 20, 51
Combustion efficiency 14, 22, 68
Combustion gases 2, 23, 50, 70
Combustion temperature 33, 73, 76
Compatibility 14, 32, 262
Composite Modified Double Base 18, 39,
Composite propellant 14, 35, 43
Compound 30
Compressibility 49
Compression 8, 34, 261
Concave 129, 173, 203
Condensed 31, 52, 245
Condition 8, 41, 102
Configuration 16, 85, 102
Consolidation 140
Consumption 10, 81, 107
Control motors 29
Conventional 31, 74, 111
Convergent 24, 61, 110
Converging 55, 61
Convex 173, 203, 212
Corrosion 20
Crack 22, 34, 38
Critical 21, 44, 149
Cross 15, 34, 45
Cross-section 34, 51, 109
Crumbling 126, 160, 276
Curvature 132, 204, 212
Cylinder 9, 41, 102
Cylindrical 102, 110, 120

D

Damping 37, 327, 329
DBP 18, 30, 41
Decoring 43, 140, 218
Deflagration 30, 101, 102
Degree of freedom 49
De-Laval 25
Demoulding 43
Dendrite 230
Density 8, 32, 55
Design 8, 26, 62
Diameter 7, 15, 24
Diesel 9, 260
Digital 226, 261, 276
Dimension 59, 64, 102
Dimensional 53, 121, 128
Discharge coefficient 68, 112,
Displacement 7, 93, 251
Distributed combustion 327, 328
Divergent 6, 24, 56
DOA 38
DOP 38
Double base propellant 18, 33, 50
Downstream 54, 59,
Dual thrust 16, 129, 135
Duration 7, 16, 39

E

EDB 18, 29, 40
Effective Exhaust Velocity 6
Efficiency 12, 22, 67
Ejection 3, 24, 95
Electrical propulsion 14, 18
Electromagnetic propulsion 14, 18,
Electrons 30
Elemental 30, 45
End burning 109, 122, 150
End face 102, 124, 135
End of burning 107, 127, 148
Energy 2, 9, 12
Energy conversion 2, 12, 15
Energy limited device 14
Energy transfer 2, 326
Engine 2, 7, 84
Entrance 10, 52, 65
Equation 53, 56, 68
Erosive burning 106, 110, 153
Ester 32
Evaluation 41, 91, 103
Exhaust 2, 5, 11
Exhaust gases 2, 14, 30
Exit 6, 11, 15
Exit Area 61, 63
Exit diameter 68, 123, 298
Exit Plane 6, 10, 15
Exothermic 17, 30, 43
Expansion 6, 8, 10

Explosion 31, 259
External 4, 66, 84
Extruded Double base 18, 29, 40
Extrusion 18, 33, 40

F

Fail 13, 308
Failure modes 258
FeAA 38,
Fillet radius 213, 217, 221
Fin 108, 119, 121
Final 19, 30, 35
Finalization 251, 272, 294
Fire 41
First Law of Thermodynamics 8
Fission 17
Flame 15, 22, 24
Flash 32
Flight 5, 7, 46
Flow turning 327, 328, 337
Force 2, 6, 80
Formulations 32, 71, 107
Forward 3, 5, 36
Freezing point 20
Friction 7, 33, 43
FRP 18,
Fuel 2, 6, 8
Fuel consumption 13
Fusion 17

G

Gamma 49, 50, 296, 301
GAP 35, 144, 172
Gas Generator 7, 17, 29
Gases 2, 4, 5
Gelled propellants 18, 21
Geometry 62 , 101, 107
Glycidyl Azide Polymer 35
Grain diameter 171, 177, 357
Gravity 4, 21, 75
GSLV 85
Gun Powder 22
Gun propellant 107, 108, 169

H

Half angle 111, 204, 218
Half width 233
Hazard 14, 25, 34
Hazard division 259
HD 141, 155, 158
Head end ignition 24, 28
Heat 7, 8, 10
Heat Addition 8, 9, 10
Heat capacity 49
Heat of combustion 33, 48, 74
Heat of Formation 14, 17, 31
Heat Rejection 9, 10,
Heat Transfer 45, 106, 268
Hepta-tubular 161, 172
Heterogeneous 18, 29, 30
Heterogeneous propellant 18, 29, 30
HMX 18
Hole 120, 128, 129
Homogenous 8, 18, 30
Homogenous propellants 18, 30, 31
Horizontal 41, 127, 218,
Horse power 12
Hot spot 268
HTPB 18, 35, 37
Hybrid propellant 21, 22
Hydrogen 21, 22, 30
Hydroxy terminated poly butadiene 35
Hygroscopic 36, 39, 44
Hypergolic 21

I

Igniter 15, 18, 22
Ignition 9, 21, 22
Ignition delay 132, 265, 266
Impact 33, 43, 300
Incremental 247
Inert 24, 32, 42
Ingredients 14. 18, 22
Inhibited 114, 121, 122
Inhibition 15, 43, 102
Initial 8, 9, 16
Inner diameter 102, 107, 111
Insensitive 22, 33
Instability 24, 35, 249
Instrumented 249, 250, 251
Insulation 14, 15, 17
Intake 8, 9, 321
Internal 4, 5, 16
Internal ballistic 48, 101, 112
Inward 9, 234, 313, 320
Isentropic 9, 49, 51,

J

JATO 20
Jet Propulsion 3, 4, 7

K

Kinetic energy 2, 6, 7
Klemmung 108
Kneading 40

L

L* combustion instability 333
Lateral 5, 63, 102
Launch Vehicle 85, 86, 87
Length 24, 42, 45
Linear 73, 82, 83
Liquid propellant 9, 16, 20
Loading 18, 37, 38
Loading density 42
Lobe radius 204, 205, 207
Lobed 202, 203, 204
Local 244, 245, 246
LOH 21
Longitudinal 5, 11, 52
LOX 21
LPCL 23, 34, 39
LVDT 269

M

Mach number 53, 55, 58
Macro-molecules 30, 32, 35
Magnesium 22, 23, 30
Magnetic propulsion 14, 18
Main motor 16, 140, 178
Mandrel 45, 109, 140
Mass 3, 6, 10
Mass discharge 112, 113, 115
Mass flow rate 6, 10, 14
Mass generation 112, 116, 246
Mass ratio 81, 83, 85
Mathematical 48, 52, 81
Maximum web 111, 124, 126
Mechanical properties 19, 33, 34
Mechanical Work 9
MESA 33, 105
Metallic fuel 35, 37
Missile 3, 16, 22
Modifiers 38, 41, 45
Modulus 19, 34
Molecular weight 31, 32, 37, 48
Molecules 12, 14, 17
Momentum 6 , 7, 10
Momentum Thrust 6, 10, 61
Monopropellant 21, 73
Motion 3, 4, 6
Motor 6, 10, 12
Mould 42, 43, 245
MTV 23
Multi-perforated 42, 108, 119
Multiple 16, 25, 35,
Multiple nozzle 321
Multi-tubular 119, 139, 140
Muraour's 104, 105, 114

N

Nag 16
NASA 87
nBD 38
NC 18, 30, 32
Negative 8, 36, 55
Neutral 108, 111, 122
Neutrality 111, 120, 145
NG 18, 30, 31
Nitramine 18, 39
Nitrate 22, 23, 35
Nitrocellulose 18, 30, 72
Nitrogen 22, 31, 32
Nitroglycerin 18, 30, 74,
Non-acoustic instability 332
Non-steady 127, 325
Normalized 142, 143, 148
Nozzle 2, 4, 6
Nozzle area ratio 26, 60, 68
Nozzle efficiency 67, 301, 302
Nozzle end ignition 24
Nuclear propulsion 14
Nuclear rocket 17
Number 1, 8, 13, 16
Numerical 1, 6, 48

O

Offset port 139, 156, 158
Opacity 32
Operation 2, 4, 5
Optimum 19, 63, 64,
Outer diameter 107, 108, 111
Outlet 54, 65, 66,
Outward 7, 9, 19
Over-expanded nozzle 11
Overhang 252, 260, 299
Oxidizer 18, 21, 22
Oxygen 14, 21, 22
Oxygen balance 31, 32
Oxygen rich 31, 32

P

Parallel 11, 42, 45
Parameter 48, 51, 58
Parametric 80, 89, 101
Particulate 123, 300, 303
Particulate damping 327, 328, 335
Paste 40, 41, 42
Patterns 148, 335
Payload 2, 4, 13, 26
Peak pressure 23, 178, 265
Peak thrust 297, 298, 360
Penner's 105
Performance 1, 6, 8, 13
Periphery 134, 158, 159
Peristalsis 3
Physical 1, 18, 20, 21
Piobert's 102, 107
Pitch 17, 172, 176, 188
Pitch circle diameter 173
Planar 233
Plasticizer 31, 32, 41
Plateau burning 33
Point 3, 7, 8, 13
Port area 101, 109, 110
Port diameter 227, 228, 313
Positive 55, 104, 213
Power 9, 12, 13, 14
Power law 127, 255, 277, 280
Practical 23, 38, 46, 49
Prediction 1, 29, 53, 63
Predictive 119, 226, 231
Pre-load 253, 254, 299
Premixed 37, 101, 105
Press 41
Pressing 19, 22, 39, 41
Pressure 2, 5, 6, 7
Pressure gauges 250, 331
Pressure Thrust 6, 7, 61, 63
Processing 1, 14, 19, 29
Profile 17, 76, 101, 108
Progress 125, 325
Progressive 33, 108, 121, 134
Progressivity ratio 170, 177, 180, 200
Projection 3, 242, 259, 313
Propellant grain 41, 107, 108
Propellant loading 109, 110, 111, 143
Propellant volume 112, 204, 207
Propellants 226, 244, 245
Properties 19, 20, 29
Propulsion 2, 4, 6,
Propulsive force 2, 4
PSLV 80, 85, 86, 96
Pyrogen 23
Pyrotechnic 22, 23, 254

Q

Quadratic 72, 127, 175
Quenching 41

R

Radial 19, 109, 140
Radius 128, 132, 172
Ramp 226, 227
Range 22, 36, 38
Ray length 204, 206, 207
RDX 18, 41, 74
Reaction 2, 3, 4, 5
Reaction device 2, 7, 63
Regressive 108, 111, 124
Resolution 262, 263
Response 262, 326, 327
Retardants 103
Revolution 5, 129, 130
RFNA 21
Robert Goddard 4, 5
Rocket 1, 3, 5
Rocket motor casing 14, 19, 45,
Rocket Performance 6, 8, 13, 75
Rocket propellant 1, 16, 22, 29
Rocket propulsion 29, 52, 120
Rocketry 6, 18, 46
Roll 17, 25,
Rolling 41

S

Saddle 260
Safety 14, 41, 43
Scarfed nozzle 25,
Section 1, 6, 7, 18
Self 15, 23, 34
Self-sustaining 15, 257
Sensitive 31, 41, 113
Sensitivity 32, 33, 41
Shape 24, 43, 48
Side thrust 299
Single grain 17, 134, 162
Skin temperature 250, 252, 268
Slag ejection 332
Slant 154, 156, 218
Sliver 108, 170, 171

Slope 98, 104, 143
Sloshing 20, 21
Slurry 37, 38, 40
Smokeless 31, 46, 336
Solar energy 18, 81,
Solar propulsion 14
Solid 1, 3, 5, 9
Solid loading 37, 38, 44
Solid propellant 1, 3, 9
Solid sustainer 107, 108, 109
Solventless extrusion 40
Sonic 53, 54, 58
Sound 53, 249, 251
Space 3, 4, 10, 12
Spark 8, 33, 260
Specific gravity 21
Specific heat 21, 48, 49, 50
Specific impulse 6, 20, 22, 34
Speed 9, 12, 13, 41
Spherical 109, 129, 130, 131
Spillage 20, 21
Stability 14, 21, 23, 32
Staging 1, 80, 81, 84
Stagnation 11, 57, 58
Star 108, 109, 110
Star angle 203
Star half angle 111
Star petals 203, 207
Star point 215, 216, 218
Star radius 204, 205, 208
Static firing 250, 251, 253
Steady state 112, 114, 116
Stem 230, 231, 232
Stick 122
Storage3, 10, 18, 20
Straight 24, 25, 28, 105
Straight nozzle 25, 28
Strain 249, 251, 261
Strain gauge 262, 263
Strand burner 103, 255, 272
Stroke 9, 10, 13
Structural 19, 20, 37, 89
Subsonic 55, 101
Summerfield 105, 113, 114
Supersonic 15, 54, 55
Surface area 23, 108, 111
Sustainer 3, 16, 106, 108
Symmetry 25, 129, 203, 226

T

T* combustion instability 333
Tactical 20, 31, 66
Tail-off 108, 112, 132, 154
Tailoring 16, 141, 209, 217
Taper 109, 139, 141
Teflon 23
Temperature 2, 7, 8, 15
Tensile strength 19, 34
Tether propulsion 14, 18
Theoretical 42, 48, 81
Thermal 2, 7, 14, 17
Thermal conductivity 21, 45, 258
Thermal insulation 14, 19, 39, 126
Thermocouple 252, 268, 269
Thermodynamic Cycle 8, 9, 10
Thermodynamics 2, 8, 326
Thermoplastic 30, 34
Thixotropic 21
Throat area 15, 60, 63, 67
Throat Diameter 15, 17, 25
Throat erosion 24, 25, 114, 124
Thrust 3, 4, 6, 11
Thrust coefficient 6, 48, 63, 64
Thrust wall 250, 252, 260, 296
Thruster 7, 18, 124
Time 8, 14, 20, 25
Tip radius 204, 207, 212
TMP 38
Top land 202, 216, 218, 220
Total impulse 76, 114, 116, 297
Toxic 20, 36
Transducer 261, 262, 263, 269
Transience 22, 23, 112, 262
Transition 36, 37, 129, 131
Transport vehicle 8, 12
Transportation 2, 7, 14, 26
Transverse 328, 329, 331, 335
Triangle 196, 232, 240
Tsiolkovosky 80, 81, 83, 93
Tubes 107, 162, 166
Tubular 102, 107, 108, 109
Turbulent 22, 106
Turning 153, 327, 329, 337

U

Under-expanded nozzle 11
Upstream 54

V

Vacuum propulsion 14
Vapour pressure 21

Variation 8, 10, 11, 21
Velocity 23, 25, 30, 39
Velocity gain 6, 11, 80, 85
Vertical 43, 93, 100, 251
Vibration 49, 69, 249, 260
Viscosity 21, 30, 37, 44
Viton 23
Volume 9, 19, 23, 31
Volumetric loading 110, 120, 121, 125

W

Water 4, 31, 34, 40
Web 34, 40, 45, 101
Web burn time 266, 267, 272
Web burnt 108, 127, 130
Web consumed 122, 124, 126, 142
Web transition 137, 157, 158
Weight 13, 24, 31, 49

Y

Yaw 17

Z

Zero 6, 8, 51, 66
Zone 101, 129, 130, 145